Data Analysis
in Astronomy

ETTORE MAJORANA INTERNATIONAL SCIENCE SERIES
Series Editor:
Antonino Zichichi
European Physical Society
Geneva, Switzerland

(PHYSICAL SCIENCES)

Recent volumes in the series:

Volume 15 **UNIFICATION OF THE FUNDAMENTAL PARTICLE INTERACTIONS II**
Edited by John Ellis and Sergio Ferrara

Volume 16 **THE SEARCH FOR CHARM, BEAUTY, AND TRUTH AT HIGH ENERGIES**
Edited by G. Bellini and S. C. C. Ting

Volume 17 **PHYSICS AT LEAR WITH LOW-ENERGY COOLED ANTIPROTONS**
Edited by Ugo Gastaldi and Robert Klapisch

Volume 18 **FREE ELECTRON LASERS**
Edited by S. Martellucci and Arthur N. Chester

Volume 19 **PROBLEMS AND METHODS FOR LITHOSPHERIC EXPLORATIONS**
Edited by Roberto Cassinis

Volume 20 **FLAVOR MIXING IN WEAK INTERACTIONS**
Edited by Ling-Lie Chau

Volume 21 **ELECTROWEAK EFFECTS AT HIGH ENERGIES**
Edited by Harvey B. Newman

Volume 22 **LASER PHOTOBIOLOGY AND PHOTOMEDICINE**
Edited by S. Martellucci and A. N. Chester

Volume 23 **FUNDAMENTAL INTERACTIONS IN LOW-ENERGY SYSTEMS**
Edited by P. Dalpiaz, G. Fiorentini, and G. Torelli

Volume 24 **DATA ANALYSIS IN ASTRONOMY**
Edited by V. Di Gesù, L. Scarsi, P. Crane, J. H. Friedman, and S. Levialdi

---

A Continuation Order Plan is available for this series. A continuation order will bring delivery of each new volume immediately upon publication. Volumes are billed only upon actual shipment. For further information please contact the publisher.

Data Analysis in Astronomy

Edited by

V. Di Gesù
University of Palermo and
Institute of Cosmic Physics and Informatics/CNR
Palermo, Italy

L. Scarsi
University of Palermo and
Institute of Cosmic Physics and Informatics/CNR
Palermo, Italy

P. Crane
European Southern Observatory
Garching/Munich, Federal Republic of Germany

J. H. Friedman
Stanford University
Stanford, California

and

S. Levialdi
University of Rome
Rome, Italy

Plenum Press • New York and London

Library of Congress Cataloging in Publication Data

International Workshop on Data Analysis in Astronomy (1984: Erice, Sicily)
Data analysis in astronomy.

(Ettore Majorana international science series. Physical sciences; v. 24)
"Proceedings of the International Workshop on Data Analysis in Astronomy, held May 28–June 4, 1984, in Erice, Sicily, Italy"—T.p. verso.
Bibliography: p.
Includes index.
1. Astronomy—Data processing—Congresses. I. Di Gesù, V. II. Title. III. Series.
QB51.3.E43I58   1984   520'.28'5   85-9436
ISBN 0-306-42018-X

Proceedings of the International Workshop on Data Analysis in Astronomy,
held May 28–June 4, 1984, in Erice, Sicily, Italy

©1985 Plenum Press, New York
A Division of Plenum Publishing Corporation
233 Spring Street, New York, N.Y. 10013

All rights reserved

No part of this book may be reproduced, stored in a retrieval system, or transmitted, in any form or by any means, electronic, mechanical, photocopying, microfilming, recording, or otherwise, without written permission from the Publisher

Printed in the United States of America

PREFACE

The international Workshop on "Data Analysis in Astronomy" was intended to give a presentation of experiences that have been acquired in data analysis and image processing, developments and applications that are steadily growing up in Astronomy. The quality and the quantity of ground and satellite observations require more sophisticated data analysis methods and better computational tools. The Workshop has reviewed the present state of the art, explored new methods and discussed a wide range of applications. The topics which have been selected have covered the main fields of interest for data analysis in Astronomy. The Workshop has been focused on the methods used and their significant applications.
Results which gave a major contribution to the physical interpretation of the data have been stressed in the presentations. Attention has been devoted to the description of operational system for data analysis in astronomy.

The success of the meeting has been the results of the coordinated effort of several people from the organizers to those who presented a contribution and/or took part in the discussion. We wish to thank the members of the Workshop scientific committee Prof. M. Capaccioli, Prof. G.De Biase, Prof. G.Sedmak, Prof. A.Zichichi and of the local organizing committee Dr.R.Buccheri and Dr. M.C.Maccarone together with Miss P.Savalli and Dr. A.Gabriele of the E.Majorana Center for their support and the unvaluable part in arranging the Workshop.

P.Crane
V.Di Gesù
J.H.Friedman
S.Levialdi
L.Scarsi

LIST OF CHAIRMEN

Session:
Data Analysis Methodologies	P. Benvenuti
	P. Mussio
	P. G. Sedmak
Systems for Data Analysis	M. Disney
	M. Capaccioli
Image Processing	S. Tanimoto
	A. Gagalowictz
	F. Rocca
Parallel Processing	J. F. Friedman

Panel discussion:
Trend in Optical and Radio Data Analysis	R. Albrecht
Data Analysis Trends in X-ray and γ-Ray Astronomy	M. E. Ozel
Systems for Data Analysis: What they are, what they should be?	P. Crane
Trend on Parallel Processing Applications	S. Levialdi

Concluding discussion:
How Can Computer Science Contribute to the Solution of the Problems posed by Astronomers	L. Scarsi

CONTENTS

DATA ANALYSIS METHODOLOGIES

The Data Analysis Facilities That Astronomers Want...... 3
   M.Disney

Time Analysis in Astronomy: Tools for Periodicity
     Searches... 15
   R.Buccheri and B.Sacco

Cluster Analysis by Mixture Identification............. 29
   P.A.Devijver

Graphical Methods of Exploratory Data Analysis.......... 45
   J.H.Friedman, J.A.McDonald and W.Stuetzle

Faint Objects in Crowded Fields: Discrimination and
     and Counting..................................... 57
   M.L.Malagnini, F.Pasian, M.Pucillo and P.Santin

Multivariate Statistics to Analyze Extraterrestial
     Particles from the Ocean Floor................... 67
   I.E.Frank, B.A.Bates and D.E.Brownlee

Statistical Analysis of Pulse Processes................. 75
   S.Frasca

Application of Bootstrap Sampling in Gamma-ray
     Astronomy: Time Variability in Pulsed Emission
     from Crab Pulsar................................. 81
   M.E.Özel and H.Mayer-Hasselwander

Binary Slice Fit Ellipticity Analysis.................... 87
  L.Rusconi and P.G.Sedmak

An Automated Method for Velocity Field Analysis......... 93
  G.Van Moorsel

The Star Observation Strategy for HIPPARCOS............. 99
  S.Vaghi

Panel discussion on: Data Analysis Trends in X-ray and
    γ-ray Astronomy.................................... 119
  M.E.Özel

Panel discussion on: Trends in Optical and Radio Data
    Analysis... 135
  R.Albrecht

SYSTEMS FOR DATA ANALYSIS

Introduction to Data Analysis Systems for Astronomy..... 157
  R.Allen

FIPS: Objectives and Achievements....................... 165
  T.D.Westrup, J.Gras and W.Kegel

MIDAS... 183
  P.Crane, K.Banse, P.Grosbol, C.Ounnas and D.Ponz

Space Telescope Science Data Analysis................... 191
  R.Albrecht

NRAO's Astronomical Image Processing System (AIPS)...... 195
  D.C.Wells

ASTRONET.. 211
  P.G.Sedmak

The CDCA Package: Past, Present and Future.............. 223
  A.Bijaoui, A.Llebaria and D.Cesarsky

Design of a Pattern-Directed System for the Interpreta-
    tion of Digitized Astronomical Images.............. 227
  P.Mussio

An Astronomical Data Analyzing Monitor.................. 235
  D.Teuber

RIAIP - A System for Interacting with Astronomical Images... 241
R.De Amicis and G.A.De Biase

Experience Report on the Transportation of Software Systems in Connection with the Creation of MOTHER, the VAX/VMS version of the Groningen HERMES....... 249
P.Steffen, C.Crezelius, P.M.W.Kalberla, R.Steube, P.T.Rayner and J.P.Terlouw

SAIA: A System for Astronomical Image Analysis.......... 257
B.Garilli, P.Mussio and A.Rampini

A System for Deep Plate Analysis: An Overview........... 263
P.Di Chio, G.Pittella and A.Vignato

The Groningen Image Processing System................... 271
R.J.Allen, R.D.Ekers and J.P.Terlouw

Panel discussion on Systems for Data Analysis: What They Are, What They Should be?........................ 303
P.Crane

IMAGE PROCESSING

Astronomical Input to Image Processing, Astronomical Output from Image Processing........................ 323
E.J.Groth

Review Methodologies in Digital Image Processing........ 343
A.Gagalowicz

2-D Photometry.. 363
M.Capaccioli

Spectrometry.. 379
R.A.E. Fosbury

Time Dependent Analysis................................. 387
R.G.Kron

Morphological Analysis of Extended Objects.............. 401
F.Pasian and P.Santin

Solar Image Processing With the Clark Lake Radioheliograph.. 411
T.E.Gergely and M.J.Mahoney

Display of Three-dimensional Data in Radioastronomy..... 417
   A.H.Rots

Data Compression Techniques in Image Processing for
      Astronomy.. 425
   C.Cafforio, I.De Lotto, F.Rocca and M.Savini

Fast Digital Image Processing Algorithms and Techniques
      for Object Recognition and Decomposition........... 431
   V.Cappellini, A.Del Bimbo and A.Mecocci

Automatic Analysis of Crowded Fields...................... 439
   M.J.Irwin

DWARF: The Dwingeloo-Westerbork Astronomical Reduction
      Facility... 449
   J.P.Hamaker, R.H.Harten, G.N.J.van Diepen and
   K.Kombrink

PARALLEL PROCESSING

Steps Toward Parallel Processing........................ 459
   S.Levialdi

New Architectures for Image Processing.................. 479
   V.Cantoni

Data Structures and Languages in Support of Parallel
      Image Processing for Astronomy..................... 497
   S.L.Tanimoto

Morphology and Probability in Image Processing........... 509
   A.G.Fabbri

Trend in Parallel Processing Applications............... 523
   S. Levialdi

CONCLUDING DISCUSSION

Panel discussion on How Can Computer Science
   Contribute to the Solution of Problems
   Posed by Astronomers................................. 527
   L.Scarsi

Index.. 539

SESSION

DATA ANALYSIS METHODOLOGIES

CHAIRMEN:

P. BENVENUTI
P. MUSSIO
P.G. SEDMAK

THE DATA ANALYSIS FACILITIES THAT ASTRONOMERS WANT

Mike Disney

Department of Astronomy and Applied Mathematics
University College
P.O.Box 78, Cardiff CF1 1XL, U.K.

A. INTRODUCTION

I am not an expert data analyst, as will become apparent, but
I have been concerned to provide astronomers with the data analysis
facilities (D.A.F.) they need, particularly in the U.K. Although
my views are coloured by my recent experiences in setting up and
running STARLINK I should point out that they are not the official
views of that project, from which I have now retired. This talk
is collected under the following headings:

   (1) The need and importance of D.A.F.'s.
   (2) What astronomers ideally want.
   (3) A very brief survey of what is available now.
   (4) Some of the main deficiencies and problems with today's
       systems.
   (5) The future.

B. THE NEED AND IMPORTANCE OF GOOD DATA ANALYSIS FACILITIES

There are certain bodily functions about which it is not fashionable to talk, although they are vital to our health and survival.
I sometimes think data analysis enjoys the same low reputation among
astronomers. It is not a fashionable subject. Look around this hall
and you won't see the directors of large observatories. I believe
the reason for this is that most directors have students or assistants who do their data analysis for them, but not their observing.
They will therefore go to hear endless repetitive talks about
telescope design but not systems design. They may be mistaken.

When we are students we think of the observational astronomer as spending his nights gazing at the heavens and his days thinking Great Thoughts about the Universe. When we become observers we find that the vast majority of our time is actually spent in the fiddling details of data analysis and calibration. If directors did a little operational research on the way that their astronomers spend their days, data analysis might become more fashionable and more time might then become available for Great Thoughts. Though I shall be preaching to the converted here it might be as well to summarise clearly why good data analysis facilities are so crucial.

(a) Recent instrumental developments have turned astronomers from a data-starved into a data-gorged community. It's not so much that we have many more pictures of the Universe, though the data-flood is enormous (see Table 1), but that these pictures now come to us in a linear, digital form, and that they have been taken through new windows on the Universe. They deserve and require the sophisticated analyses which can reveal so much more these days than an old fashioned glance at the data. Note in particular the enormous flood of data CCD's and plate measuring machines like the APM and COSMOS will confront us with. A CCD fitted to quite a modest telescope can easily produce 150 astrophysically useful frames a week, as well as 100 calibrations. Without sophisticated D.A.F.'s the money spent on building the satellites and telescopes may partially go to waste.

Table 1. Main Sources of Astronomical Data.

Instrument.	Pixels/Frame	Bits/Pixel	Frames/day	Bits/day
Optical photographic.	$(10^4)^2$	8	10	8 G-bits
Optical TV/CCD	$(512)^2$	16	10	40 M
VLA	$(1000)^2$	16	15	200 M
Optical Spectros.	10^4	10	10	1 M
Imaging X-ray satellite.	$>(1000)^2$	16	10	>160 M
Satellite Planetary Camera.	$>(512)^2$	16	channel capacity	~ 1 G

(b) If, as I have suggested, astronomer-time is at a premium, then it can no longer be wasted.

(c) A good scientist needs to play with his data, to indulge in the "I wonder if..." syndrome, easily and quickly. If not he may remain unaware of some of the systematic errors, be blind to the most significant results, or be forced into approximations which disguise the truth. For instance, lacking a computer in 1952, Thornton Page (1952) was forced to reduce his classical survey of double galaxies using the simplifying assumption of circular orbits. For twenty years thereafter responsible astronomers found it hard to believe in large amounts of missing mass in spiral galaxies. Repeating this work, but with a computer, which allowed him to play with the assumptions, Turner could show that Page's assumptions lead to masses that are too low by at least an order of magnitude.

(d) Astronomy, because it is observational rather than experimental, is above all a statistical subject. It is often impossible to disentangle a significant law without first analysing the data from a very large number of objects. Table 2 is a crude attempt to estimate the number of separate objects N(obs) that had to be studied before my capriciously chosen list of landmark discoveries were made. It must be easy to study data statistically, and to do it objectively using modern statistical techniques. For instance we have been pleasantly surprised to find how effective multivariate analysis is in turning up unseen and unsuspected correlations between the properties of a large number of elliptical galaxies (Sparks et al 1984). Without sophisticated D.A.F.'s astronomers will be unwilling or unable to undertake and properly analyse the large statistical surveys that are necessary.

Table 2. Landmark Discoveries.

Discovery:	1	1-10	N(obs) 10-100	>100	V.Large.
1. Our galaxy as an island					x
2. Distance to stars	x			x	
3. H-R diagram (stellar structure)				x	
4. Cepheid Period -Lum.Reln.				x	
5. Size of our galaxy				x	
6. Distance to galaxies				x	
7. Expansion of the Universe			x		
8. Rotation of our galaxy				x	
9. Measurement of Ho				x	
10. Interstellar obscuration					x
11. Stellar luminosity function					x
12. Missing mass in clusters				x	
13. Evolution of radio sources					x
14. Discovery of Q.S.O.'s	x				
15. Distance of Q.S.O.'s				x	
16. Galaxy Correlation Function					x
17. H-Rdiag.(Stellar evolution)					x
18. Flat galaxy-rotation curves			x		

(e) Astronomy can be a highly competitive subject, particularly on large telescopes of various kinds. Astronomers with inefficient D.A.F.'s will rarely be in a position to reduce their data in a competitive timescale. Serious competitors for time on world-class instruments like the Space Telescope and the V.L.A. will find sophisticated D.A.F.'s a sine-qua-non.

C. THE FACILITIES ASTRONOMERS WOULD LIKE TO HAVE

(i) The astronomer wants <u>flexibility</u> and <u>informality</u> in the reduction process. In practice this means a <u>local</u> terminal which doesn't require much booking ahead. Centralized national facilities or systems provided locally at a remote telescope will properly serve only those living on site. Since the data-transfer rates required for interactive image-processing certainly ought to run at >1 Mbit/sec (Disney 1982) the local terminal presently requires a local mega-mini with a 1 MIPS performance and something like a G-Byte of fast store. Though such a system might presently cost 0.5M$ I am told that within 5 years time this will fall to 50k$. At the same time plans are going ahead for very high speed transmission networks based on optical fibres and satellite links. Clearly, 10 years from now we will have all the <u>hardware</u> we need to provide a superb image processing capability.

(ii) The astronomer may wish to interact with his data on a more or less real time basis because (a) one needs to use one's judgement and experience in dealing with marginal and complex data, often affected by spurious foreground and background sources; (b) the eye is much better than the machine at pattern recognition; (c) one wishes to inspect immediately the effect of various operations carried out on the data to see how effective they have been; and (d) a batch-processing algorithm may fail entirely to spot the unforeseen and perhaps most exciting discoveries.

While interaction <u>may</u> be essential it should not be forced on the astronomer when it <u>is</u> unnecessary, for it may then become time-consuming and extremely irritating. One needs to switch back and forth between interactive and batch modes on demand. Typically the observer may wish to spend a great deal of time interacting with the first few frames of a run, until his analysis is running smoothly, and then switch gradually over to batch mode.

Interaction will generally require a sophisticated image display system (I.D.S.) with computing power and storage of its own under command of the background C.P.U. The symbiosis between I.D.S. and background C.P.U. should be such as to facilitate batch/interactive switching. A good I.D.S. like the Sigma ARGS can now be had for less than 20k$ and such machines are bound to get cheaper.

(iii) Good <u>software support</u> is absolutely essential for if it is not all the rapidly improving hardware will be wasted to say nothing of the scarce talents of young astronomers who will become astronomical

programmers rather than vice-versa. Whereas we can rely on the
immense resources of the big I.D.P.firms to provide our hardware it
is up to us to organize our own specialized software and I see this
as our main task in the decade to come. The user requires easy-to-
use well documented application packages which cover many of the main
instruments and tasks that he has to perform. He wants to know what
is available, how to get hold of it, and what its limitations are.
He wants on-tap advice when things go wrong and he needs to communicate with other astronomers who have used the packages before. He
needs to modify packages to his own particular needs. And when he
writes new software himself it must be easy to translate it into a
form where it can be widely and easily used by his colleagues. Above
all he wants to work in a familiar environment or 'marketplace' where
it is easy to find and exchange ideas and where there is a minimum of
new tricks and jargon to be learned. And when he travels to a new
site or acquires a new computer he wants to find himself in familiar
territory.

I have no doubt that most of these goals are technically achievable provided, (a) that their importance is recognized at the highest
levels; (b) that strong management decisions are taken to see that
software proliferation is avoided and (c) that good communications
are provided between the various national and international centres
probably at this stage by 5 k-baud dial up lines and through various
international conventions on the transport of data.

(iv) Access to various archives and data bases will become increasingly vital. One only has to recognize how much valuable research
has been made possible by the availability of archives like I.U.E.,
The Palomar Sky Survey, the 3C Catalogue and the Reference Catalogue
of Bright Galaxies, to recognize how much more will be possible when
at the touch of the keyboard one can call up say all the stored data
on an individual QSO or on a class of X-ray binaries. To do the job
properly will cost a great deal of effort and money but it is my
impression that this will pay off handsomely. But it is not easy to
see who will take the initiative here.

The astronomers' requirements for D.A.F's do not differ in any
fundamental way from those of other tight-knit scientific communities
which operate internationally, for instance geophysicists, crystal-
lographers, radiographers, particle and atmospheric physicists,
oceanographers and remote sensing scientists.

D.   EXISTING DATA ANALYSIS FACILITIES

(a) Existing University main-frames seldom have either the resources
for greedy interactive users or the sophisticated I.D.S.'s needed,
but they can do batch processing cheaply.
(b) Much useful work has been done on the <u>departmental mini</u>, but for
financial reasons it is usually oldish (16-bit), underpowered, and
weak on the I.D.S.side. The software is usually locally specific,

and the installation will suck up an enormous local effort to keep it going. The falling costs of 32-bit machines and I.D.S.'s offer succour here in the near future, provided software can be shared widely.

(c) At national observatory sites or headquarters, for instance the VLA,KPNO,ESO, the AAO etc. one usually finds lavish up-to-date D.A.Fs. The hardware will be excellent but too expensive for wider use. The software is good, but usually custom-built in a hurry and and rapidly changing with the instruments it is designed to serve. The system is not usually designed to be portable and one generally finds that it is highly dependent on a local computer 'king' or 'kings'.

The outside user will generally find such systems excellent for removing the instrumental signature, less good for deep analysis, and often overbooked.

(d) Interactive computing networks like STARLINK and ASTRONET (Disney and Wallace 1982) are beginning to spread. These consist of several mega-minis from the same manufacturing range joined by slow-speed (< 10k bits/sec) links for the exchange and coordination of software.

Judging from STARLINK, such a system does indeed result in software sharing. Because the whole system will be highly visible, with a larger user community, it is possible to buy good equipment and attract excellent staff. The links provide very useful communication channels between astronomers for many purposes beyond software. The homogeneity of the system makes it possible to adopt and enforce network-wide software and data-handling conventions.

On the other hand such a network may become large enough to become insular and so may be tempted into grandiloquent software enterprises. It is often difficult to find someone available to give you help when you urgently need it. It turns out to be amazingly difficult at times to get a sensible degree of cooperation between nodes. And it is all too easy to oversell the system. As a result financial bodies may be reluctant to fund the necessary people as well as the machines, whilst the astronomers, instead of rolling up their sleeves may sit back and wait for the software to arrive from elsewhere.

Nevertheless such networks provide an exhilerating standard of data analysis at the national level from which users would be reluctant to retreat.

(e) Those communities which either cannot afford or do not wish to set up such networks may have national plans. For instance the Groth subcommittee (1983) have recommended in the U.S.: (i) 4 new VAX-equivalent systems be funded each year, each provided with 1 person, the machines to be replaced every 6 years; (ii) that the National Centres take a lead in distributing and documenting software and (iii) that software be machine-independent or portable.

Such a strategy, if it can be executed, posesses the great virtue of machine independence, which I feel will be important with the hardware developments to come. But keeping momentum up in hard financial times will be difficult.

E. OUR MAIN PROBLEMS AT PRESENT

(a) <u>Lack</u> of coordinated effort and central <u>planning</u>. Because data analysis is unromantic and because it does not require machines or sums of heroic size, each group tends to feel it can settle its problems de novo. The problem here is making visible the hidden costs, both in terms of money and opportunities lost, of such piecemeal cowboy development.

(b) <u>Differences</u> in <u>hardware</u>, which is ever improving, make people despair of cooperating. This is especially true at the moment with I.D.S.'s. If indeed the goal of machine-independent software is not a chimera then we must aim hard for it. If that proves impossible, most people will be tempted to remain with a single manufacturer - which in the astronomical case is DEC.

(c) <u>Measuring performance.</u> With a telescope or instrument it is usually possible to define and measure the performance, to identify the weaknesses and usually improve the situation. But a D.A.F. is an open-ended project; its final potential is albeit impossible to quantify and the evaluation of its performance tends to be subjective. This makes management very difficult.

(d) Providing D.A.F.'s means <u>funding people</u> as well as machines, and this can be very difficult. Funding agencies naturally don't like long term personnel commitments while the salaries we can offer do not compare with industry. Certainly in the U.K. this is a very grave problem.

(e) Making it <u>easier to borrow</u> software than to write it is not an easy task. A scientists natural inclination when faced with a difficulty is very often to get on and solve it for himself. To him software is usually only part of a much larger enterprise and he may feel uneasy about relying on someone else's efforts. He will pay for a piece of hardware off the shelf because its specification is exact, its performance is guaranteed, and the manufacturer will send someone along to mend it if it goes wrong. He will be rightly suspicious of a software package, even if it is free, whose specification is ever-changing, whose validation is unclear, whose final delivery date is not guaranteed and whose support is nominal. Moreover compulsive computer programming can become an all but incurable disease.

(f) Being seduced into <u>grandiloquent software projects</u>. It is the easiest and most delightful past-time to plan and outline software castles in the air. It is so easy to assure ourselves that whereas everyone else has got it wrong, we can do the job properly ourselves. However, outlining brilliant plans is very different from writing and testing hard lines of code and submitting them to the public's criticism. I bet we'll hear of some more grand schemes at this meeting.

We have learned our own hard lesson from the STARLINK software "Environment", which still isn't ready 4 years and 100,000 lines of code after we started. Based on that experience I <u>think</u> I would generally recommend:
(i) Sticking to commercially available operating systems, with all their limitations for astronomy, like VMS and UNIX;
(ii) Providing specific application programs (expandable) within these systems and giving the best of them manpower support;
(iii) Funding futuristic software schemes as separate projects, not to be relied on in advance, but to be accepted only if and when they are proven and tested.

(g) Being overrun by '<u>computer nuts</u>', and in particular by software egotists who are stupid enough to believe that no-one but themselves can write a sensible program. My advice would be to put the management of D.A.F's securely in the hands of astronomers who have no interest in writing their names in software lights, sacking the prima-donnas and employing, wherever possible, only people trained in astronomy, preferably PhD's.

(h) Rewarding and supporting properly those people who are willing to write or share their software with the rest of us. I myself believe you need formal agreements to see that young people especially, get their names on papers which make extensive use of their software.

F. WHAT WE ARE CHIEFLY LACKING

(a) Some sort of <u>international committee</u> which meets occasionally to decide on, (i) the adoption of conventions - e.g. FITS, graphics packages, and high level command languages; (ii) the adoption of operating systems like UNIX/VMS etc., (iii) how to share out the effort of providing major software packages and subsystems e.g. S.T., I.U.E., archiving and (iv) the organization of international line-links and other communications. If only representatives from the big organizations like S.T.ScI., E.S.O., K.P.N.O., N.R.A.O., WESTERBORK, STARLINK and ASTRONET could get together quickly and informally on some of these questions I believe most of us would be content to follow their lead. We are less interested in a perfect scheme than in ensuring that convergence gradually takes place.
(b) A lack of <u>international line-links</u> between major astronomy centres.

Although many people were sceptical at the outset, we are now very conscious within STARLINK of the great importance of our low-speed (<10k bit/sec) line links between nodes. They contribute a nervous system which is used for all sorts of purposes including software exchange, up-dates, management, astronomer to astronomer mail, documentation, the notice of telescope and instrument changes, data exchange, advertisements, warnings and so on and so on. It would be unthinkable to operate without these links. They are as much as the soul as the nervous system of the network. They are not expensive to install or run and I am certain that once regular line-links are set up between the major international centres they will act as a forceful and indispensable catalyst for cooperation.

Once again somebody has to take the lead. Since the S.T. is likely to be the first major world astronomical instrument, I would urge the S.T.ScI to lead the way.

(c) A clear idea of the amount of <u>manpower support</u> needed for data analysis.

It is hard to convince the funding authorities of the support needed if they look on the D.A.F's simply as computer centres. In fact if the D.A.F. supplies astronomers with user-friendly software capable of dealing with data from a wide variety of sources it is taking over many of the roles traditionally filled by research assistants of various kinds. I have the feeling there needs to be a largish temporary hump in our manpower support, a hump which can die away if and when flexible machine-transportable software exists to carry out many of the mainline reduction tasks. In the meantime it would help if people made and <u>published</u> estimates of the manpower costs of their various software systems. This would help us to make realistic estimates of what we will need in future.

(d) A clear idea of the <u>degree of sophistication of our users</u>.

Decisions have to be made about the level of expertise we can presuppose in our users. Traditionally national centres have had to deal with users who may know little or nothing of the computer system in operation. Coping with such users can slow the whole system down and be very expensive, especially when the majority of users may be highly sophisticated. I believe this is a luxury we cannot generally afford and training programs must exist to provide a reasonable degree of expertise.

(e) Plans for proper <u>archives and data-bases</u>.

Optical disc technology will allow for rapid strides in this area, and of all subjects astronomy, being exploratory and statistical, could reap enormous benefits. Unique archives such as I.U.E., IRAS and S.T. could become major research tools provided they are practically and easily accessible. The costs of organizing all this could be considerable but astronomy will only be one of the

subject areas clamouring for such facilities. Once again I hope
that the S.T.Sci.Inst. and the S.T. European Coordinating Facility
will take a decisive initiative.

(f) Channels for scientific communication.

For two scientists to interact naturally with one another they
need to talk, to show each other rough diagrams and calculations,
and to exchange papers and occasionally pictures. Given a low-speed
(<10k bit/sec) link (including a voice line) between two observatories say, with an analogue T.V. + D/A converter at each end (vide
Groningen or the E.S.A.FITS system) it is entirely possible for full
scientific conversations to take place. I believe they will
revolutionize the way science - and a great deal else will be done
(e.g. business and management). We can either wait for the commercial product or get on with it ourselves now. There's no need to
wait.

(g) Insufficient emphasis on standard theoretical as well as
observational packages being mounted for general use. I have been
very impressed for instance with the astronomical use that can be
made of the 'Statistical Package for the Social Sciences' which is
now widely distributed. Programs for particle dynamics, stellar
atmospheres, abundance analysis, gas-dynamics, clustering analysis
and so on can and should be mounted and documented for general use
as part of the general progress. Once again the reward system has
to be worked out at the management level.

G. THE FUTURE

(1) So far as hardware is concerned we are in the hands of very
capable industrial suppliers. Since our needs are not unique in
hardware terms I see the relatively large market being provided with
wondrously cheap and powerful tools. I think we can probably assume
that by 1990 the average small department will be able to purchase
for less than 50,000 current dollars a D.A.F. of the type we need,
superior to anything on the market today at 10 x the price. With
such decentralized hardware the whole burden will fall upon the
provision, sharing and management of machine-independent software.
No commercial firm is going to produce that for us, and we have to
plan to see that what we have done already will not be wasted in
future.

(2) Optical discs will store 10^{12} bits each - and they are not the
end of the line. Such very cheap fast storage will, I hope, lead to
the provision of DATA bases and ARCHIVES on a scale and with an ease
unheard of before. But who is going to organize it all? I hope that
centres like Strasbourg or the S.T.Sc.I. will take the initiative.

(3) The greatest changes by far will however not come through

computing but through communications technology. Widely and cheaply
available communications networks of high bit-rate (1M bit/sec or
more) based either on optic fibre or satellite links (e.g.project
UNIVERSE between CERN and some of its international supporting
institutions at 1M bit/sec) will so revolutionize the ways we do
science that it is hard to be clairvoyant. I like Fig.1 which shows
the relationship between an astronomer and the main resources of his
working environment. The way astronomy is done in any epoch depends
very much on the speed and convenience of the various channels
between one resource and another. One only has to consider the impact
of the jet airliner on the subject over the past 20 years, to realize
this. If you consider how technology is going to sophisticate and
speed up many of the links in Fig.1. it is not hard to foresee that:
(a) Observing, even on ground-based telescopes will be done remotely
in many though not all cases (Disney et al 1983), and this will lead
to flexible scheduling.
(b) Face to face astronomer interactions which I believe are very
important for the science, will be augmented and partially superseded
by console-to-console communication. Electronic interactions,
collaborations and even seminars will be a regular possibility and I
foresee that most of the advantages presently enjoyed by large
institutes will spread to a wider community. Astronomy may then
become comparatively decentralized.

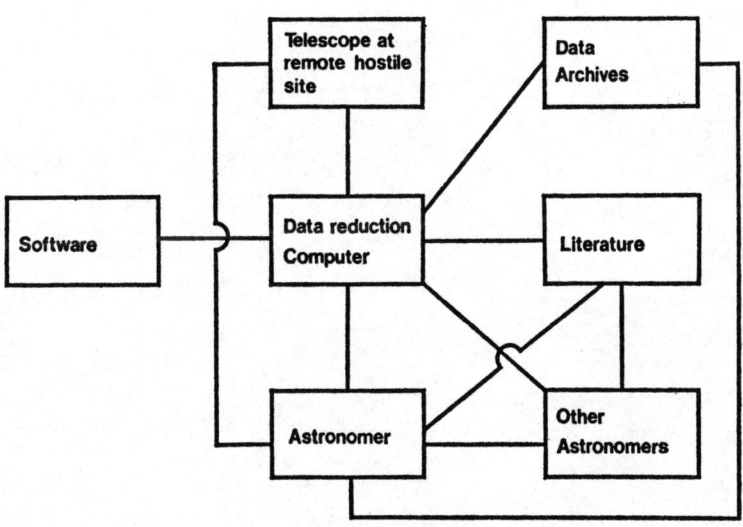

Fig.1.

(c) Data in different degrees of preparedness will whistle around
more freely. Journals will be gradually superseded by electronic
preprints to which the reader's comments will attach themselves,

providing an almost instant peer-group review.

The facilities to do all these things will come about not because of astronomers but because they will be of great use and profit to a much wider class of mankind. We must be ready to exploit them.

REFERENCES

Disney M.J., and Wallace P.T., 1982, Q.Jl.R.astr.Soc., 23, 485.
Disney M.J. et al, 1983, Review of Observing Time Allocations, SERC
    publication (to appear in Q.J. R.Astron.Soc.)
Groth et al, 1983, Astronomy and Astrophysics for the 1980's,
    National Academy Press, Vol.2., p.302.
Page T., 1952, Ap.J., 116, 63.
Sparks W.B. et al, 1984, Mon.N.R.Astr.Soc., 207, 445.
Turner E., 1976, Ap.J., 208, 304.

TIME ANALYSIS IN ASTRONOMY: TOOLS FOR PERIODICITY SEARCHES

Rosolino Buccheri and Bruno Sacco

IFCAI-CNR, Palermo, Italy

SUMMARY

The paper deals with periodicity searches in radio and gamma--ray Astronomy with special considerations for pulsar searches. The basic methodologies (Fast Fourier Transform, Rayleigh Test, Epoch Folding) are reviewed with the main objective to compare cost and sensitivities in different applications. One of the main results of the discussion is that FFT procedures are convenient in unbiased searches for periodicity in radio astronomy, while in spark chamber gamma-ray astronomy, where the measurements are spread over a long integration time, unbiased searches are very difficult with the existing computing facilities and analyses with a-priori knowledge on the period values to look for are better done using the Raleigh test with harmonics folding (Z_n test).

1. INTRODUCTION

The analysis of time series is an important subject in Astronomy because of the largo class of time-variable celestial phenomena of both periodical and irregular type.In the course of last twenty years in which Astrophysics has rapidly evolved in all its branches, it has become necessary to improve and specialize the existing time analysis techniques in order to reach always better sensitivities also in relation to their cost.

These two items,cost and sensitivity, will be specifically addressed in this paper because of the long integration times required to reach lower and lower visibility thresholds mainly in those classes of astronomical problems where time invariability is of periodical nature.

In the case of irregular variability the problem of the optimization of the ratio cost/sensitivity is mainly dependent on the shape and intensity of the studied signal and is more of statistical nature. We will not approach this problem here leaving it to further studies. We will instead deal with time periodical signals as detected, for example, in the pulsar searches both in radio astronomy and in high energy astrophysics. It will be seen that the optimization of the analysis requires fairly different approaches in the two cases because of the different data base obtainable in the radio and in the high energy domain.

The procedures here described and the considerations derived are based on well-known Fourier and Folding techniques and do not aim to be exhaustive of the great amount of work that has been made in this field. The computational methods used to evaluate the Fast Fourier Transform are described by Cochran et al. (1967) and Cooley et al. (1970) whereas the basic algorithms are given in Cooley and Tukey (1965). We will assume the reader confident with the basics of these techniques. We believe however that the effort of summarizing their main characteristics and their comparitive evaluation concerning cost and sensitivity in different applications is of some utility to whom has to deal with periodical phenomena in Astronomy.

2. SEARCH FOR PULSARS IN RADIOASTRONOMY

2.1. THE USE OF FFT

The experimental characteristics of radiotelescopes and the range of period values searched, going from the msec level to few seconds, make the use of FFT techniques highly advisable. In fact the detected radiation, in a given bandwidth Δf, is generally sampled at few msec intervals (S) such to make possible the detection of periodicities down to 2xS (for the Nyquist theorem). As a result of the Fourier Transformation to the frequency domain, the Power Spectrum Density (PSD) is obtained:

$$z(f_n) = \frac{1}{\Delta t} \left[\left(\sum_{i,1}^{N} x_i \cos 2\pi i f_n S \right)^2 + \left(\sum_{i,1}^{N} x_i \sin 2\pi i f_n S \right)^2 \right] \quad (1)$$

at all frequencies $f_n = n/\Delta t$

where N is the total number of sampling intervals,
   n = 1,2,...,N/2   and
   Δt = N·S is the total observing time.

The CPU time of the computation, as can be seen from eq. (1), is proportional to the square of the number N of samples

$$\text{cost} \propto N^2$$

The use of the FFT technique removes much of the redundancy in the calculation of repeated products and enables a large reduction in the calculation time.

The CPU time in this case is

$$\text{cost} \propto N \log N$$

which is especially convenient for large N.

The necessity to have a large number N of samples and therefore a long integration time Δt is particularly important in radio pulsar searches due to sensitivity considerations: a pulsar with period P and pulse width W (equal to the sampling intervals S) shows up in the Power Spectrum with an intensity x (signal-to-noise) given by

$$x = \frac{y}{P} \sqrt{\frac{W \Delta t}{2 + \frac{P^2}{W \Delta t y^2}}} \qquad (2)$$

where y (<< 1) is the input signal-to-noise and W << P (Burns and Clark 1969). It is clear from (2) that low intensity pulsars will be detectable above a certain threshold on x, after the FFT, only if the integration time is sufficiently long.

2.2. LIMITATION ON THE INTEGRATION TIME

A limitation on the integration time is however necessary for several reasons. One of them is constituted by the intrinsic variations of the pulsar period as, for example, the slowing down or the Doppler shift in the case of binary systems.

The maximum integration time necessary to avoid the shift of the signal from one PSD channel to the next is of the order of

$$\Delta t_{max} \sim \frac{P}{\sqrt{2 \dot{P}}} \qquad (3)$$

where $\dot{P}$ is the intrinsec change of the pulsar period.

This limitation, in the worse case of the msec binary radio pulsar (PSR1953+29) with 6.1 ms period and 52×10^{-15} s/s of apparent $\dot{P}$ (due to the binary motion over $\sim$ 60 days, see Boriakoff et al. 1984) is $\Delta t_{max} \sim 5$ hours.

Such an integration time is not possible to achieve for cost considerations. If one wants for instance, sample the data at 1 msec interval in order to be able to observe periodicities down to 2 ms, one would collect something like 20 million datapoints so requiring a computer memory too large for the existing computing systems and a CPU⌡ computing time not realistic; in the past pulsar surveys where more or less extended regions of the sky have been investingated, the integration time was less then 10 minutes with sampling intervals of 10 to 20 ms (requiring N $\sim$ 60000) thus, allowing fast analysis of many observations.

The discovery of millisecond pulsars (Backer et al., 1982, Boriakoff et al. 1983) render now necessary to use sampling intervals down to the ms level which increases the number of data points N necessary to achieve the same sensitivity. Considering that an FFT of 1 million data points takes about 25 minutes of CPU computing time at the VAX 11/750, it is understood that N $\sim$ 1 million is a limit, reachable only when having available a good computing system, especially in the case of a deep pulsar survey.

2.3. SENSITIVITY CONSIDERATIONS

The limitation on the observing time and therefore on the number of data points N imply, according to equation (2), a limitation on the minimum pulsar flux $\emptyset_{min}$ detectable, which is one of the main problems of a pulsar search. We can compute from (2) the valute of $\emptyset_{min}$. The input signal y to the FFT is given by (in terms of ratio signal-to-noise).

$$y = \frac{\varepsilon \emptyset}{T_G} \sqrt{\Delta f \cdot \tau} \qquad (4)$$

where T_G is the antenna temperature (°K) ε the antenna sensitivity (°K/Jy), $\emptyset$ the pulsar flux in mJy, τ a time constant roughly equal to the sample interval S and Δf is expressed in MHz.

Let's recall the measured power z, as given by equation 1 and call $\bar{z}$ the expected power in absence of signal; the "normalized" power $2z/\bar{z}$, being the sum of the square modulus of real and imaginary parts of a Fourier Transform which are in turn sums of a large number of random variables (Bendat and Pearsol, 1971), behaves for the Central Limit Theorem as a χ^2 with 2 dof. The equation 2 can therefore be written as

$$x = \frac{z - \bar{z}}{\sigma(z)} = \frac{y}{P}\sqrt{\frac{w\Delta t}{2 + \frac{P^2}{w\Delta t y^2}}} = \frac{2z/\bar{z} - 2}{2}$$

where $\sigma(z) = \bar{z}$. It follows

$$z = \frac{y}{P}\sqrt{\frac{w\Delta t}{2 + \frac{P^2}{w\Delta t y^2}}}\, \bar{z} + \bar{z}$$

If the final PSD is obtained by summing the signals of M subdivisions of the bandwidth Δf, we have

$$z_M = \sum_{i,1}^{M} z_i = \frac{My}{P}\sqrt{\frac{w\Delta t}{2 + \frac{P^2}{w\Delta t y^2}}}\, \bar{z} + M\bar{z}$$

where z_M is a random variable with mean $\mu = M\bar{z}$ and standard deviation $\sigma = \bar{z}\sqrt{M}$. Its significance, in terms on number of standard deviations is given by

$$n_\sigma = \frac{z_M - \bar{z}M}{\bar{z}\sqrt{M}} = \sqrt{M}\,\frac{y}{P}\sqrt{\frac{w\Delta t}{2 + \frac{P^2}{w\Delta t y^2}}}$$

Taking into account equation (4) and inverting, it results

$$\emptyset_A = \emptyset\,\frac{w}{P} = \frac{T_G}{\epsilon}\frac{1}{\sqrt{\Delta f \Delta t}}\sqrt{n_\sigma^2 + n_\sigma^2\sqrt{n_\sigma^2 + M}}\;\; \text{mJy}$$

where $\emptyset_A$ is the pulsar flux $\emptyset$ averaged over the period P. In the more general case when the observation is made up of n groups of indipendent observations each lasting $\Delta t = NS = N \times P_{min}/2$

$$\emptyset_{n\sigma} = \frac{T_G}{\epsilon}\frac{\sqrt{2}}{\sqrt{L\Delta f n\, N\, P_{min}}}\sqrt{n_\sigma^2 + n_\sigma\sqrt{n_\sigma^2 + Mn}}\;\; \text{mJy} \qquad (5)$$

where P_{min} is the minimum period detectable because of the sampling interval used and the term L has been included to take into account the possibility to add the signal of L different polarizations. If we put a threshold on the FFT output n_σ we get for $\emptyset_{n_\sigma}$ the visibility threshold $\emptyset_{min}$.

2.4. SENSITIVITY DEGRADATION DUE TO INTERSTELLAR DISPERSION

Eq. (5) is valid in the conditions of a small pulsar duty cycle (W<<P) and a train pulse equal to the sampling interval.

These conditions an not achieved when the interstellar dispersion is not negligible. In fact the smearing of the pulsar pulse due to interstellar dispersion is given by

$$\delta t \simeq 8.3 \times 10^3 \frac{\Delta f}{f^3} \text{DM s} \qquad (6)$$

where DM (in pc/cm^3) is the value of the pulsar dispersion measure (which is proportional to the pulsar distance from the observing site) and f is the central observing frequency in MHz. At large values of the pulsar distance (and therefore of DM) the pulsar pulse is broadened of an amount δt which, in some cases, can become a substantial fraction of the period. In this last cases (large DM and/or small P) the harmonics content of the pulse spectral component is decreased and the pulsar drops below detectability threshold. The subdivision of the whole bandwidth in M sub-bandwidths reduces the problem. In general, given a number of subdivisions M of the bandwidth Δf, the value of the ratio DM/P will be limited by

$$\frac{DM}{P} \lesssim M \frac{f^3}{\Delta f} \frac{\alpha \times 10^{-3}}{8.3} \qquad (7)$$

where $\alpha << 1$ in order that the pulse broadening be negligible with respect to the pulsar period.

From the catalog of Manchester and Taylor (1981) is appears that the weakest pulsar detected (PSR 1919+20) has a flux of 1 mJy (at 400 MHz), a dispersion measure DM $\sim$ 70 pc/cm^3 and a period of $\sim$ 761 ms. According to equation (7) the pulse broadening is equal to 0.0024 in the standard conditions of the Arecibo Radiotelescope facilities (M=30, f=430 MHz, Δf=7.5 MHz). From eq. 5 it derives that the pulsar can be detected (at a confidence level of 6 sigmas) using for example N = 64K data points sampled at 15 ms interval for a total observing time of 17 minutes (with the Arecibo 430 MHz receiver at 2 polarizations, with T_G = 120 °K and ε = 10 °K/Jy).

In the cases where the ratio DM/P is much higher than for PSR 1919+
+20 case, α can reach values comparable with unity and the pulsar
becomes undetectable. This is the case, in the standard Arecibo conditions, when DM/P $\sim$ 40000, which for P = 1 ms becomes DM $\sim$ 40 pc/
/cm^3. In the same conditions, to detect a pulsar with 1 ms period
down to 1 mJy flux (provided DM be less than 4 pc/cm^2, corresponding
in average to a maximum distance of 130 pc) we would need for example an observation time of 17 minutes and 30 FFT's each of 2 million data points sampled at 0.5 msec. The cost of such an analysis
would be exceedingly high and future surveys aiming to discover
new superfast pulsars have to address new improved techniques both
in the data acquisition systems and in the data analysis facilities
(memory and computing speed).

3. PERIODICITY SEARCHES IN GAMMA-RAY ASTRONOMY

3.1. UNBIASED SEARCHES

The experimental situation in gamma-ray Astronomy is very much
different than in Radioastronomy.

While in this last case the detected radiation, as obtained by
the receiving antenna, is a continuous function of the time giving
the antenna temperature and sampled out at a constant rate, the radiation in the gamma-ray energy domain is detected under form of
single photon arrival times at a very low counting rate and irregularly spaced. This is due to the low gamma-ray fluxes arriving at
the earth and to the particular physical process involved in the
detection by the present gamma-ray detectors. The low counting statistics require long integration times in order to collect a significant number of source photons. In a typical gamma-ray experiment
like the European satellite COS-B (Scarsi et al., 1977) an observing time of about one month was requested to collect few hundred
photons from the Crab pulsar out of a total of $\sim$ 1000 photons.
This implies serious limitations in the pulsar searches in gamma-
-ray astronomy.

In fact from eq. 3 it can be seen that the class of pulsars
that could be searched with a COS-B-like experiment is such that

$$P \gtrsim 3.7 \times 10^6 \sqrt{\dot{P}} \qquad (8)$$

i.e. a limited class of old pulsars.

A second limitation comes again from the long integration times
and relates to the necessity to avoid an exceedingly high number N of data points. In the COS-B case (observations of one month

time) a maximum value of N = 4M data points implies a sampling S ≃ 0.6 s and therefore a minimum detectable period of 1.2 s. This limitation, together with that expressed by eq. 8 are represented in fig. 1 showing the measured values of period and period derivative for the actual catalog of pulsars. It remains to see which sensitivity levels can be reached in gamma-ray astronomy for periodicities limited as in fig. 1.

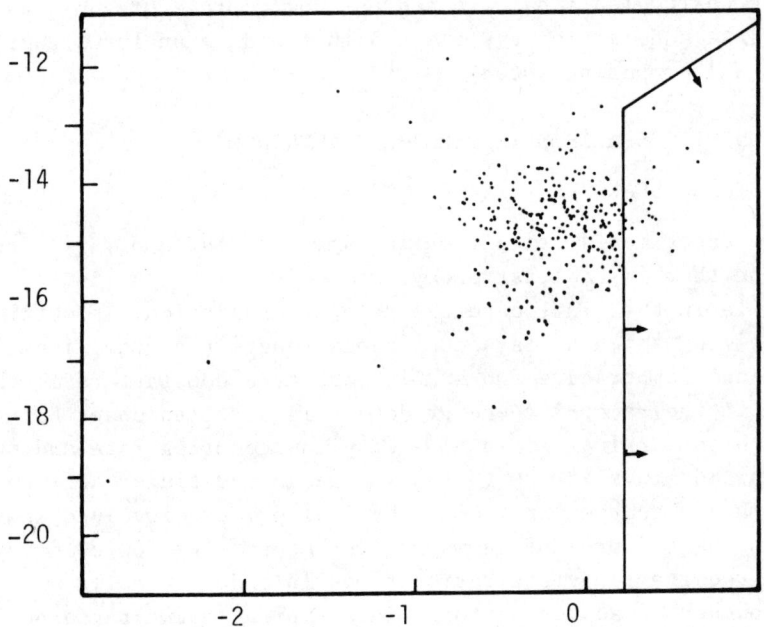

Fig. 1 - Log(P) versus log(P) for the pulsars of the Manchester and Taylor catalog

To do this let us go back to the PSD. In the case of single photons arrival times t_i eq. 1 can be written as

$$z(f_n) = \frac{1}{\Delta t}\left[(\sum_{i,1}^{N_T} \cos 2\pi t_i f_n)^2 + (\sum_{i,1}^{N_T} \sin 2\pi t_i f_n)^2\right] \quad (9)$$

where N_T is the total number of photons detected. Eq. 9 can be written as

$$z(f_n) = \frac{1}{\Delta t}(N_T + 2\sum_{i \neq j} \cos 2\pi f_n(t_i - t_j))$$

In the case of N_T arrival times uniformly distributed in the interval Δt, the variable z has an expected value $\bar{z} = N_T/\Delta t$ and a standard deviation $\sigma(z) = N_T/\Delta t$.

If the total number of photons N_T is due to the sum of N_P "pulsed" photons and N_B background photons ($N_T = N_P + N_B$)

$$z(f_n) = (N_T + \alpha N_P (N_P - 1))\frac{1}{\Delta t} \qquad (10)$$

where $\alpha < 1$ depends on the shape of the pulsed signal (in the extreme case when the "pulsed" photons have all the same phase $\alpha = 1$).

The statistical significance of z, as given by eq. 10 and expressed in no. of standard deviations from the expected average is

$$n_\sigma = \frac{z - \bar{z}}{\sigma(z)} = \alpha \frac{N_P(N_P-1)}{N_T} \qquad (11)$$

Re-examining the COS-B type example given above, we would need a value for n_σ greter than 20 in order to reach a probability level of the order of 10^{-9} in order to take into account the 2 Million values of f_n investigated in an unbiased FFT search. This corresponds to a reasonable probability for the effect to be real (99.85%) and requires, for eqs. 11

$$N_P \gtrsim \sqrt{\frac{20 \, N_T}{\alpha}}$$

In the typical COS-B case $N_T \sim 500$ and therefore the requested number of "pulsed" photons for an unbiased search is N_P about $\sim 100/\alpha \sim 140$ for an average value $\alpha = 0.5$.

This request is a very drastic one and can only be achieved (in the COS-B case) only for three of the 25 gamma-ray sources of the 2CG catalogue (Swanenburg et al., 1981). In the general case where the number of source photons is around 100 an unbiased search for periodicities, even with the limitation given by fig. 1, is not possible.

3.2 ORIENTED SEARCHES

3.2.1. THE RAYLEIGH TEST

The cost and sensitivity of a periodicity search in gamma-ray astronomy can be significantly improved if it is possible to limit the range of periods to search for making the analysis feasible.

This is because one can use eq. 9, instead of eq. 1, as a statistical variable to test (Rayleigh test; Mardia, 1972) for presence of periodicity effects on a range of K values of the period.

If the a priori knowledge on the values of the period to search is good enough (low K), the use of Raleigh test is preferable to the use of FFT both because of the lower cost ($KN_T < NlogN$) and the better sensitivity achievable due to the low trial periods K looked for. In the example of the last section, $NlogN \sim 6 \times 10^7$ while $KN_T \sim 5.10^4$, even looking for K = 100 values of the period. Moreover for 100 trials, a probability level of the order of 10^{-5} (9σ) is enough for considering statistically significant an effect and therefore sources down to $N_p = \sqrt{9\ N_T/\alpha}$ can be detected.

3.2.2. EPOCH FOLDING AND Z_n^2

In this section we want to make some further steps to the improvement of the analysis for periodicity searches in gamma-ray astronomy by describing two other tests used in literature: the epoch folding and the Z_n^2.

In the Epoch folding it is used the Pearson's χ^2 test for uniformity with the use of the statistical variable

$$X^2 = \sum_{i,1}^{m} \frac{(x_i - \overline{x})^2}{\overline{x}} \qquad (12)$$

here the total integration time Δt is subdivided into M intervals, each of lenght P (the period searched), which are folded together and the resulting phase interval is in turn subdivided into m phase bins. x_i is the number of photon arrival times falling into the i-th bin and $\overline{x}$ is the average of the x_i's. In the general case of arrival times uniformly distributed in Δt with N_T large enough to satisfy the Central Limit Theorem conditions, the variable X^2 is described by a χ^2 with m-1 degrees of freedom.

The convenience of using the Epoch folding with respect to the Rayleigh test (see for example Leahy et al., 1983) base on the fact that the Rayleigh test does not use the power content of the higher harmonics while the Epoch folding does; this is especially important in the case where the shape of the function f(t) describing the pulsation effect results from a sum of non-negligible harmonic components other than the fundamental with period P. This can easily be seen by recalling that the significance of a pulsation effect, expressed in number of standard deviations from the expected average in case of noise only, is given by eq. 11.

In the case of a very narrow pulse (in which the content of harmonics higher than the fundamental is relevant) α tends to unity for the Rayleigh test (see sec. 3.2.2) while in the case of the Epoch folding $\alpha = (1-w)/w > 1$ (see for ex Buccheri et al., 1977) where w is the pulsar duty cycle. In these cases the significance of the effect is greater with the Pearson's test.

The Epoch Folding has two disadvantages:
) reduces the time resolution of each event to the width of the histogram bin. All information within the bin is lost thus decreasing the significance of the pulsed structure.

2) In the case of low counting statistics, where the Central Limit Theorem ceases to be valid, the variance of the probability distribution of the X^2 variable is greater than that of the corresponding χ^2. The use of this last distribution can produce in these cases an overestimate of the statistical significance of effects with high values of X^2 (Gerardi et al., 1982).

This difficulties are overruled by the use of the statistical variable Z_n^2, defined by

$$Z_n = \frac{2}{N_T} \sum_{K,1}^{n} \left(\sum_{j,1}^{N_T} \cos 2\pi K\theta_j \right) + \left(\sum_{j,1}^{N_T} \sin 2\pi K\theta_j \right)$$

where $\theta_j = t_j/P$ or
including, if known, the derivative $\dot{P}$ of the period

$$\theta_j = t_j/P - \dot{P}t_j/P$$

and K is the Kth harmonics of the fundamental frequency $f = 1/P$.

The variable Z_n is distributed as the χ^2_{2n} up to reasonably high values of N_T.

Its convenience with respect to the Epoch Folding is strengthen also by the possibility to choose the harmonics to use in the search, in function of the necessity. In the gamma-ray pulsar search made by Buccheri et al., 1983, the number n was put to 2 in order to select pulsar shapes similar to the shape of the well known gamma-ray pulsars PSR0531+21 and PSR0833-45.

REFERENCES

- Becker D.C., Kulkarni S.R., Heiles C., Davis M.M. Gass W.M.
Nature 300, 615, 1982

- Bendat J.S., Pearsol A.G.
"Random data: analysis and measurement procedures"
Wiley Interscience, J.Wiley and sons, 1971

- Boriakoff V., Buccheri R., Fauci F.
Nature, 304, 417, 1983

- Buccheri et al.
Proc. 12th ESLAB Symp. "Recent Advances in gamma-ray Astronomy"
ESA SP-124, p.309, 1977

- Buccheri R. et al.
Astron.Astrophys., 128, 245, 1983

- Boriakoff V., Buccheri R., Fauci F., Turner K., Davies M.
Proc.of the "Millisecond Pulsar Meeting" Green Bank, June, 1984

- Burns V.R., Clark B.B.
Astron. Astrophys. 2, 280, 1969

- Cochran W.T. et al.
IEEE Trans.Audio Electroacustics. Special Issue on Fast Transform and its application to Digital Filtering and Spectral Analysis.
AV-15, 2,45, 1967

- Cooley J.W., Lewis P.A.W., Welch P.D.
J.Sound Vib. (12(3), 315, 1970

- Cooley J.W., Tukey J.W.
Math. Comp., 19, 297, 1965

- Gerardi G., Buccheri R., Sacco B.
COMSTAT 82, p. 111, 1982

- Leahy D.A. et al.
Ap.J., 272, 256, 1983

- Manchester R.N., Taylor J.H.
Astron. J., 86(12) 1952, 1981

- Mardia K.V.
"Statistics of directional data", New York, Academic Press, 1972

- Scarsi L. et al.
Proc. 12th ESLAB Symp. "Recent Advances in Gamma-Ray Astronomy"
ESA SP-124, p.3, 1977

- Swanenburg B.N. et al.
Ap. J., 243, L69, 1981

CLUSTER ANALYSIS BY MIXTURE IDENTIFICATION

Pierre A. Devijver

Philips Research Laboratory
Ave. Em. Van Becelaere 2, Box 8
Brussels,   Belgium

1. Introduction

Clusters analysis is frequently defined as the problem of partitioning a collection of objects into groups of similar objects according to some numerical measure of similarity. A wide variety of methods for doing this have been available for quite some time. The field has been—and remains—very well documented in the open literature [see, e.g., the books by Anderberg (1973), Diday (1979), Jambu (1978), Jambu and Lebeau (1983), Sokal and Sneath (1963), and the surveys by Diday and Simon (1976), Duda and Hart (1973), Redner and Walker (1984), to cite a few]. However, permanent emergence of new technical requirements as well as recognition of the limitations of existing techniques continue fostering intensive research activity worldwide which is mirrored by a permanent flow of publications proposing new ideas and algorithms.

In this paper, attention is focused on a well-defined family of statistical, clustering techniques in which similar objects in the data are assumed to constitute a random sample drawn from some unknown probability distribution. In other words, being given data drawn from a *mixture* of statistical distributions, we shall address the problem of *indentifying* the distributions the mixture is made of, hence the term "mixture identification".

This paper has been written with two kinds of readers in mind. Accordingly, our motivation for selecting to discuss this particular approach to cluster analysis is twofold. In the first place, it is our belief that much of the literature can lull those who do not have an immediate involvement with the theory into a false sense of security, while we know, in fact, very little about the behavior of clustering algorithms. With the newcomer in mind, we define mixture indentification as a problem in the realm of constrained optimization and every effort is made to lay bare the numerous obstacles hindering the route to the otpimal solution. In the second place, we would like to call the attention of data analysts to some techniques which have recently become quite popular and successful in engineering circles concerned with automatic speech recognition. These techniques, based on the so-called hidden Markov models (see Sec. 5), are of particular interest in that they permit to recover *contextual* information inherent in the data.

2. Identifiability of Finite Mixtures

We assume that we have a sample $\mathcal{X}_n = X_1, \cdots, X_n$ that
i) comes from c classes (components, clusters) $\omega_1, \cdots, \omega_c$ with
ii) a priori probability (or *mixing parameters*) $P(\omega_i)$, $i = 1, \cdots, c$, and
iii) class conditional densities (or *component densities*) $p(X \mid \omega_i, \theta_i)$, $i = 1, \cdots, c$, of known functional form (e.g., Gaussian), where θ_i is a parameter vector (e.g., mean vector and variance-covariance matrix).

We further assume that the mixing parameters $\{P(\omega_i)\}_{i=1}^c$, the parameter vectors $\{\theta_i\}_{i=1}^c$, and perhaps even c, the number of components, are unknown to us and we address the problem of using the samples from the mixture density $p(X \mid \Theta) = \sum_{i=1}^c P(\omega_i) p(X \mid \omega_i, \theta_i)$, where $\Theta = (\theta_1, \cdots, \theta_c)$, for the purpose of estimating all the unknowns.

The analysis of necessary and sufficient conditions for the existence of a solution to our problem has led statisticians to the concept of *identifiability*. Following Yakovitz (1970), let $\mathcal{M} = \{p(X \mid \omega, \theta)\}$ be a family of (possibly multidimensional) probability distributions parameterized by θ (whose range is some known parameter space). Let $\mathcal{P}$ be any probability distribution, defined on the parameter space, and such that $\mathcal{P}$ assigns probability 1 to finitely many mass points $\{\theta_1, \cdots, \theta_c\}$, $P(\omega_i)$ being the mass at θ_i, $i = 1, \cdots, c$. $p(X \mid \Theta)$ can be interpreted as a mapping $Q : \mathcal{P} \mapsto H(X) = p(X \mid \theta)$, i.e.,

$$Q(\mathcal{P}) = \sum_{i=1}^c P(\omega_i) p(X \mid \omega_i, \theta_i) = H(X).$$

The set of all finite mixtures on $\mathcal{M}$ is said to be identifiable if Q is one-to-one, i.e., $(\mathcal{P}_1 \neq \mathcal{P}_2) \Longrightarrow Q(\mathcal{P}_1) \neq Q(\mathcal{P}_2)$.

Duda and Hart (1973) provide a simple example of a mixture of two discrete distributions with known mixing parameters which fails to be identifiable, namely $P(X \mid \Theta) = (1/2)[\theta_1^X(1-\theta_1)^{1-X} + \theta_2^X(1-\theta_2)^{1-X}]$ where $X = 0, 1$. Readily, $P(X \mid \Theta) = \frac{1}{2}(\theta_1 + \theta_2)$ if $X = 1$, and $P(X \mid \Theta) = 1 - \frac{1}{2}(\theta_1 + \theta_2)$ if $X = 0$. In such a situation, the most we can do is to estimate $(\theta_1 + \theta_2)$, but there is no way to estimate θ_1 and θ_2 separately. The mixture is not identifiable.

Several families are known to be identifiable, e.g., mixtures of univariate Poisson and Gamma distributions [Teicher (1961), (1963)], multivariate normal and exponential distributions as well as unions of these families [Yakovitz and Spragins (1968)], the family of ν-products of any identifiable one-dimensional family [Teicher (1967)], as well as some other families less frequently encountered in practical applications [Yakovitz (1970)]. In what follows, we shall assume that the densities we are talking about are identifiable.

3. Identification by Maximum Likelihood Estimation

Let us assume that our sample $\mathcal{X}_n$ comes from a *known* number c of classes. (See Sec. 6 for the estimation of the number of components in a mixture of multivariate normal densities.) By definition, the likelihood of the observed sample $\mathcal{X}_n$ is given by

$$p(\mathcal{X}_n \mid \Theta, P) = \prod_{j=1}^{n} p(X_j \mid \Theta, P),$$

where $p(X_j \mid \Theta, P) = \sum_{i=1}^{c} P(\omega_i) p(X_j \mid \omega_i, \theta_i)$. Let us further assume that *i)* $p(\mathcal{X}_n \mid \Theta, P)$ is a differentiable function of Θ, *ii)* elements of θ_i and θ_j are functionally independent if $i \neq j$, and *iii)* $\hat{P}(\omega_i) \neq 0$ for any i, where $\hat{P}(\omega_i)$ is the maximum likelihood estimate of $P(\omega_i)$. Then, it is not difficult to show (by equating the gradient of the log-likelihood to zero) that, for any i, the maximum likelihood estimates $\hat{P}(\omega_i)$ and $\hat{\theta}_i$ for $P(\omega_i)$ and θ_i must satisfy [Duda and Hart (1973) p.193]

(1)
$$\hat{P}(\omega_i) = \frac{1}{n} \sum_{j=1}^{n} \hat{P}(\omega_i \mid X_j, \hat{\theta}_i)$$

$$\sum_{j=1}^{n} \hat{P}(\omega_i \mid X_j, \hat{\theta}_i) \nabla_{\theta_i} \log p(X_j \mid \omega_i, \hat{\theta}_i) = 0$$

$$\hat{P}(\omega_i \mid X_j, \hat{\theta}_i) = \frac{p(X_j \mid \omega_i, \hat{\theta}_i) \hat{P}(\omega_i)}{\sum_{k=1}^{c} p(X_j \mid \omega_k, \hat{\theta}_k) \hat{P}(\omega_k)}$$

where $\hat{P}(\omega_i \mid X_j, \hat{\theta}_i)$ is the maximum likelihood estimate of the a posteriori probability of sample X_j belonging to class (component, cluster) ω_i.

It is interesting to consider the specialization of these general results in the case where the component densities are multivariate normal (with mean vector μ and variance-covariance matrix Σ) $X \sim N(\mu_i, \Sigma_i)$ if $\omega(X) = \omega_i$ for $i = 1, \cdots, c$, that is

$$p(X \mid \omega_i, \theta_i) = (2\pi)^{-\frac{\nu}{2}} |\Sigma|^{-\frac{1}{2}} \exp\{-\frac{1}{2}(X - \mu_i)'\Sigma_i^{-1}(X - \mu_i)\},$$

where ν is the dimension of X, X' denotes the transpose of vector X, and $|\Sigma|$ is the determinant of Σ.

Our problem can be further specialized in various ways, e.g.,

α) known mixing parameters and (common) variance-covariance matrix, unknown mean vectors;

β) unknown mixing parameters, mean vectors, and *common* variance-covariance matrix;

γ) all parameters unknowns.

The solutions to cases α and β are obvious simplifications of the general solution to case γ which will be given hereafter.

An excellent account of the computation of the maximum likelihood estimate of μ_i in case α (for $c = 2$ and $\nu = 1$) has been given by Duda and Hart (1973). They convincingly demonstrate the existence of multiple solutions (with the values for the likelihood at the extrema failing to give a clear indication of which solution should be preferred) as well as the possibility of converging to a useless saddle-point solution. (Their example problem is also discussed by Diday and Simon (1976) using the "Dynamic Clusters Method") The problem of multiple solutions is, naturally, magnified in case β, [see Day (1969), Wolfe (1970), McLachlan (1982)] and the maximum likelihood estimation method breaks down in case γ because each data point gives rise to a singularity in the likelihood on the edge of the parameter space [Day (1969), McLachlan (1982)]. Nevertheless, Kiefer (1978) has shown that the likelihood equations (1) have a root (corresponding to a finite extremum) which is a consistent, asymptotically normal, and efficient estimator of $P(\omega_i)$ and θ_i, $i = 1, \ldots, c$. However, even if we restrict our attention to the largest of the finite local maxima of the likelihood function, we may fail to locate the root corresponding to the consistent estimator. In spite of all these problems, let us proceed with the derivation of the solution to the likelihood equations.

By some simple algebraic manipulations, it can be shown that the maximum likelihood estimators of $P(\omega_i)$ and θ_i must satisfy

$$\hat{P}(\omega_i) = \frac{1}{n} \sum_{j=1}^{n} \hat{P}(\omega_i \mid X_j, \hat{\theta}_i)$$

(2)
$$\hat{\mu}_i = \frac{\sum_{j=1}^{n} \hat{P}(\omega_i \mid X_j, \hat{\theta}_i) X_j}{\sum_{j=1}^{n} \hat{P}(\omega_i \mid X_j, \hat{\theta}_i)}$$

$$\hat{\Sigma}_i = \frac{\sum_{j=1}^n \hat{P}(\omega_i \mid X_j, \hat{\theta}_i)(X_j - \hat{\mu}_i)(X_j - \hat{\mu}_i)'}{\sum_{j=1}^n \hat{P}(\omega_i \mid X_j, \hat{\theta}_i)}$$

where

(3)
$$\hat{P}(\omega_i \mid X_j, \hat{\theta}_i) = \frac{p(X_j \mid \omega_i, \hat{\theta}_i)\hat{P}(\omega_i)}{\sum_{k=1}^c p(X_j \mid \omega_k, \hat{\theta}_k)\hat{P}(\omega_k)}$$
$$= \frac{\left|\hat{\Sigma}_i\right|^{-\frac{1}{2}} \exp\{-\frac{1}{2}(X_j - \hat{\mu}_i)'\hat{\Sigma}_i^{-1}(X_j - \hat{\mu}_i)\}\hat{P}(\omega_i)}{\sum_{k=1}^c \left|\hat{\Sigma}_k\right|^{-\frac{1}{2}} \exp\{-\frac{1}{2}(X_j - \hat{\mu}_k)'\hat{\Sigma}_k^{-1}(X_j - \hat{\mu}_k)\}\hat{P}(\omega_k)}.$$

These implicit equations can be solved iteratively, provided good initial estimates for $\hat{\mu}_i$ and $\hat{\Sigma}_i$, $i = 1, \cdots, c$ are at hand. They are used in (3) to obtain $\hat{P}(\omega_i \mid X_j, \hat{\theta}_i)$. In turn, these estimates are used in (2) to yield updated values for $\hat{\mu}_i$ and $\hat{\Sigma}_i$. This hill-climbing or gradient ascent procedure is iterated until convergence at a local maximum of the likelihood function. It is an empirical fact that this procedure may give "meaningful results". It may be regarded as the seed from which several hierarchical clustering procedures have germinated. The algorithm prescribed by (2–3) is a special instance of the so-called EM algorithm of Dempster, Laird and Rubin (1977). An exhaustive review of the EM algorithm is offered by Redner and Walker (1984) [together with 162 references].

In view of our above remarks about the existence of multiple local maxima of the likelihood function, it should be clear that the choice of initial values is the crucial step of the procedure. These could be "pre-estimated" from a large set of correctly labelled samples (provided such a set is at hand) [Duda and Hart (1973)]. Quant and Ramsey (1978) proposed to "pre-estimate" the initial values using the moment generating function of a mixture of normal distributions, in an attempt to locate the extremum corresponding to the consistent estimator. Bryant (1978) suggested taking the *decision directed* estimate (of the next section) as a starting value in the likelihood equations.

4. The Decision Directed Method

In the preceding section, we have ignored the fact that each sample $X_j \in \mathcal{X}_n$ was actually drawn from one of the subpopulations of the mixture. Therefore, in addition to estimating the mixture-parameters, one might attempt to allocate each sample to the subpopulation to which it belongs. To formalize this idea, we need some new notation : Let $\mathcal{X}^{(i)}$ be the subset of $\mathcal{X}_n$ consisting of samples coming from class ω_i, and $n_i = $ card $\mathcal{X}^{(i)}$. Then for a mixture of multivariate normal densities, the likelihood of the data is given by

$$\prod_{i=1}^{c} P(\omega_i)^{n_i} (2\pi)^{-\frac{n_i \nu}{2}} |\Sigma_i|^{-\frac{n_i}{2}} \exp\left\{-\frac{1}{2} \sum_{i=1}^{c} \sum_{X_j \in \mathcal{X}^{(i)}} (X_j - \mu_i)' \Sigma_i^{-1} (X_j - \mu_i)\right\}.$$

This likelihood is to be maximized over all possible values of the parameters, and all possible assignments of sample classes. However, there are

$$\frac{1}{c!} \sum_{i=1}^{c} \binom{c}{i} (-1)^{c-i} i^n \approx c^n/c!$$

ways of partitioning a set of n elements into c nonempty subsets thus, unless n is quite small, (in which case any hope to get meaningful results is doomed to failure, anyway) searching over all possible partitions is prohibitive. The alternative is to assign the sample X_j to class ω_i if

(4) $$\hat{P}(\omega_i \mid X_j, \hat{\theta}_i) = \max_k \hat{P}(\omega_k \mid X_j, \hat{\theta}_k).$$

Using an obvious notation, an equivalent formulation is

(5) $$\hat{\omega}(X_j) = \omega_i \text{ if } \hat{P}(\omega_i) p(X_j \mid \omega_i, \hat{\theta}_i) = \max_k \hat{P}(\omega_k) p(X_j \mid \omega_k, \hat{\theta}_k)$$

in which case X_j is assigned to $\hat{\mathcal{X}}^{(i)}$. The maximum likelihood estimates of the parameters in (5) are the standard ones, namely:

$$\hat{P}(\omega_i) = \hat{n}_i/n, \qquad (\hat{n}_i = \text{card } \hat{\mathcal{X}}^{(i)})$$

(6) $$\hat{\mu}_i = \frac{1}{\hat{n}_i} \sum_{X_j \in \hat{\mathcal{X}}^{(i)}} X_j,$$

$$\hat{\Sigma}_i = \frac{1}{\hat{n}_i} \sum_{X_j \in \hat{\mathcal{X}}^{(i)}} (X_j - \hat{\mu}_i)(X_j - \hat{\mu}_i)'.$$

It is fairly obvious that this procedure is a simplified version of the procedure of the preceding section and is obtained by making the approximation $\hat{P}(\omega_i \mid X_j, \hat{\theta}_i) = 1$ in (2–3) whenever (4) is satisfied. The optimization should be carried out under the restriction that $\hat{n}_i \geq \nu + 1$, $\forall i$, to avoid the degenerate case of infinite likelihood (resulting from the singularity of $\nu \times \nu$ sample variance-covariance matrices).

As for the procedure of the previous section, the decision directed procedure is applied iteratively. Starting from initial values for $\hat{\mu}_i$ and $\hat{\Sigma}_i$, the computation merely consists in assigning each sample to the nearest cluster center in terms of estimated Mahalanobis distance weighted by the logarithm of the estimated a priori probability. The estimates of the mixture parameters are then updated accordingly.

As pointed out by Symons (1981), it is intuitively clear that the decision directed procedure has a tendency to underestimate the variances and overestimate the distance between means whenever there is overlap between the mixture components. (In fact, Marriot (1975), and Bryant and Williamson (1978) showed that the method gives definitely inconsistent and asymptotically biased estimates quite generally.) Thus, if the primary interest is the estimation of a good allocation of the n samples to c groups (clusters), then the decision directed procedure is a viable alternative, while if the major concern is the estimation of the mixture parameters, then the procedure of the previous section should definitely be preferred.

Let W_i designate the within subpopulation sum of squares matrix, i.e., $W_i = \hat{n}_i \hat{\Sigma}_i$. Symons (1981) has shown that the optimal (decision directed) allocation is that partition which minimizes the criterion function

$$(7) \qquad \sum_{i=1}^{c} \hat{n}_i \log |W_i| - 2 \sum_{i=1}^{c} \hat{n}_i \log \hat{n}_i.$$

On the other hand, suppose the priors were ignored in the expression for the likelihood of the data [see McLachlan (1982)]. In this case the reader should have no difficulty to show that the corresponding decision directed procedure would be prescribed by

$$\hat{\omega}(X_j) = \omega_i \text{ if } p(X_j \mid \omega_i, \hat{\theta}_i) = \max_k p(X_j \mid \omega_k, \hat{\theta}_k)$$

with $\hat{\mu}_i$, $\hat{\Sigma}_i$ and $\hat{n}_i$ as in (6). The optimal allocation for the modified likelihood would then be that partition which minimizes the determinant of the pooled within subpopulation sum of squares matrix, a criterion proposed by Friedman and Rubin (1967), namely $W = \sum_{i=1}^{c} W_i$. McLachlan (1982) points out that the minimization of $|W|$ would appear to be rather robust against departure from normality. It does have a tendency to produce clusters of roughly equal size, (ignoring the priors is somehow equivalent to assuming they are all known to be equal to $1/c$) while the criterion of (7) would appear to gome some way to overcome this drawback.

For more reading on applications of decision directed procedures to pattern classification and signal detection, see e.g., Patrick et. al. (1970), and

Kazakos and Davisson (1980), and Sclove (1983) for an application to image segmentation.

5. Maximum Likelihood Identification in Hidden Markov Chains

Recent work on speech recognition has been concerned with the problem of identifying (normal) mixtures whose components labels form a finite-state, stationary Markov chain with unknown transition probabilities. In other words, part of the information in the data is assumed to be encoded in the temporal sequence of cluster identities. [In this problem domain, the mixture identification problem is sometimes going by the name of "vector quantization" see Gray (1984) and Rabiner et. al. (1983)] Thus, we shall assume that the underlying Markov process has c states $\omega_1, \cdots, \omega_c$, and, for the purpose of the present discussion, we shall make the simplifying assumption that the observations are drawn from a finite alphabet $Y_1, \cdots, Y_N$ (we shall briefly return to the case of normal mixtures at the end of this section). The Markov chain is specified in terms of an *initial* state distribution $P_i = P(\omega^1 = \omega_i)$—where for any $1 \leq \tau \leq T$, $\omega^\tau = \omega_i$ denotes that the process is in *state* ω_i at *time* τ—and a matrix of state transition probabilities $P_{ij} = P(\omega^{\tau+1} = \omega_j \mid \omega^\tau = \omega_i)$, $1 \leq i, j \leq c$, $\forall \tau, 1 \leq \tau \leq T-1$. The random process associated with the states is represented by c probability distributions $p_j(Y_k) = P(X^\tau = Y_k \mid \omega^\tau = \omega_j)$, $1 \leq j \leq c$, $1 \leq k \leq N$, $1 \leq \tau \leq T$, which mirror the assumption that, given the state at time τ, X^τ is independent of past and future states and observations.†

The likelihood L of a T-sequence of observations $X^1, \cdots, X^T$ is given by

$$L = \sum_{\omega^1, \cdots, \omega^T = \omega_1}^{\omega_c} P(\omega^1) p(X^1 \mid \omega^1) \prod_{\tau=1}^{T-1} P(\omega^{\tau+1} \mid \omega^\tau) p(X^\tau \mid \omega^\tau)$$

(8)
$$= \sum_{i_1, \cdots, i_T = 1}^{c} P_{i_1} p_{i_1}(X^1) \prod_{\tau=1}^{T-1} P_{i_\tau i_{\tau+1}} p_{i_{\tau+1}}(X^{\tau+1}).$$

It should be noted that, according to (8), the computation of L involves $2Tc^{T+1}$ multiplications. For most applications, this would prove intractable. Fortunately, Baum and his coworkers [Baum (1972), Baum et. al. (1970)] were able to devise a clever method for computing L with a work factor linear in T. Let $\mathcal{F}_\tau(i) \doteq P(X^1, \cdots, X^\tau, \omega^\tau = \omega_i)$. Thus, $\mathcal{F}_1(i) = P_i p_i(X^1)$, and $\mathcal{F}_\tau(i)$

---

† Every X^τ is a probabilistic function of an unseen Markov process (and is by no way Markov). Hence the term hidden Markov chains.

is given by the following recursive relationship for computing the "*Forward* probabilities"

$$\mathcal{F}_{\tau+1}(j) = \sum_{i=1}^{c} \mathcal{F}_\tau(i) P_{ij} p_j(X^{\tau+1}), \qquad 1 \leq \tau \leq T-1.$$

Likewise, let $\mathcal{B}_\tau(i) \doteq P(X^{\tau+1},\cdots,X^T \mid \omega^\tau = \omega_i)$, and $\mathcal{B}_T(j) = 1 \; \forall j$. Then, $\mathcal{B}_\tau(i)$ is given by the following recursive relationship for computing the "*Backward* probabilities"

$$\mathcal{B}_\tau(i) = \sum_{j=1}^{c} P_{ij} p_j(X^{\tau+1}) \mathcal{B}_{\tau+1}(j), \qquad T-1 \geq \tau \geq 1.$$

Now, the remarkable thing about these relationships is that

(9) $$L = \sum_{i=1}^{c} \mathcal{F}_\tau(i)\mathcal{B}_\tau(i) = \sum_{i=1}^{c}\sum_{j=1}^{c} \mathcal{F}_\tau(i) P_{ij} p_j(X^\tau) \mathcal{B}_{\tau+1}(j),$$

identically in τ. This shows that L can be computed with $4c^2 T$ multiplications.

With these preliminaries, let us return to our indentification problem of finding maximum likelihood estimates for the parameters $\{P_i, P_{ij}, p_j(Y_k)\}$, given the sample $X^1,\cdots,X^T$. This problem can be cast in the mould of constrained optimization, and the solution is obtained in terms of implicit equations

$$P_i = \frac{P_i \partial L/\partial P_i}{\sum_j P_j \partial L/\partial P_j}, \qquad P_{ij} = \frac{P_{ij}\partial L/\partial P_{ij}}{\sum_j P_{ij}\partial L/\partial P_{ij}},$$

(10) $$p_j(Y_k) = \frac{p_j(Y_k)\partial L/\partial p_j(Y_k)}{\sum_k p_j(Y_k)\partial L/\partial p_j(Y_k)}.$$

By substituting for the partials froms (9), we obtain equations whose probabilistic interpretation bears a close resemblance with that of (2), viz.,

(11) $$P_i = \frac{\mathcal{F}_1(i)\mathcal{B}_1(i)}{\sum_j \mathcal{F}_1(j)\mathcal{B}_1(j)} = \frac{P(X^1,\cdots,X^T,\omega^1=\omega_i)}{\sum_j P(X^1,\cdots,X^T,\omega^1=\omega_j)},$$

$$P_{ij} = \frac{\sum_\tau \mathcal{F}_\tau(i) P_{ij} p_j(X^{\tau+1}) \mathcal{B}_{\tau+1}(j)}{\sum_\tau \mathcal{F}_\tau(i)\mathcal{B}_\tau(i)} = \frac{\sum_\tau P(X^1,\cdots,X^T,\omega^\tau=\omega_i,\omega^{\tau+1}=\omega_j)}{\sum_\tau P(X^1,\cdots,X^T,\omega^\tau=\omega_i)},$$

$$p_j(Y_k) = \frac{\sum_{\tau \mid X^\tau=Y_k} \mathcal{F}_\tau(j)\mathcal{B}_\tau(j)}{\sum_\tau \mathcal{F}_\tau(j)\mathcal{B}_\tau(j)} = \frac{\sum_{\tau \mid X^\tau=Y_k} P(X^1,\cdots,X^\tau,\cdots,X^T,\omega^\tau=\omega_j)}{\sum_\tau P(X^1,\cdots,X^T,\omega^\tau=\omega_j)}.$$

It is essentially these equations which serve as a basis for iteratively updating initial parameter estimates. $\mathcal{F}$'s and $\mathcal{B}$'s are computed using the forward and backward recurrences respectively for the current values of the estimators. They are then used in (11) to compute updated values of the estimates. Again, this gradient ascent procedure is continued until convergence at a local maximum of the likelihood function. See Baum (1972) and Levinson et. al. (1983) for exhaustive discussions of convergence properties.

The analogy with the material in Sec. 3 becomes even more obvious when we return to mixtures of multivariate normal distributions with distinct variance-covariance matrices. The first equation in (11) remains essentially unchanged. Using an obvious notation, the next two ones become

$$\hat{P}_{ij} = \frac{\sum_\tau \hat{\mathcal{F}}_\tau(i)\hat{P}_{ij}p(X^{\tau+1}|\omega_j,\hat{\theta}_j)\hat{\mathcal{B}}_{\tau+1}(j)}{\sum_j \sum_\tau \hat{\mathcal{F}}_\tau(i)\hat{P}_{ij}p(X^{\tau+1}|\omega_j,\hat{\theta}_j)\hat{\mathcal{B}}_{\tau+1}(j)}$$

where $p(X^{\tau+1}|\omega_j,\hat{\theta}_j) \sim N(X^{\tau+1};\hat{\mu}_j,\hat{\Sigma}_j)$ with

(12) $$\hat{\mu}_j = \frac{\sum_\tau \hat{\mathcal{F}}_\tau(j)\hat{\mathcal{B}}_\tau(j)X^\tau}{\sum_\tau \hat{\mathcal{F}}_\tau(j)\hat{\mathcal{B}}_\tau(j)} = \frac{\sum_\tau \hat{P}(\omega^\tau = \omega_j|X^1,\cdots,X^T)X^T}{\sum_\tau \hat{P}(\omega^\tau = \omega_j|X^1,\cdots,X^T)}$$

and

$$\hat{\Sigma}_j = \frac{\sum_\tau \hat{\mathcal{F}}_\tau(j)\hat{\mathcal{B}}_\tau(j)(X^\tau - \hat{\mu}_j)(X^\tau - \hat{\mu}_j)'}{\sum_\tau \hat{\mathcal{F}}(j)\hat{\mathcal{B}}_\tau(j)}$$

with a similar interpretation. It should be noted that the rightmost part of (12) emphasizes the a posteriori (Bayesian) nature of this re-estimation scheme.

As with the standard maximum likelihood method of Sec. 3, convergence to the (finite) global maximum of the likelihood function is not guaranteed. A local extremum may be attained as well as a singular solution on the edge of the parameter space. See Liporace (1982) for further details. Baum et. al. (1970) have investigated the behavior of the algorithm for mixtures of univariate Poisson distributions on the non-negative integers, binomial distributions on the integers, and univariate normal and gamma densities. Liporace (1982) extended their results to multivariate normal and Cauchy densities. Experiments by Levinson et. al. (1983) have shown that skewness of the underlying Markov model is detrimental to convergence rate and that the choice of initial values is again quite critical. See Rabiner et. al. (1983) for an application to speaker-independent, isolated words recognition, and Jelinek and Mercer (1982) for an investigation of the sparse-data problem.

We have seen, in Sec. 4, that some simplification of the maximum likelihood method arose from attempting to make deterministic assignments of each sample to the class (clusters) to which it is currently assumed to belong. The very same idea can be pursued in the present context of hidden Markov chains. The technique consists in determining the sequence $\tilde{\omega}^1, \cdots, \tilde{\omega}^T$ that maximizes the joint probability $P(\omega^1, \cdots, \omega^T; X^1, \cdots, X^T)$ given the current values of the parameters, and to update the estimators according to the class (clusters) assignments prescribed by $\tilde{\omega}^1, \cdots, \tilde{\omega}^T$. Readily, this turns out to be a maximization problem over all possible T-sequences of states which can be solved quite efficently by the dynamic programming algorithm of Viterbi [Forney (1973)]. Space limitations do not allow us to give the method the attention it deserves, and we must content ourselves to refer the reader to the literature, see e.g., Jelinek et. al. (1983). Presently, we shall bring to an end our brief incursion into the province of hidden Markov models. The next section will address—among other things—the important problem of estimating the number of clusters that are actually present in a sample drawn from the basic model described in Sec. 2.

5. The Method of Moments

Since the pionneering work of K. Pearson it has been known that the decomposition of a mixture into its components can be performed on the basis of *moment equations*. Our presentation is based essentially on the recent work of Fukunaga and Flick (1983). [See also the interesting papers by Cooper and Cooper (1964), and Cooper (1967).] To illustrate the use of the method of moments, let us assume that we are dealing with a mixture consisting of an *unknown* number c of ν-dimensional normal densities with identical variance-covariance matrices, $\Sigma_i = \Sigma, \forall i$, (e.g., different signals corrupted by the same noise) and let us contemplate estimating c. We first assume that our sample is normalized with respect to its mean and variance-covariance matrix as follows:

$$(13) \quad E\{X\} = \sum_{i=1}^{c} P(\omega_i)\mu_i = 0, \quad \text{and} \quad E\{XX'\} = \sum_{i=1}^{c} P(\omega_i)S_i = I,$$

where E denotes expectation, S_i is the autocorrelation matrix of component density $p(X \mid \omega_i, \theta_i)$, viz., $S_i = \Sigma_i + \mu_i\mu_i'$, and I is the $\nu \times \nu$ identity matrix. No generality is lost since the normalization requires a simple translation and linear transformation of the coordinate system.

The kurtosis matrix (i.e., a matrix of fourth order moments), is given by

$$(14) \quad K = E\{(X'X)XX'\} - (\nu + 2)I$$

and for a distribution normalized by (13), K becomes

$$(15) \quad K = \sum_{i=1}^{c} P(\omega_i)[\text{tr}(\Sigma + \mu_i\mu_i') - (\nu + 2)]\mu_i\mu_i',$$

where $\text{tr}(\cdot)$ is the trace of matrix $(\cdot)$. This is a weighted sum of direct products (which are rank one matrices). Thus, if $c \leq \nu$, $\text{rank}(K) = c - 1$. We get $c - 1$ instead of c because, by (13), only $c - 1$ μ_i's are linearly independent. The kurtosis matrix can be estimated as

$$\hat{K} = \frac{1}{n}\sum_{j=1}^{n}(X'X)XX' - (\nu + 2)I.$$

Now, it is well known that higher order moments do not always measure what they are supposed to do. So, unless n is fairly large, $\hat{K}$ may be expected to be of full rank. However, if $c \leq \nu$, an eigenvalue analysis may be expected to reveal $\hat{c} - 1$ dominant eigenvalues indicating the presence of $\hat{c}$ clusters in the data.†

Let us now briefly address the problem of using moment equations for estimating the parameters of a Gaussian mixture for the case when it is known that $c = 2$ and $P(\omega_1) = P(\omega_2)$. [See Fukunaga and Flick (1983) for more details and more general cases.] For a bimodal distribution normalized by (13), we have $\mu_2 = -\mu_1$, and $S_2 = 2I - S_1$. Thus, all the parameters are completely determined by μ_1 and S_1. Let x_i, m_i, and s_{ij} designate the ith components of X and μ_1, and the ijth component of S_1 respectively. For a Gaussian mixture, it is known that third and higher unconditional moments can be expressed as functions of second and lower conditional ones. For instance, with $P(\omega_1) = P(\omega_2)$, we have

(16) $$t_i = E\{x_i^3\} = 3(s_{ii} - 1)m_i,$$

(17) $$u_i = E\{x_i^4\} = 3(s_{ii} - 1)^2 - 2m_i^4 + 3,$$

where, clearly, left hand sides are unconditional moments that can be estimated from the data while the right hand sides involve some of the desired conditional ones. From (16) and (17),

$$\pm t_i = 3(s_{ii} - 1)\left[\frac{3(s_{ii} - 1)^2 - (u_i - 3)}{2}\right]^{1/4}$$

This equation can be solved for s_{ii} for known (estimated) t_i and u_i. On the other hand, it is easy to show that the region containing allowed solutions of

---

† In the framework of the decision directed procedures of Sec. 4, Marriot (1971) has suggested taking $\hat{c}$ to be the number which minimizes $c^2|W|$ where $W = \sum_{i=1}^{\hat{c}} \hat{n}_i \hat{\Sigma}_i$. It is an empirical fact that this approach does in no way provide a more clear-cut estimate of the actual value of c.

s_{ii} is determined by $-(1-m_i^2) \le s_{ii} - 1 \le 1 - m_i^2$. Ordinarily, there are two solutions of s_{ii} for each $i = 1, \cdots, \nu$, hence 2^ν possible solutions for μ_1 and S_1. The selection of the best one (in a sense defined hereafter) can be combined with the estimation of the off-diagonal elements of S_1. Let

$$(18) \qquad \begin{aligned} v_{ij} &= E\{x_i^2 x_j\} = 2s_{ij}m_i + (s_{ii} - 1)m_j, \\ v_{ji} &= E\{x_j^2 x_i\} = 2s_{ji}m_j + (s_{jj} - 1)m_i, \end{aligned}$$

for $i \ne j$. In principle these two equations should be interchangeable since $s_{ij} = s_{ji}$. However, in practice, they fail to give identical results because estimates must serve as "knowns". The (heuristic) solution proposed by Fukunaga and Flick (1983) is to use a least square error approach. Let the error ϵ_{ij} be defined as

$$\begin{aligned} \epsilon_{ij} = &\{q_{ij} - 2[s_{ij}m_i + (s_{ii} - 1)m_j]\}^2 \\ &+ \{q_{ji} - 2[s_{ji}m_j + (s_{jj} - 1)m_i]\}^2. \end{aligned}$$

Then, it is a simple matter to show that ϵ_{ij} is minimized with respect to s_{ij} when

$$s_{ij} = \frac{[q_{ij} - (s_{ii} - 1)m_j]m_i + [q_{ji} - (s_{jj} - 1)m_i]m_j}{2(m_i^2 + m_j^2)}.$$

To determine the best solution, the accumulated error $\epsilon_0 = \sum_{i \ne j} \epsilon_{ij}$ is minimized over all 2^ν possible solutions.

In summary, with the above method, each dimension is first treated separately; the results are combined and multiple solutions are eliminated using the third order moments in (18).

6. Alternative Approaches

The unsupervised Bayesian learning method extend the maximum likelihood technique of Sec. 2. In this method, it is assumed that the parameter vector to be estimated is itself a random variable, and that our prior knowledge about this parameter vector can be embodied in an a priori distribution $p(\Theta)$. Unsupervised Bayesian learning originated in a (hardly accessible) report by Daly, and is very well discussed in Duda an Hart (1973). See also Young and Calvert (1974), Symmons (1981), and Makov and Smith (1977) for the estimation of mixing parameters using a simple quasi-Bayes approach.

Stochastic approximation was suggested by Young and Coraluppi (1970) for one-dimensional normal mixtures, and extended to the multivariate case by Mizoguchi and Shimura (1975) for bimodal normal mixtures with known variance-covariance matrices or known mean vectors. A fast stochastic approximation algorithm was proposed by Kazakos (1977) for the estimation of the mixing parameters. Stanat (1968) and Sammon (1968) used Fourrier transforms for the decomposition of mutlivariate normal and multivariate Bernoulli mixtures, while Postaire and Vasseur (1981), suggested an approach based on searching the mode of the mixture distribution.

References

Anderberg, M.R., (1973). *Clusters Analysis for Application*, New York: Academic Press.
Baum, L.E., T. Petrie, G. Soules, and **N. Weiss**, (1972). A maximization technique occurring in the statistical analysis of probabilistic functions of Markov chains. *Ann. Math. Statist.*, **41**, 164–171.
Baum, L.E., (1972). An inequality and associated maximization technique in statistical estimation for probabilistic functions of Markov processes. *Inequalities*, **3**, 1–8.
Bryant, P., (1978). Contributions to the discussion of the paper by R.E, Quandt and J.B. Ramsey. *J. Amer. Statist. Assoc.*, **73**, 748–749.
Bryant, P., and **J.A. Williamson**, (1978). Asymptotic behavior of classification maximum likelihood estimates. *Biometrika*, **65**, 273–281.
Cooper, D.B., and **P.W. Cooper**, (1964). Non supervised adaptive signal detection and pattern recognition. *Inform. Contr.*, **7**, 416–444.
Cooper, P.W., (1967). Some topics on nonsupervised adaptive detection for multivariate normal distributions. In *Computer and Information Sciences, II*, 123–146, J.T. Tou ed., New York: Academic Press.
Day, N.E., (1969). Estimating the components of a mixture of normal distributions. *Biometrika*, **56**, 463–474.
Dempster, A.P., N.M. Laird, and **D.B. Rubin**, (1977). Maximum likelihood estimation from incomplete data via the EM algorithm. *J. Royal Statist. Soc.* Ser. B, **39**, 1–38.
Diday, E., and **J–C. Simon**, (1976). Clustering analysis. In *Digital Pattern Recognition*, 47–94, K.S. Fu ed. Berlin: Springer Verlag.
Diday, E., (1979). *Optimisation en Classification Automatique*, Rocquencourt: INRIA.
Duda, R.O., and **P.E. Hart**, (1973). *Pattern Classification and Scene Analysis*, New York: Wiley.
Forney, G.D., Jr. (1973). The Viterbi Algorithm. *Proc. IEEE*, **61**, 268–278.
Friedman, H.P., and **J. Rubin**, (1967). On some invariant criterion for grouping. *J. Amer. Statist. Assoc.*, **62**, 1159–1178.
Fukunaga, K., and **T.E. Flick**, (1983). Estimation of the parameters of a Gaussian mixture using the method of moments. *IEEE Trans. Pattern Anal. Machine Intell.*, **PAMI-4**, 410–416.
Gray, R.M., (1984). Vector quantization. *IEEE ASSP Magazine*, **1**, 4–29.
Jambu, M., (1978). *Classification Automatique pour l'Analyse des Données*, Paris: Dunod.

Jambu, M., and M.-O. Lebeau, (1983). *Cluster Analysis and Data Analysis*, Amsterdam: North-Holland.

Jelinek, F., and R.L. Mercer, (1980). Interpolated estimation of Markov source parameters from sparse data. In *Pattern Recognition in Practice*, 381-397, E.S. Gelsema and L.N. Kanal, eds., Amsterdam: North-Holland.

Jelinek, F., R.L. Mercer, and L.R. Bahl, (1982). Continuous speech recognition: Statistical methods. In *Handbook of Statistics 2*, 549-573, P.R. Krishnaiah and L.N. Kanal, eds. Amsterdam: North-Holland.

Kazakos, D., (1977), Recursive estimation of prior probability using a mixture. *IEEE Trans. Inform. Theory*, IT-23, 203-211.

Kazakos, D., and L.D. Davisson, (1980). An improved decision directed detector. *IEEE Trans. Inform. Theory*, IT-26, 113-115.

Kiefer, N., (1978). Discrete parameter variation: efficient estimation of a switching regression model. *Econometrika*, 46, 427-434.

Levinson, S.E., L.R. Rabiner, and M.M. Sondhi, (1983). An introduction to the application of the theory of probabilistic functions of a Markov process to automatic speech recognition. *B.S.T.J.*, 62, 1035-1074.

Liporace, L.A., (1982). Maximum likelihood estimation for multivariate observations of Markov sources. *IEEE Trans. Inform. Theory*, IT-28, 729-734.

Makov U.E., and A.F.M. Smith, (1977). A quasi-Bayes unsupervised learning procedure for priors. *IEEE Trans. Inform. Theory*, IT-16, 761-764.

Marriot, F.H.C., (1971). Practical problems in a method of cluster analysis. *Biometrics*, 27, 501-514.

Marriot, F.H.C., (1975). Separating mixtures of normal distributions. *Biometrics*, 31, 767-769.

McLachlan, G.M., (1982). The classification and mixture maximum likelihood approaches to cluster analysis. In *Handbook of Statistics 2*, 199-208, P.R. Krishnaiah and L.N. Kanal, eds. Amsterdam: North-Holland.

Mizoguchi, R., and M. Shimura, (1975). An approach to unsupervised learning pattern classification. *IEEE Trans. Comput.*, C-24, 979-983.

Patrick, E.A., J.P. Costello, and F.C. Monds, (1970). Decision directed estimation of a two-class boundary. *IEEE Trans. Comput.*, C-19, 197-205.

Postaire J-G, and C.P.A. Vasseur, (1981). An approximate solution to normal mixture identification with application to unsupervised pattern classification. *IEEE Trans. Pattern Anal. Machine Intell.*, PAMI-3, 163-179.

Quandt, R.E., and J.B. Ramsey, (1978). Estimating mixtures of normal distributions and switching regressions. *J. Amer. Statist. Assoc.*, 73, 730-738.

Rabiner, L.R., S.E. Levinson, and M.M. Sondhi, (1982). On the application of vector quantization and hidden Markov models to speaker-independent, isolated word recognition. *B.S.T.J.*, 62, 1075-1105.

Redner, R.A., and H.F. Walker, (1984). Mixture densities, maximum likelihood and the EM algorithm. *Siam Review*, 26, 195-239.

Sammon, J.W., (1968). An adaptive technique for multiple signal detection and identification. In *Pattern Recognition*, 409-439, L.N. Kanal ed. Washington: Thomson Book Cy.

Sclove, S.L., (1983). Application of the conditional population-mixture model to image segmentation. *IEEE Trans. Pattern Anal. Machine Intell.*, **PAMI-5**, 428-433.

Scott, A.J., and M.J. Symons, (1971). Clustering methods based on likelihood ratio criteria. *Biometrics*, **27**, 387-398.

Sokal, R.R., and P.H.A. Sneath, (1963). *Principles of Numerical Taxonomy*, San Francisco: W.H. Freeman & Cy.

Stanat, D.F., (1968). Unsupervised learning of mixtures of probability functions. In *Pattern Recognition*, 357-389, L.N. Kanal ed. Washington: Thomson Book Cy.

Symons, M.J., (1981). Clustering criteria and multivariate normal mixtures. *Biometrics*, **37**, 35-43.

Teicher, H., (1961). Identifiability of mixtures. *Ann. Math. Stat.*, **32**, 244-248.

Teicher, H., (1963). Identifiability of finite mixtures. *Ann. Math. Stat.*, **34**, 1265-1269.

Teicher, H., (1967). Identifiability of mixtures of product measures. *Ann. Math. Stat.*, **38**, 1300-1302.

Wolfe, J.H., (1970). Pattern clustering by multivariate cluster analysis. *Multivariate Behavioral Research*, **5**, 329-350.

Yakovitz, S., and J. Spragins, (1968). On the identifiability of finite mixtures. *Ann. Math. Stat.*, **39**, 209-214.

Yakovitz, S., (1970). Unsupervised learning and the identification of finite mixtures. *IEEE Trans. Inform. Theory*, **IT-16**, 330-338.

Young T.Y., and C. Coraluppi, (1970). Stochastic estimation of a function of normal density functions using an information criterion. *IEEE Trans. Inform. Theory*, **IT-16**, 258-263.

Young T.Y., and T.W. Calvert, (1974). *Classification, Estimation and Pattern Recognition*. New York: American Elsevier.

GRAPHICAL METHODS OF EXPLORATORY DATA ANALYSIS

Jerome H. Friedman, John A. McDonald and Werner Stuetzle

Stanford Linear Accelerator Center and Stanford University
Stanford, California 94305

ABSTRACT

Orion I is a graphic system used to study applications of computer graphics – especially interactive motion graphics– in statistics. Orion I is the newest of a family of "Prim" systems whose most striking common feature is the use of real-time motion graphics to display three-dimensional scatterplots. Orion I differs from earlier Prim systems through the use of modern and relatively inexpensive raster graphics and microprocessor technology. It also delivers more computing power to its user; Orion I can perform more sophisticated real-time computations than were possible on previous such systems. We demonstrate some of Orion I's capabilities in our film: "Exploring data with Orion I".

PREFACE

This paper accompanies a film that demonstrates some programs written for the Orion I workstation. Orion I is an experimental computer graphics system built by the Computation Research Group at the Stanford Linear Accelerator Center (SLAC) in 1980-81. It is used to develop applications of interactive graphics to data analysis.

We intend this paper for an audience familiar with computer graphics, but not necessarily with statistics. We hope that it provides some perspective on the place of interactive graphics in statistics and an introduction to several areas of current research.

GOALS

Description and Inference

The goal of statistics is to analyze data. A data set consists of, say, n observations. For each observation, we measure p variables. It will be a useful idealization to assume that the values each variable takes on are reasonably represented by real or floating point numbers. Then each observation can be thought of as a vector in a p-dimensional real vector space. A data set is then a set of vectors. This type of data is typical of the field of statistics known as 'multivariate analysis'.

How we analyze a given data set is very dependent on context. Sometimes we are given data that arises from an experiment designed to answer a particular question. In this case, we may have a great deal of prior knowledge about our data, and have confidence that we know what to expect in it. This is a situation in which statistical inference may be appropriate. An inference is a generalization from a given data set to some larger population (real or hypothetical) from which the data set is presumed to be a sample.

Often, however, we are given data to analyze about which we have very little prior knowledge. The goal then is to "just look at the data and see what is going on". Statisticians call this description or exploratory data analysis [12,10]. We aim to describe features of the particular data set at hand and not draw (formal) conclusions about larger populations. There are two parts to description: exploration, to discover interesting features of our data, and summary, to report what we have discovered.

In order to make inferences, we need a probability model of the process by which our data set is sampled from the larger population. Our inferences will only be reasonable if the probability model is a reasonable approximation of the way the data is generated. In order to construct a reasonable probability model, we must have a great deal of prior knowledge about the data. A common and serious flaw in standard statistical practice is the use of inappropriate models.

In classical multivariate analysis, one typically assumes that observations are independent samples from a multivariate Gaussian distribution. The Gaussian distribution models only ellipsoidal shaped point clouds; it is inappropriate when the data set seems to come from a distribution that has a more complicated shape. In our experience, most real data sets, especially those with many (more than three) variables, have some features that are clearly not Gaussian. The Gaussian distribution, and other very similar models that characterize classical multivariate analysis, have received a great deal of attention from theoretical statisticians. This interest is not because they are reasonable models of the generation of any data. Rather, the Gaussian distribution is favored because it is easy to analyze mathematically.

Description is a more primitive and fundamental problem than inference. If inference is to be done successfully, then we must use good models. In order to get enough prior knowledge about a problem to construct a reasonable model, we must, at some time, look at data.

The Value of Interactive Graphics

In statistics, graphical methods are used primarily for description. Our research, which emphasizes interaction and real-time motion, concentrates on the exploratory part of description. We want to develop methods that allow us to discover, understand, and summarize the multivariate "structure" of our data set. Since, in general, we have little reliable prior knowledge about our data, we are especially interested in methods that allow us to discover <u>unanticipated kinds</u> of structure in data. We also want to be able to respond to unexpected circumstances in an intelligent way.

This is where interactive graphics has much to contribute. First of all, computer graphics can quickly expose different views of data to human perception so that the data analyst can detect many kinds of patterns in the data. The analyst can interpret apparent patterns using his knowledge of the context in which the data arose. Also, the analyst can select directions for further exploration using results so far.

Thus, interactive graphical methods combine human talents for perception of patterns and judgment using the full context of a problem with a machine's ability to do rapid and accurate computation.

An Example of Structure

We have intentionally left "structure" undefined. What we mean by "structure" is any apparent pattern or interesting feature in a graphical (or numerical) description of data. Structure is a subjective perception. Quantitative measures of specific aspects of structure can be very useful, but we need to avoid too restrictive definitions.

In developing new graphical methods, it is useful to have some examples of structure in mind. We discuss next a simple kind of structure that is illustrated in our film: "Exploring Data with Orion I".

A typical and simple thing that we would like to know about a set of data is the following: Does the data set naturally separate into groups or clusters?

In the film, we look at a data set presented by Harrison and Rubinfeld

[9,1]. They measured 14 variables for each of 506 census tracts in the Boston Standard Metropolitan Statistical Area. They were interested in examining the dependence of housing price (represented by the median value of owner-occupied houses) on air pollution (represented by nitrogen oxide concentration). The remaining 12 variables measured other quantities thought to influence housing prices, such as crime rate, average number of rooms per house, etc.

Harrison and Rubinfeld tried to determine the dependence of housing price on pollution by forming a prediction rule (a linear regression) for median housing value as a function of nitrogen oxide concentration and the other 12 variables. The effect of pollution alone on housing price was presumed to be reflected in the partial dependence of the prediction rule on nitrogen oxide.

In the exploration movie, we do not consider explicitly the dependence of housing price on pollution. We concentrate instead on clustering.

Before fitting any prediction rule, it is natural to consider whether it is appropriate to fit one rule for all the data. If the data set separated in a natural way into distinct and internally homogeneous groups (or clusters), then we would want to consider fitting different rules for each group.

Statisticians have developed many algorithms for clustering data. These algorithms usually rely on a notion of distance in the data space to partition a data set into isolated clumps. Observations are considered similar if they are close together and dissimilar if they are far apart. A cluster is a group of points that are close to each other and far from any other points.

With Orion I, we can use other criteria besides separation to partition a data set. In particular, we can use subjective perception of patterns to define natural groupings. A data set may naturally divide into two groups which clearly follow distinct patterns. Yet the difference between the groups may not be easily summarized by any natural measure of distance.

In the film, we show how the Harrison-Rubinfeld data can be divided naturally into several groups. The major division turns out to be between urban and suburban-rural census tracts. However, the urban and suburban tracts do not form isolated clumps. Instead, in certain views, they lie in intersecting, perpendicular planes. Because the planes intersect, the groups are not isolated in any distance measure. But the separation in the two groups is obvious when seen.

METHODS

The really challenging problem in multivariate data analysis is to discover and understand structure involving more than three variables at once.

Suppose our data set involves only a single variable. Then we can see most of what we want with a histogram.

We look at two-dimensional data with a conventional scatterplot. In conventional scatterplots, observations are represented by points which are plotted at horizontal and vertical positions corresponding to the values of two variables.

To look at three-dimensional data, we draw a three-dimensional version of the scatterplot. Real-time motion graphics makes it possible to draw pictures that appear three-dimensional. We subject the data to repeated small rotations and display the projection of the rotated data as points on the two-dimensional screen. If we can compute and display rotations fast enough (>10 per second), then we get an illusion of continuous motion. Apparent parallax in the motion of the points provides a convincing and accurate perception of the shape of the point cloud in the three-dimensional space.

In a conventional scatterplot, it is easy to read off approximate values of the two variables for any particular point. In moving, three-dimensional scatterplots estimating the values of the three variables for any particular point is difficult, if not impossible. However, perceiving the shape of the point cloud is easy. Because we are interested in detecting patterns, we are much more interested in overall shape than in individual points.

There is no completely satisfactory method that lets us look at more than three variables at a time. Three basic approaches to many (more than three) dimensional structure are being studied with Orion I. They are:

1. <u>Higher dimensional views</u>: we try to represent as many variables as possible in a single picture.

2. <u>Projection Pursuit</u>: we try to find a low dimensional picture that captures the structure in the many dimensional data space.

3. <u>Multiple Views</u>: we look at several low dimensional views simultaneously; by making connections between the low dimensional views, we hope to see higher dimensional structure.

<u>Higher Dimensional Views</u>

One way to see high-dimensional structure is to try to invent pictures that show as many dimensions at a time as possible.

One of the simplest ways to add dimensions to a picture is through color. We start with a three-dimensional scatterplot. We can add a fourth variable to the picture by giving each point in the scatterplot a color that depends on the value of a fourth variable. With an appropriately chosen color spectrum, we can easily see simple or gross dependence of the fourth variable on position in the three-dimension space. Our ability to perceive distinctions in color does not compare to our ability to perceive position in space; we should expect to miss subtle or complicated relationships between a color variable and three position variables. Color works best for a variable that takes on only a small number of discrete values.

There are many more tricks that let us add dimension to a picture. For example, we can represent each observation in the scatterplot by a circle, rather than by a simple point. The radius of the circle can depend on the value of a fifth variable.

In a simple scatterplot, observations are represented by featureless points. We add dimension to the picture by replacing points with objects that have features, such as color, size, and shape. These features can be used to represent variables in addition to those represented by a point's position in the scatterplot. These "featurefull" objects, sometime called <u>glyphs</u>, are well known [8].

With each new dimension, the glyphs become more complicated and the picture becomes more difficult to interpret. We need more experience to determine which are good ways of adding dimension to glyphs and to understand the limitations of each method.

<u>Projection Pursuit</u>

The basic problem with adding dimension to a single view is that the picture quickly becomes impossible to understand. An alternative is to restrict our picture to a few (3 or less) dimensions and then try to find low-dimensional pictures that capture interesting aspects of the multivariate structure in our data. This is the basic idea of projection pursuit [4,5].

In general, we could consider any mapping from the many-dimensional data space to a low-dimensional picture. The projection pursuit methods that have been developed so far restrict the mappings considered to orthogonal projections of the data onto 1, 2 or 3-dimensional subspaces of the data space. More general versions of projection pursuit would include methods similar to multidimensional scaling [8].

The original projection pursuit algorithm [7] used a numerical optimizer to search for 1 or 2-dimensional projections that maximized a "clottedness" index which was intended as a measure of interesting structure.

Automatic projection pursuit methods have been developed for more well defined problems: non-parametric regression, non-parametric classification, and non-parametric density estimation. In these problems, we build up a model that summarizes the apparent dependence in our data set of a response on some predictors. Interesting views are those in which the summary best fits or explains the response.

We are, of course, not restricted to a single low-dimensional view. Models of a response will usually be constructed from several views.

Interactive versions of the automatic projection pursuit methods are being developed for Orion I. The system allows a user to manually imitate the Rosenbrock search strategy [11] used by the numerical optimizer in the automatic versions. However, a human being can search to optimize subjective criteria, using perception and judgment, instead of having to rely on a single, numerical measure of what constitutes an interesting view.

Multiple Views

This approach is inspired by *M and N Plots* of Diaconis and Friedman [2]. A two and two plot is one kind of M and N plot; two and two plots are used to display four-dimensional data. To make a two and two plot, we draw two two-dimensional scatterplots side by side; the scatterplot show different pairs of variables. We then connect corresponding points in the two scatterplots by lines. To get a picture that is not confusing, we will often not draw all the lines. Diaconis and Friedman give an algorithm for deciding which lines to draw.

Our idea is a modification of the above. Instead of connecting corresponding points by lines, we draw corresponding points in the same color.

Briefly, this is how the program works. On the screen there are two scatterplots, side by side, showing four variables. Thee is also a cursor on the screen in one of the two scatterplots. The scatterplot that the cursor is in is the active scatterplot. We can move the cursor by moving the trackerball. Points near the cursor in the active scatterplot are red. Points at an intermediate distance from the cursor are purple. Points far from the cursor are blue. Points in the non-active scatterplot are given the same color as the corresponding point in the active scatterplot. The colors are continuously updated as the cursor is moved. We can also move the cursor from one scatterplot to the other, changing which scatterplot is active.

Using color instead of line segments to connect points has a disadvantage; it is not as precise at showing us the connection between a pair of points representing the same observation. However, we are not usually very interested in single observations. More often, we want to see how a region in one scatterplot maps into the other scatterplot; the combination of local coloring and the moveable cursor is a good way of seeing regional relationships in the two scatterplots.

Another advantage of color over line segments is that it is possible to look at more than two scatterplots at once. Connecting corresponding points with line segments in more than two scatterplots at a time would produce a hopelessly confusing picture. With color, on the other hand, it is no more difficult to look at three or more scatterplots at once than it is to look at two.

MACHINERY

Prim-9

Orion I is the youngest descendent of a graphics system called Prim-9, which was built at SLAC in 1974 [4]. Prim-9 was used to explore up to 9-dimensional data. It used real time motion to display three-dimensional scatterplots. Through a combination of projection and rotation, a user of Prim-9 could view an arbitrary three-dimensional subspace of the 9-dimensional data. Isolation and masking were used to divide a data set into subsets.

The computing for Prim-9 was done in a large mainframe (IBM 360/91) and used a significant part of the mainframe's capacity. A Varian minicomputer was kept busy transferring data to the IDIIOM vector drawing display. The whole system, including the 360/91, cost millions of dollars. The part devoted exclusively to graphics cost several hundreds of thousands of dollars.

Other Prims

Successors to Prim-9 were built at the Swiss Federal Institute of Technology in 1978 (Prim-S) and at Harvard in 1979-80 (Prim-H) [3].

Prim-H improves on Prim-9 in both hardware and software. It is based on a VAX 11/780 "midi-" computer and an Evans and Sutherland Picture System 2 (a vector drawing display). It incorporates a flexible statistical package (ISP). However, the system has a price tag still over several hundreds of thousands of dollars. The VAX is shared with perhaps two dozen other users. Computation for rotations is done by hardware in the Evans and Sutherland. No demanding computation can be done on the VAX in real-time.

Orion I

There are two ways in which the Orion I hardware is a substantial improvement over previous Prim systems: price and computing power. The total cost for hardware in Orion I is less than $30,000. The computing power is equivalent to that of a large mainframe computer (say one-half of an IBM 370/168) and is devoted to a single user. The hardware is described in detail by Friedman and Stuetzle [6].

The basic requirement for real-time motion graphics is the ability to compute and draw new pictures fast enough to give the illusion of continuous motion. Five pictures per second is a barely acceptable rate. Ten to twenty times per second gives smoother motion and more natural response for interaction with a user.

Orion I and the earlier Prims use real-time motion to display three-dimensional scatterplots. We can view three-dimensional objects on a two-dimensional display by continuously rotating the objects in the three-dimensional space and displaying the moving projection of the object onto the screen. In a scatterplot, the object we want to look at is a cloud of points. A typical point cloud will contain from 100 to 1000 points. So our hardware must be able to execute the viewing transformation, which is basically a multiplication by a 3x3 rotation matrix, on up to 1000 3-vectors ten times per second. The system must also be able to erase and draw 1000 points ten times per second.

The important parts of the hardware are:

1. A SUN microcomputer, based on the Motorola MC68000 microprocessor. The SUN microcomputer is a MULTIBUS board developed by the Stanford Computer Science Department for Stanford University Network. It is the <u>master processor</u> of the Orion I system. It controls the action of the other parts of the system and handles the interaction with the user. The SUN is programmed mostly in Pascal; a few critical routines for picture drawing are in 68000 assembly language.

2. A Lexidata 3400 <u>raster graphics frame buffer</u>, which stores and displays the current picture. The picture is determined by a "raster" of 1280 (horizontal) by 1024 (vertical) colored dots, called <u>pixels</u>. There are 8 bits of memory for each pixel which, through a look-up table, determine the setting (from 0 to 255) of each of the three color guns (red, green, blue). Thus, at any time, there may be 256 different colors on the screen, from a potential palette of 2**24. The Lexidata 3400 contains a microprocessor that is used for drawing vectors, circles, and characters. Color raster graphics devices like the Lexidata are cheaper and more

flexible than the black and white line drawing displays used in earlier Prim systems.

3. An <u>arithmetic processor</u> called a 168/E. The 168/E is basically an IBM 370/168 cpu without channels and interrupt capabilities. It was developed by SLAC engineers for the processing of particle physics data and has about half the speed of the true 370/168. Because it has no input/output facilities, it is strictly a slave processor. The 168/E serves us as a flexible floating point or array processor. It is programmed in FORTRAN.

A programmer of Orion I works on:

4. The <u>host</u>, an IBM 3081 mainframe. The host is used only for software development and long term data storage. Programs are edited and cross-compiled or cross-assembled on the 3081. Data sets are also prepared on the 3081. Programs and data are downloaded to the SUN board and the 168/E through a high speed serial interface that connects the MULTI-BUS and the 3081 (by emulating an IBM 3277 local terminal). Strictly speaking, the 3081 is not considered a part of the Orion I workstation; it does not have an active role in any of the interactive graphics on the Orion I.

A user of Orion I deals most of the time with:

5. a 19 inch <u>color monitor</u> that displays the current picture.

6. A <u>trackerball</u>, which is a hard plastic ball about three inches in diameter set into a metal box so that the top of the ball sticks out. The ball can be easily rotated by hand. The ball sends two coordinates to the SUN computer. In Orion I, the trackerball provides the angles of a rotation; the apparent motion of a point cloud on the screen mimics the motion of the trackerball under a user's hand. The trackerball has six switches which are used for discrete input to programs.

The part of the system that currently limits what can be done in real time is the Lexidata frame buffer. Its speed of erasing and redrawing pictures does not keep up with the computational speed of the SUNboard and the 168/E. In fact, the 168/E can execute our most demanding real time computation, a sophisticated smoothing algorithm, at much more than ten times a second. This smoothing algorithm [5] is much more demanding than simple rotation.

It is also worth noting that rotations can be done in real time by the SUNboard alone. Thus, a system could be built with all the capabilities of earlier Prim systems using only a SUNboard or some other 68000 based microcomputer and without a 168/E. Such systems are now commercially available for about $25,000 for basic hardware and about $40,000 for a complete, stand-alone system with hard disk, UNIX-like operating system, resident high level language compilers, etc.

REFERENCES

1. Belsey, D.A., Kuh, E., and Welsch, R.E.   Regression Diagnostics, 1980.

2. Diaconis, P. and Friedman, J.H.   "M and N Plots," Tech. Report 151, Dept. of Statistics, Stanford University, April 1980.

3. Donoho, D., Huber, P.J. and Thoma, H.   The Use of Kinematic Displays to Represent High Dimensional Data, Research Report PJH-5, Dept. of Statistics, Harvard University, March 1981.

4. Fisherkeller, M.A., Friedman, J.H. and Tukey, J.W.   PRIM-9, An Interactive Multidimensional Data Display and Analysis System. Proc. 4th International Congress for Stereology, 1975.

5. Friedman, J.H. and Stuetzle, W.   "Projection Pursuit Regression," JASA, Vol. 76, 1981.

6. Friedman, J.H. and Stuetzle, W.   "'Hardware for Kinematic Statistical Graphics," Tech. Rep. Orion-00, Dept. of Statistics, Stanford University, Feb. 1982.

7. Friedman, J.H. and Tukey, J.W.   "'A projection pursuit algorithm for exploratory data analysis," IEEE Trans. Comput., C-23, pp 881-890, 1974.

8. Gnanadesikan, R. Methods for Statistical Data Analysis of Multivariate Observations, 1977.

9. Harrison, D. and Rubinfeld, D.L.   "Hedonic Prices and the Demand for Clean Air," Journal of Environmental Economics and Management, 5, 1978.

10. Mosteller, F. and Tukey, J.W.   Data Analysis and Regression, 1977.

11. Rosenbrock, H.H.   "An automatic method for finding the greatest or least value of a function," Comput. J., 3, 1960.

12. Tukey, J.W.   Exploratory Data Analysis, 1977.

FAINT OBJECTS IN CROWDED FIELDS:
DISCRIMINATION AND COUNTING

M.L. Malagnini*, F. Pasian,
M. Pucillo and P. Santin

Osservatorio Astronomico
* and Universita' degli Studi
Trieste, Italy

INTRODUCTION

    The analysis and classification of faint objects measured on photographic plates (Malagnini et al.,1984) involve a number of problems, which are particularly relevant in the case of crowded regions. The density of the object images can be evaluated by comparing the distribution of the object dimensions with the distribution of the distances between any object and its nearest neighbor. We refer to the case in which the modes of the two distributions are comparable, thus asking for particular caution in the automatic analysis of individual objects. The aspects we will discuss refer to: i) local background estimates and automatic detection, ii) preliminary screening and constraints on the image characteristics and on environment relationships to pass the object to the classifier. The examples given here refer to photographic material processed at the ASTRONET center of Trieste, Osservatorio Astronomico, and fully referenced in the papers by Malagnini et al. (1984) and Santin (1984).

LOCAL BACKGROUND ESTIMATES AND AUTOMATIC DETECTION

When working in crowded regions, object detection becomes critical. The choice of the detection algorithm depends on the characteristics of the image and primarily of the background from which objects are to be extracted. Moreover, since the digitization process may introduce spurious signals (spikes), care must be taken in order to remove such kind of instrumental noise. In the case of stationary background, as in the one shown in Fig. 1a, a constant thresholding can be applied, since analyzing the histogram of density values for the whole image (Fig. 1b), a suitable threshold can be identified. Problems arise only when selecting an unappropriate threshold value: if a high figure is chosen, faint objects are missed, while if the threshold value is too low, a single detected object may contain some close companions, and spurious pixel aggregates may be interpreted as objects.

When dealing with images having non-stationary background, as in the case depicted in Figs. 1c and 1d, more refined methods must be used. Adaptive thresholding may help overcoming the variable-

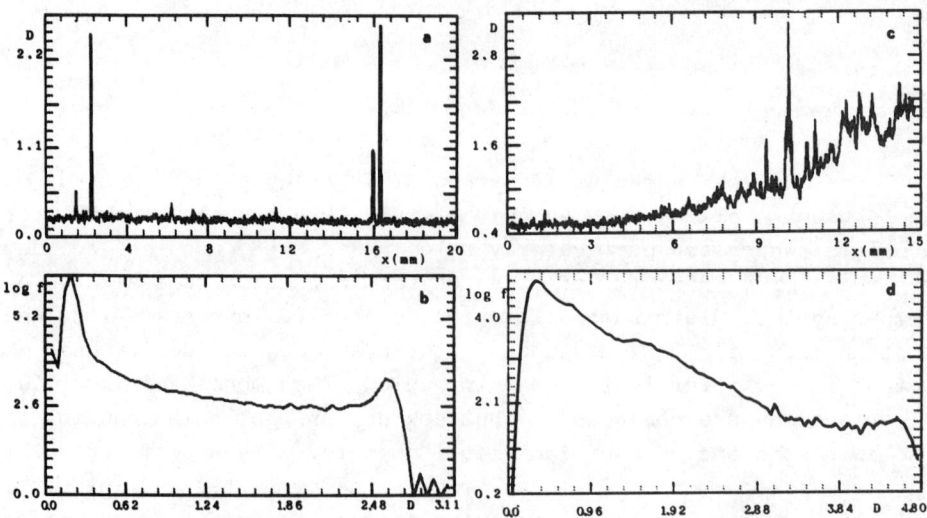

Fig. 1. Example of stationary background : (a) density profile along a fixed direction; (b) histogram of density values for the whole image. Example of non-stationary background : (c) density profile along a fixed direction; (d) histogram of density values for the whole image.

background problem. It can be applied either to original images, or to transformed images or to filtered images. In the following, the general characteristics of the adaptive thresholding algorithm, the transformation devised for object detection, and a class of non-linear digital filters for pre-processing will be described.

Adaptive Thresholding

This method consists in a local background computation; from the computed value, the local threshold value for detection is derived. In a moving window, of suitably chosen size, the average and the standard deviation of pixel values are computed at each step, after rejection of all values above the threshold computed at the preceding step. The threshold is usually assumed at the 3 RMS level above the average. Objects are defined as groups of contiguous pixels above the local threshold. This technique can be applied successfully to raw data in the case of smoothly-varying backgrounds.

Information Transform

A further improvement is achieved through the information transform method, described by Santin (1984). Each pixel of the original image is mapped into a conjugate pixel, whose value represents the measure of local information content in a predefined neighborhood of the original pixel. In the conjugate domain the background features are smoothed and the object features are enhanced, thus allowing the use of a constant threshold for detection. This fact is evident when comparing Fig. 1d, where the histogram of density values for a test image is shown, with Fig. 2,

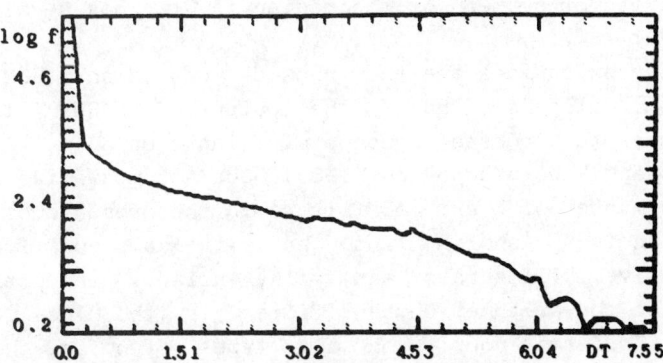

Fig. 2.   Levels histogram in the transformed domain for the same image referred to in Fig. 1d.

where the level histogram in the transformed domain for the same image is shown. Furthermore, the barycenters positions are invariant with respect to the information transform, thus allowing the definition of the objects centroids directly in the transformed domain. This method demonstrated to be particularly efficient when dealing with variable or complex backgrounds, due to its object-enhancing characteristics.

Non-linear Digital Filtering

Filtering operations can be performed on an image, both for the computation of the background, which is to be subtracted from the original image before object detection, and for the removal of isolated spikes which are not to be detected as objects. There is a wide variety of filters among which to choose the most suitable to the specific problem. Linear techniques possess mathematical simplicity, but do not perform well in the presence of signal-dependent noise and are not suitable for the removal of spiky (impulsive) noise. Recently (Kundu et al.,1983; Pitas and Venetsanopoulos,1984), a class of non-linear filters for image processing have been introduced; their properties look quite promising for our kind of problem. These filters are defined as non-linear statistical means; their computation require only arithmetic operations, and not logical ones.

When the definition is applied to an ordered sequence of data, it may include, as special cases, the alpha-trimmed means and the median filter. Given an input sequence of N ordered data, the output of the alpha-trimmed mean filter (Bednar and Watt,1984) is the average of the central values, provided that a certain number of values (α N) have been removed from the top and from the bottom of the ordered sequence. In particular, the output of the filter is the arithmetic mean or the median if α has a value of 0 or 0.5, respectively.

Further extensions are the ranked non-median (RNM) filters (Nodes and Gallagher,1982): the output of an RNM filter is the r-th value of the ordered sequence of N input data.

Pitas and Venetsanopoulos (1984) made a comparative study of the performance of various filters in the presence of different types of noise (short- and heavy-tailed additive noise, multiplicative, film-grain and spiky noise). In particular, it results that the arithmetic mean filter is the less efficient, since it performs poorly on all types of noise, short-tailed additive noise excluded. On the other hand, from this comparison

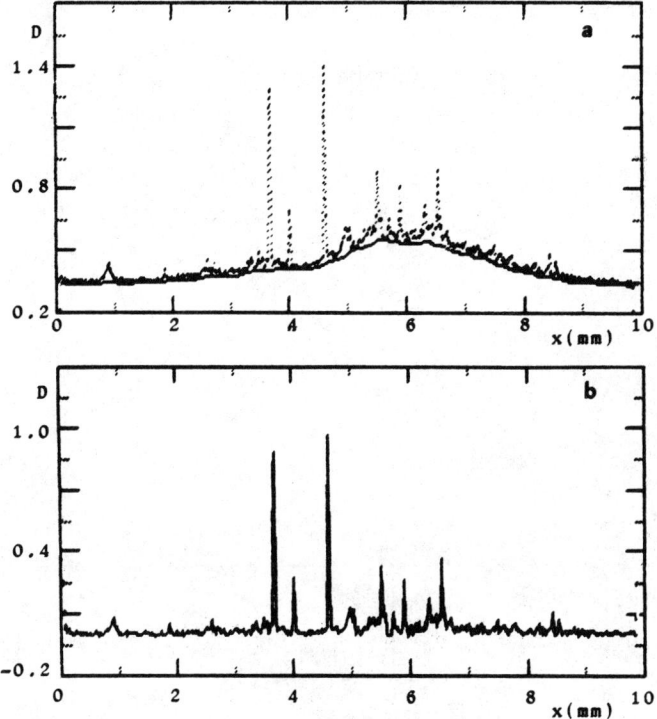

Fig. 3. Effects of RNM filtering : (a) original density profile (dashed line) and estimated background; (b) filtered density profile.

and from previous extensive analyses of statistical (Justusson,1981) and deterministic (Tyan,1981) properties of median filters, it results that such kind of filters are quite efficient for the removal of spiky and heavy-tailed additive noise while preserving edges and slopes. Thus median filtering appears to be the best suited to avoid spurious detections.

Non-linear filtering can be applied also to solve the problem of object detection in the presence of non-stationary backgrounds. The approach used is to compute the image background through filtering; the objects are extracted from the image obtained after background subtraction. In particular, the performance of an RNM filter appears to be quite satisfactory in the case of strongly-varying background, as shown in Figs. 3a and 3b.

OBJECT SCREENING

After object detection and before the classification phase, an intermediate step of object screening is required. Constraints on the object characteristics are needed to select those for which further analysis is not feasible or to select objects having peculiar features, so to perform further processing on this restricted sample, possibly asking for individual rather than standard analysis.

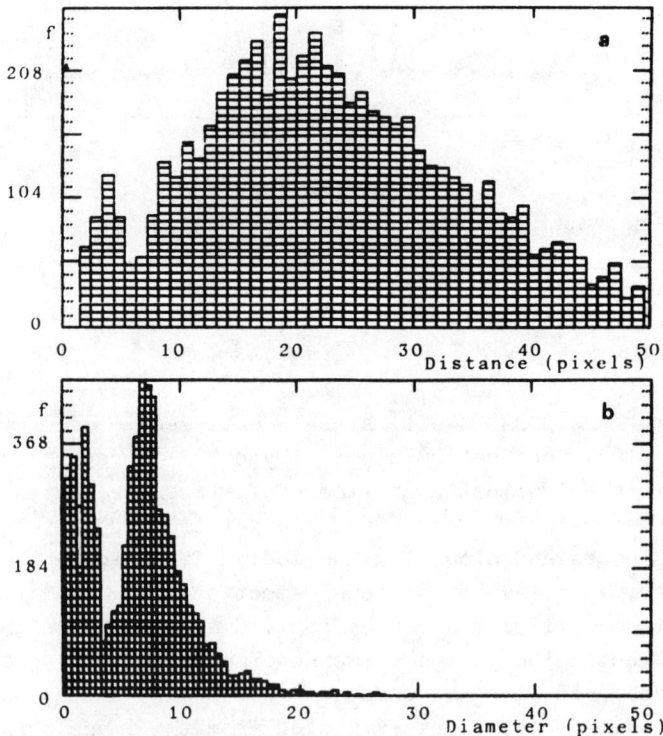

Fig. 4. (a) Histogram of distances between each object and its nearest neighbor; (b) histogram of estimated object diameters. Both in (a) and (b), the first modes refer to spurious detections.

At first, a preliminary screening, based on the object area, is performed. As a matter of fact, the detector may identify aggregates of pixels too small to be classified as physical objects. These detections are disregarded, together with those too close to image edges.

Furthermore, features related to position with respect to other objects, multiplicity and symmetry are analyzed.
For each object, the nearest neighbor is identified; the label of the neighbor and the distance between the two objects are recorded. In Fig. 4a the distribution of such distances is shown for the objects detected on the test image referred to in Figs. 1a and 1b. When comparing Fig. 4a with Fig. 4b, where the object diameter distribution is depicted, it can be seen that a great number of objects has comparable values for diameter and distance from the nearest neighbor. The modes at low values in both figures represent aggregates of pixels too small to permit reliable analysis and rejected on the basis of the area value.

The structure of each detected object is analyzed by means of its projections (marginal distributions) in the X, Y, +45 and -45 degrees directions on a variable size window. From these projections, modified with respect to those defined by Pavlidis (1982) to allow normalization, two asymmetry indexes are derived. The differences in absolute value between the widths at half-height (in pixels) of each couple of orthogonal projections are computed; the maximum of these two values is taken as the first index and gives an estimate of the ellipticity of the object. The second index is computed as the maximum of the four distances (in pixels) between the centroid of each projection and its mode; this value takes into account asymmetries of pixel values distribution with respect to the barycenter of the object.

From the normalized projections, a further information may be extracted on possible complex structures. If for at least in one of the projections more then one maximum is found above a given threshold, the object is flagged as complex-structured, thus requiring further individual analysis for classification. Two examples of such cases are shown in Figs. 5 and 6: the isolevel description of the object is given together with its projections in the X, Y, +45 and -45 degrees directions. Fig. 5 represents an object defined as single by the detection algorithm but classified as double by the analysis of its projections. Fig. 6 shows clearly the complexity of an object defined as single in the detection phase.

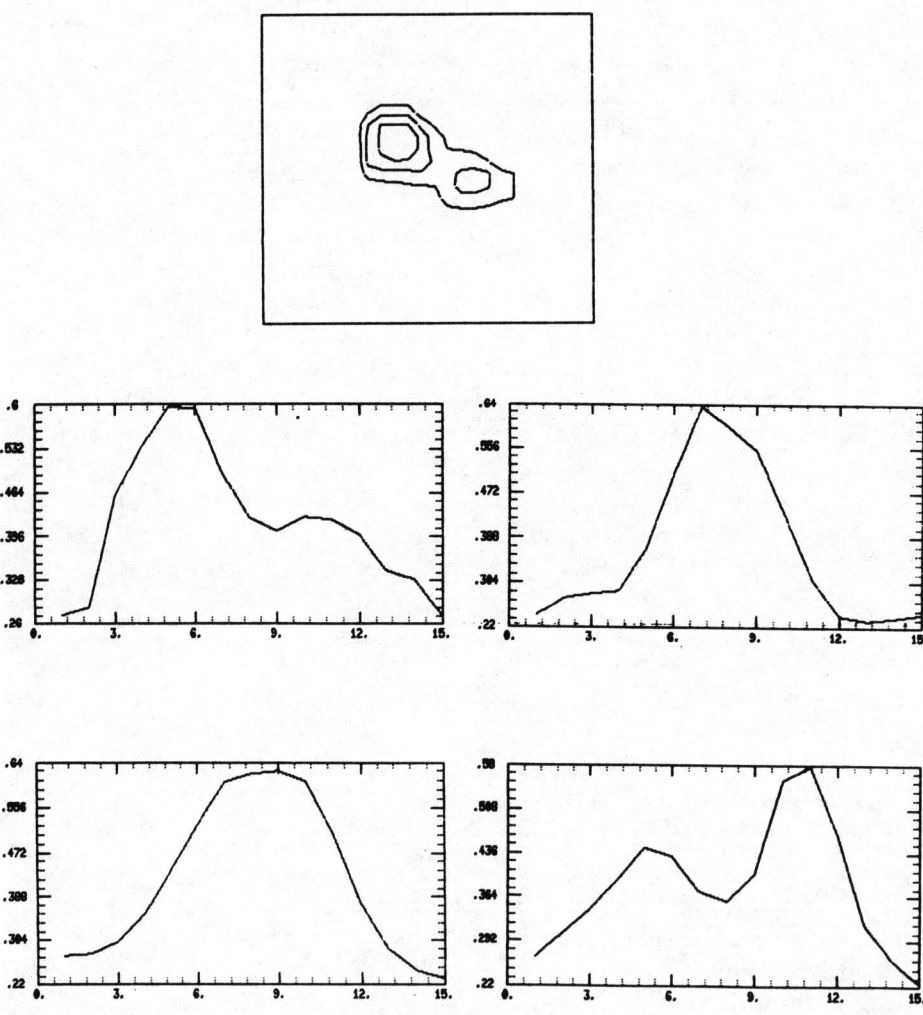

Fig. 5. On top, isodensity representation of an object defined as double. The projections along the four main directions are shown below.

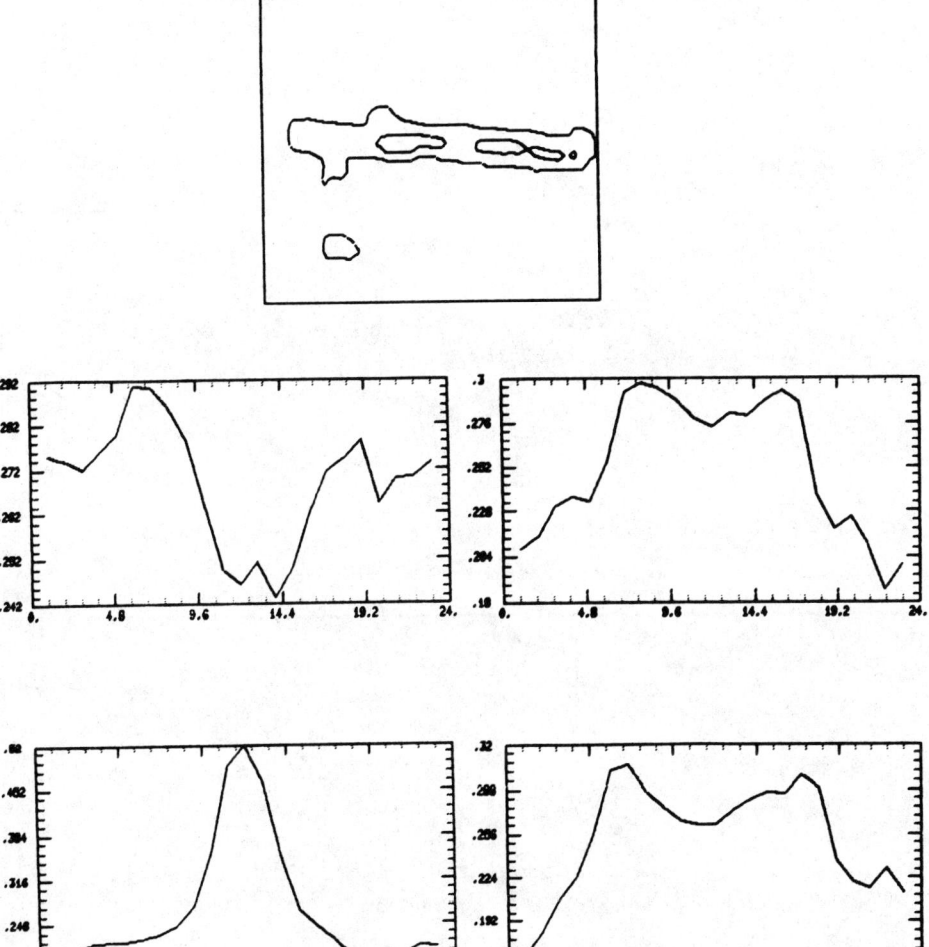

Fig. 6. On top, isodensity representation of an object defined as complex. The projections along the four main directions are shown below.

REFERENCES

1. Bednar J.B., Watt L.T., 1984, Alpha-trimmed means and their relationship to median filters, IEEE Trans. ASSP 32, 145:152.

2. Chang S.K., 1984, Image information measure and encoding techniques, in: "Digital Image Analysis", S. Levialdi, ed., Pitman, London.

3. Kundu A., Mitra S.K., Vaidyanathan P.P., Generalized mean filters: a new class of nonlinear filters for image processing, 1983, Proc. 6th Symp. on Circuit Theory and Design, 185:187.

4. Justusson B.I., 1981, Median filtering: statistical properties, in: "Two-dimensional Digital Signal Processing", Vol. II, T.S. Huang, ed., Springer, Berlin.

5. Malagnini M.L., Pucillo M., Santin P., Sedmak G., Sicuranza G.L., 1984, A system for object detection and image classification on photographic plates, in: "Astronomy with Schmidt-type telescopes", M. Capaccioli, ed., Reidel, Dordrecht, The Netherlands.

6. Nodes T.A., Gallagher N.C. Jr., 1982, Median filters: some modifications and their properties, IEEE Trans. ASSP 30, 739:746.

7. Pavlidis T., 1982, "Graphics and Image Processing", Springer, Berlin.

8. Pitas I., Venetsanopoulos A., 1984, Non-linear statistical means in image processing, (preprint).

9. Santin P., 1984, Object detection with the information transform, Mem. SAIt (in press).

10. Tyan S.G., 1981, Median filtering: deterministic properties, in: "Two-dimensional Digital Signal Processing", Vol. II, T.S. Huang, ed., Springer, Berlin.

MULTIVARIATE STATISTICS TO ANALYZE EXTRATERRESTRIAL PARTICLES

FROM THE OCEAN FLOOR

Ildiko E. Frank**, Bernard A. Bates* and Donald E. Brownlee*

** Department of Chemistry, Laboratory for Chemometrics
   University of Washington
   Seattle, WA. 98195
 * Department of Astronomy
   University of Washington
   Seattle, WA. 98195

INTRODUCTION

The elemental composition of cosmic particles found deep on the ocean floor is of astronomical interest, because of its representativeness of small meteoroids, that melt during high velocity entry into the atmosphere. The properties of these deep sea spheres (DSS) give information about the composition of the earth crossing population of small meteoroids. Meteor trajectory analysis has shown that most small meteoroids are debris from short period comets. Their elemental and Sr isotopic abundances and their content of the cosmogenic isotopes are proof of their extraterrestrial origin. Studying DSS chemical compositions unaffected by atmospheric entry is a vehicle to learn about meteoroids. With this goal in mind, chemical analysis of about 300 polished sections of 250-800 μm DSS, collected 1000 km east of Hawaii was made by electron microprobe technique. Each sample was characterized by the following elemental concentrations: Al, Ca, Cr, Fe, Mg, Mn, Na, Ni, P, Si, Ti. In general, it was found that the relative abundances of these elements in each particle is close to solar (chondritic) proportions. In a previous study[1] conclusions about the composition of the major part and some peculiar sub-groups of the DSS samples were drawn on the basis of the univariate (one variable at a time) distributions of the elemental concentrations.

In this paper various multivariate statistical methods are used to explore the multivariate nature of the data in order to reveal underlying structure, find natural groupings among the samples, determine the importance and similarity among the elemental concentrations. Our intention is to provide an example and suggest steps for exploring data that could be followed in other similar multivariate studies.

STATISTICAL METHODS

1. Data Representation

In order to apply multivariate statistical technics to a problem, mathematical representation of the data has to be given. Therefore, the first step of each exploratory data analysis is to organize the acquired information in a computer compatible form. In our case each of the 300 samples characterized by 11 different elemental concentration measurements can be thought of as a point in an 11 dimensional coordinate system. Each point is described by an eleven dimensional row vector, the entries of which contain the 11 coordinates, i.e. the 11 chemical measurements. Putting these vectors one after each other results in a 300 x 11 matrix where the samples are associated with rows and the elemental concentrations with columns. The sequence of the elements (dimensions) must be the same through all the samples. In a statistical sense, the rows represent the observations of the 11 random variables, each described by a column.

2. Univariate Distributions

As a pre-examination, it is useful to look at the distribution of each variable separately by calculating univariate histograms (frequency vs. concentration range plots). The information is twofold: not only the location of the largest frequency and the spread of the frequency along the concentration ranges can be observed, but also anomalies of the distribution can be revealed, like asymmetry, heavy tail, outlier samples. These anomalies might be corrected by appropriate transformations.

3. Scaling

The next step in the analysis is to determine the apriori importance of the dimensions relative to each other, i.e. to define the scale of each coordinate system. In our case because of lack of information, the dimensions are given equal weight in the analysis, and each coordinate is centered to the origin. This transformation is done by subtracting from each concentration entry of a column, the mean of that column (dimension) and dividing each centered concentration entry of a column by the standard deviation of that column. This way each variable is scaled to zero mean and unit variance and has equal importance.

4. Looking at Multidimensional Structures

The human being is the best pattern recognizer in two or three dimensions, but it is above the human capability to see any structure in higher than three dimensional space. Therefore, to be able to visualize the relative position of points in high (in our case 11) dimensional space a two (or three) dimensional representation of the samples is needed, which reflects faithfully the high dimensional structure. Two basic principles can be used: either the two new coordinates are linear combinations of the original (11) coordinates so that the maximum variance of the high dimensional data is retained on the plane; or find any curved surface in the high dimensions, on which the mapping of the points preserve the best of the distances among the points. The first principle is realized by a linear projection of the high dimensional points on the first two principal componenent axes, also called Karhunen-Loève projection. There are several algorithms available for the second principle. Non-linear mapping finds the two dimensional plane iteratively, minimizing the difference between the distance calculated in the high dimensional space and the distance calculated in the two dimensional space

If the distances (or the reciprocal quantities, the similarities) among the points are represented on an ordinal scale rather than on an interval scale, in other words, if only the order of the distances is important to preserve, but not their ratios, then multidimensional scaling gives a solution similar to non-linear mapping.

Visualization and comparison of these two dimensional plots reveal groups of samples, underlying structures, outlier samples, inhomogenity in the high dimensional distribution.

5. Cluster Analysis

Another group of multivariate methods based on the distances among the samples is called clustering. These technics, similar to the previous group, also give an insight to the structure of the data, by exploring the natural clusters and indicating single samples lying far away from the main body of the data. Two hierarchical techniques were applied in this study: the minimal spanning tree (MST) and the agglomerative dendrogram (AD) method. The first step in each cluster analysis algorithm is to calculate the distances among the samples. The most often used metric is the Euclidean distance, the high dimensional version of which is

a natural extension of the well known two dimensional formula, the
sum of the squared differences of the coordinates over all
dimensions.

The MST algorithm is a graph-theoretical method, each sample
is represented by a node and the nodes are connected by edges.
The goal is to connect each node so that the sum of the connecting
edges is minimal. The samples connected by the shorter edges form
the natural clusters.

The AD method also creates a tree aggregating into common
cluster samples or sample groups for which the interpoint distance
is below a preset level.

7. Principal Component Analysis

This method is used to discover underlying factors in the
data set. Usually the measurements represented by the coordinate
axes of the high dimensional space are correlated, which means
that the data structure can be approximately represented in a
lower dimensional space. The goal of the principal component
analysis is to find the axes in a lower dimensional space by
linear combination of the original axes so that most data variance
is retained. These new lower dimensional axes are the representations of the underlying factors, so the interpretation of the
coefficients of the linear combinations resulting from these axes
provide information about the nature of these factors. These
coefficients also indicate the importance of each original
measurement (dimension).

8. Classification

So far, in the analysis only unsupervised techniques were used,
which means that the structure of the data was explored without
any apriori information about samples belonging to distinct
subgroups, so called categories. Classification methods calculate
models (decision rules for separating categories) on the basis
of a training set. The training set samples are not only described
in terms of the measurements, but also assigned to a particular
category. The goal of the classification algorithms is to model
the relationship among the various measurements and the category
membership of a sample. The classification model fulfills two
requirements: it gives a quantitative description of the relationship and, also, provides a rule to classify samples with
unknown categories.

In this study three classification models were calculated
to confirm the results from the cluster analysis and from the two

dimensional plots: linear discrimination, K-nearest neighbor and SIMCA. The linear discriminator algorithm calculates hyperplanes, which separate the categories defined by the training samples. The K-nearest neighbor algorithm assigns a sample to the same category to which the majority of its nearest neighbors belong. Here K is the (preset) number of the nearest neighbors included in the calculation. This algorithm is similar to the clustering methods and the non-linear mapping techniques in that it is based on the distances among the samples, and its success depends on the choice of the distance metric. The SIMCA model describes each category by a disjoint principal component model and the distance of the samples from each category model is calculated using these principal components as a basis for a metric. The samples are classified into the closest category. This algorithm not only gives a classification rule but also gives a quantification of how much a sample belongs to each category in terms of how much of its variance is described by each principal component model. As additional information, the modeling and discriminating power of each variable is given.

A more detailed description and discussion of these multivariate methods especially addressed to chemical data analysis is available in the literature[2,3,4].

RESULTS

There were four questions to be answered by the multivariate analysis of the DSS composition data:
- Which elements are important and similar to each other?
- Is the data homogeneous or are there natural clusters among the samples?
- In which elemental concentrations do the clusters differ?
- Can these clusters be assigned to well known elemental composition categories?

The study performed on this data set included all the steps described above in the previous section. The computations were done by two software packages, ARTHUR and DISPLAY, developed at the Laboratory for Chemometrics (Department of Chemistry, University of Washington, Seattle). They are interactive FORTRAN programs containing (among others) the above discussed algorithms.

To explore the importance and similarities of the 11 elements, principal components were calculated. The first component covering 30% of the variance mainly composed of Fe, Si and Mg, the second (20%) of Al, Ti and Ca, the third (14%) of P, Cr, Ni and the fourth (11%) of Na. The first principal component contains the major chemical components, the elements in the second and third components are close in the periodic table, therefore, show similar characteristics. These variable groups were confirmed also by hierarchical cluster analysis performed on the transposed data matrix, i.e. the dendrogram was calculated from the distance matrix among chemical elements. The hierarchy is shown on Figure 1. A similar picture is obtained on the two-dimensional non-linear plot of the chemical elements given in Figure 2.

The cluster analysis of the sample distance matrix showed that the data set is not homogenous, there are five distinct clusters. The same sample groups appeared on the Karhunen-Loève projection. The contours of these clusters are shown in Figure 3. The separability of these clusters was verified by three classification methods, the goodness of classification of samples into the five categories revealed by the cluster analysis and the two dimensional plots were between 85% and 95% with category overlap between 1:4 and 2:3. The elements of the highest discriminating power according to the SIMCA method are: Si, Cr, Fe, Al. The histograms of these four elements calculated in each category separately have their peak at the concentration range shown in Table 1.

Finally, an attempt to assign the five clusters to known elemental compositions was made. Each sample was classified to one of the seven categories represented by a prototype sample of the known seven major chondritic meteorite groups. The classification was made by K-nearest neighbor technique with K=1. It turned out that samples from both our clusters 1 and 4 were

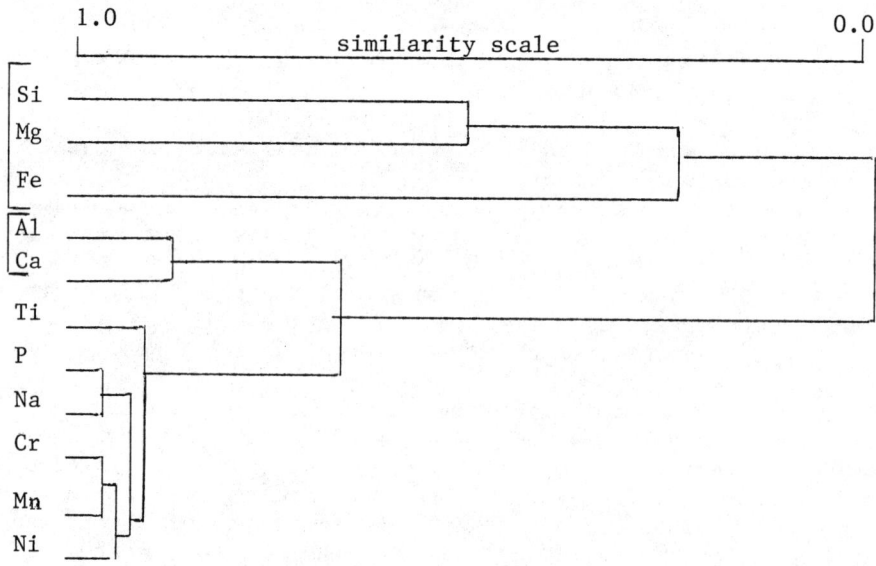

Figure 1. Hierarchical Clustering Among Chemical Elements.

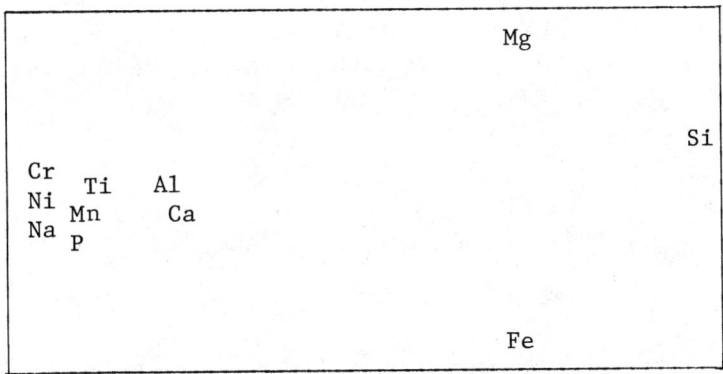

Figure 2. Two-dimensional Non-linear Plot of the Chemical Elements

classified to chrondritic composition C_{II} (as denoted in the literature), samples from clusters 2 and 3 to composition C_{III} and samples from cluster 5 to composition denoted LL. The whole analysis was repeated with elemental concentrations normalized to the Si concentrations to see whether the inhomogenity of the data set was not merely the result of the different conditions of the particles entering in the atmosphere. The results from the normalized data set were similar to the previous ones, so the above hypothesis was rejected.

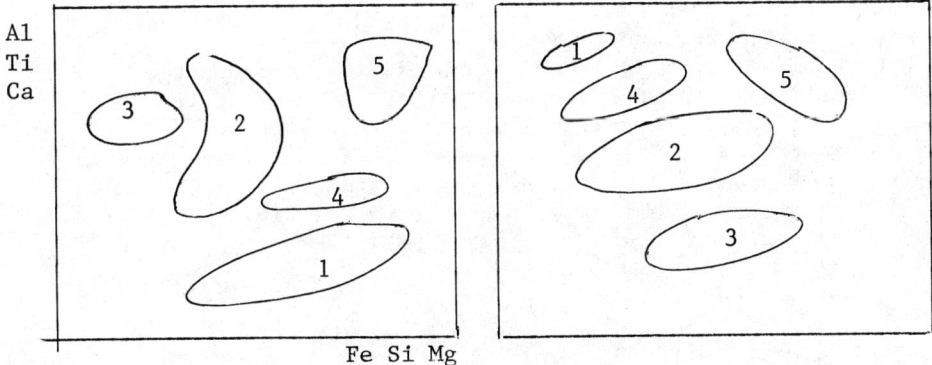

Figure 3. Karhunen-Loève Projection and Non-linear Mapping of Samples

Table 1. Concentration Range of the Histogram Peaks of Four Elements in Five Categories

element \ category	1	2	3	4	5
Si	medium	medium	low	medium	high
Cr	low	high	high	low	low
Fe	medium	medium	high	medium	low
Al	medium	low	medium	medium	low

CONCLUSION

Our goal in this paper was to illustrate the application and power of multivariate statistical technics. The various steps and levels of a typical data analysis are described in Section 2. The performance of these methods was demonstrated on DSS data set. Information was obtained about the grouping of chemical elements and their relative importance. Examination of the structure of the data set resulted in five distinct clusters. Their separability was proved by classification methods, and these were compared with various typical compositions well known in the literature. Our multivariate analysis discovered that there are two subgroups in both C_{II} and C_{III} categories.

REFERENCES

1. B. A. Bates and D. E. Brownlee, Lunar and Planetary Science, XV:40 (1984).
2. K. Varmuza, "Pattern Recognition in Chemistry," Springer-Verlag, New York (1980).
3. D. L. Massart, A. Dijkstra and L. Kaufman, "Evaluation and Optimization of Laboratory Methods and Analytical Procedures," Elsevier Scientific Publishing Co., Amsterdam (1978).
4. D. L. Massart and L. Kaufman, "The Interpretation of Analytical Chemical Data by the Use of Cluster Analysis," John Wiley & Sons, New York (1983).

STATISTICAL ANALYSIS OF PULSE PROCESSES

Sergio Frasca

Dipartimento di Fisica Universita' di Roma
Rome, Italy

Sometimes astrophysical or geophysical data consist of sequences of pulses, in which the time structure, the amplitude, or other features of every single pulse are not significant or of preminent interest, but we consider only the time occurrences of the pulses and we can look for periodicities, swarming, or correlation with other physical variables. This is the case of, e.g., the large pulses of gravitational antennas, the periodicities of which are now investigated by the group of Rome [1 - 4]. Other examples are the detection of gamma rays from weak sources and the problem of earthquakes, in which the correlation with the tide generating potential is interesting in order to witness the importance of local stresses as a cause of the earthquakes .
A theory for these cases is not deeply developed and often empirical statistical methods are used. In this approach the phenomenon is modeled as a "pulse process" called "non-uniform Poisson process" [5 - 6]

1) $$\{x(t)\} = \{\sum_k \delta(t - t_k)\}$$

and the probability of having n pulses in the time interval (t_1, t_2) is

2) $$P(n; t_1, t_2) = e^{-\int_{t_1}^{t_2} \lambda(t) dt} \cdot \frac{\left(\int_{t_1}^{t_2} \lambda(t) dt\right)^n}{n!}$$

and $\lambda(t)$ is defined by

3) $$\int_{t_1}^{t_2} \lambda(t)\,dt = E\left[\int_{t_1}^{t_2} x(t)\,dt\right]$$

for every (t_1, t_2). So $\lambda(t)$ is "the expected number of pulses (or "events") per unit time at time t" and to study the pulse process means to study the $\lambda(t)$.

Often the main problem is to look for periodicities in the $\lambda(t)$ (i.e. in the events). This can be done by estimating the power spectrum of the $\lambda(t)$

4) $$S_\lambda(\omega) = \int_{-\infty}^{\infty} R_{\lambda\lambda}(\tau) e^{-j\omega\tau}\,d\tau = \lim_{T\to\infty} \frac{1}{2T}\left|\int_{-T}^{T} \lambda(t) e^{-j\omega t}\,dt\right|^2$$

where $R_{\lambda\lambda}(\tau)$ is the autocorrelation of λ (+++); using 3) we obtain

5) $$\hat{S}_\lambda(\omega) = \frac{1}{T_0}\left|\sum_k e^{-j\omega t_k}\right|^2 = \frac{|X(\omega)|^2}{T_0}$$

where T_0 is the observation time and $X(\omega)$ is the Fourier transform of $x(t)$.

In practice the function

6) $$F(\omega) = \frac{1}{N}\left|\sum_i e^{-j\omega t_i}\right|^2 = \frac{T_0}{N}\hat{S}_\lambda(\omega)$$

(where N is the number of events) is more convenient. In fact, in the case of $\lambda(t)$=const, it is easy to demonstrate that the amplitude of F (if $T_0 \gg 1$ and $N > 10$) has probability density

7) $$p_F(\xi) = U(\xi) e^{-\xi}$$

We call $F(\omega)$ the "pulse spectrum" of the $x(t)$. If the $\lambda(t)$ is of the form

8) $$\lambda(t) = \lambda_0 + \lambda_1 \sin(\omega_1 t + \varphi_1)$$

(+++) For a discussion on the second relation of 4), see [5].

where $\lambda_0 = N/T_0$, then the estimated value of the peak at pulsation ω_1 is

9) $$E[F(\omega_1)] \simeq \frac{T_0}{N} \cdot \frac{1}{T_0} \cdot \left| \int_0^{T_0} [\lambda_0 + \lambda_1 \sin(\omega_1 t + \varphi_1)] e^{-j\omega_1 t} dt \right|^2 \simeq$$
$$\simeq 1 + \frac{T_0^2}{4N} \lambda_1^2 = 1 + \frac{N}{4} \left(\frac{\lambda_1}{\lambda_0}\right)^2$$

and so, from $F(\omega_1)$ we can estimate λ_1

10) $$\lambda_1 \simeq \frac{2}{T_0} \sqrt{N \cdot F(\omega_1)}$$

Often observations are not continuous and this can introduce in the $F(\omega)$ spurious periodicities or hide real ones. A method to circumvent this, very effective in practice, consists in modifying the $F(\omega)$ in

11) $$F_0(\omega) = \frac{|X(\omega) - \lambda_0 \cdot O(\omega)|^2}{N}$$

where $O(\omega)$ is the Fourier transform of $o(t)$, which is equal to 1 during observations and null elsewhere, and

12) $$\lambda_0 = \frac{N}{\int_{-\infty}^{\infty} o(t) dt}$$

To understand the meaning of 11), it is easy to see that if $\lambda(t)$ is constant and $o(t)$ is periodical, $F(\omega)$ shows the periodicity of $o(t)$ and $F_0(\omega)$ doesn't.

If the periodicities of $\lambda(t)$ are not coherent, i.e. if the lines in the spectrum of $\lambda(t)$ are wider than the resolution, we can use an analogue of the "windowed" autocorrelation function, that we call "pulse autocorrelation function"

13) $$R_{xx}(\tau) = \sum_{i,j} \delta(\tau - t_i + t_j) \cdot W(\tau)$$

or an hystogram of it, and obtain the estimation of the spectrum of λ as

14) $$\hat{S}_{xx}(\omega) = k \int_{-T_0}^{T_0} R_{xx}(\tau) e^{-j\omega\tau} d\tau$$

k depends on the window function W, which can be chosen in different ways, for example as

15) $$W(\tau) = \begin{cases} 1 - a|\tau| & \text{for } |\tau| < 1/a \\ 0 & \text{elsewhere} \end{cases}$$

depending on the coherence time of the periodicities; in the case of $W(\tau)=1$, 14) is equivalent to 5).

The pulse autocorrelation function, in a suitable form, can be used to obtain a maximum entropy estimation of the power spectrum.

Another interesting case is that of the estimation of faint spectral lines in presence of stronger ones. A good method in this case is to "weight" the events, obtaining

16) $$F_1(\omega) = \frac{\left| \sum_{k}^{N} a_k e^{-j\omega t_k} \right|^2}{\sum_k a_k^2}$$

If, for example, a strong periodicity of the type

17) $$\lambda_1 \sin(\omega_1 t + \varphi_1)$$

is present, the weights are, with $\lambda_0 = N/T_0$,

18) $$a_i = 1 - \frac{\lambda_1}{\lambda_0} \sin(\omega_1 t_i + \varphi_1)$$

The application of these methods at the case of large pulses of gravitational antennas was very fruitful.

Bibliography

1. E. Amaldi, E. Coccia, S. Frasca, I. Modena, P. Rapagnani, F. Ricci, G. V. Pallottino, G. Pizzella, P. Bonifazi, C. Cosmelli, U. Giovanardi, V. Iafolla,

S. Ugazio, G. Vannaroni "Background of Gravitational-Wave Antennas of Possible Terrestrial Origin - I", Il Nuovo Cimento vol. 4 C, pag. 295

2. E. Amaldi, S. Frasca, G. V. Pallottino, G. Pizzella, P. Bonifazi "Background of Gravitational-Wave Antennas of Possible Terrestrial Origin - II", Il Nuovo Cimento, vol. 4 C, pag. 309

3. E. Amaldi, E. Coccia, S. Frasca, F. Ricci, P. Bonifazi, V. Iafolla, G. Natali, G.V. Pallottino, G. Pizzella "Background of Gravitational-Wave Antennas of Possible Terrestrial Origin - III", Il Nuovo Cimento, vol. 4 C, pag. 441

4. E.Amaldi, P.Bonifazi, G.Cavallari, S.Frasca, G.V.Pallottino, G.Pizzella, P.Rapagnani, F.Ricci "Analysis of the Frascati and Geneva Gravitational Wave Antenna Background Exhibiting the One Half Sidereal Day Period", Ist. di Fisica "G.Marconi", Univ. di Roma "La Sapienza", Nota Interna n. 801, 18 Aprile 1983.

5. A.Papoulis "Probability, Random Variables and Stochastic Processes", McGraw-Hill (1965).

6. S.Frasca "Pulse Process Analysis: Theory and Applications", Signal Processing II, Proceedings of EUSIPCO-83, pag. 475, North Holland 1983

APPLICATION OF BOOTSTRAP SAMPLING IN γ-RAY ASTRONOMY:

TIME VARIABILITY IN PULSED EMISSION FROM CRAB PULSAR

M.E. Özel[*] and H. Mayer-Haßelwander

Max-Planck-Institut für Extraterrestrische Physik
8046 Garching bei München, FRG

1. INTRODUCTION

New statistical and computational methods that aim to make full use of today's cheap and fast computing facilities become more and more widespread. One such scheme is called the bootstrap and fits very well for many astronomical applications. It is based on the well-known sampling plan called "sampling with replacement". Digital computers make the method very practical for the investigation of various trends present in a limited set of data which is usually a small fraction of the total population. A wide presentation of the method for the "outsiders" can be found in Diaconis and Efron (1983) and Efron (1982). Also see Simpson (1984, this volume).

The specific problem we attempt to apply the method and demonstrate its feasibility is that of the following:

Crab pulsar (PSR 0521+21) is a strong high-energy γ-ray ($\gtrsim$ 100 MeV) photon emitter and was viewed by the ESA's γ-ray satellite experiment COS-B 6 times during its 6 1/2 years life-time. Observation epochs are widely separated ($\lesssim$ 2 yr) and about 1 month of duration. γ-ray light curves of the pulsar obtained from each of these observations are given in Fig. 1. An immediate cursary observation in this figure is the apparent change in the relative strengths of the 1st (primary) and the 2nd (secondary) pulse. As a measure of this variability, the ratio of the number of γ-ray photons in the 2nd pulse, N(P2), to the 1st pulse N(P1),

$$R_i = \frac{N_i(P2)}{N_i(P1)} \quad (i = 1,6 \text{ COS-B observations}) \tag{1}$$

---

[*]Present address: Max-Planck-Institut für Radioastronomie
Auf dem Hügel 69, 5300 Bonn 1, FRG

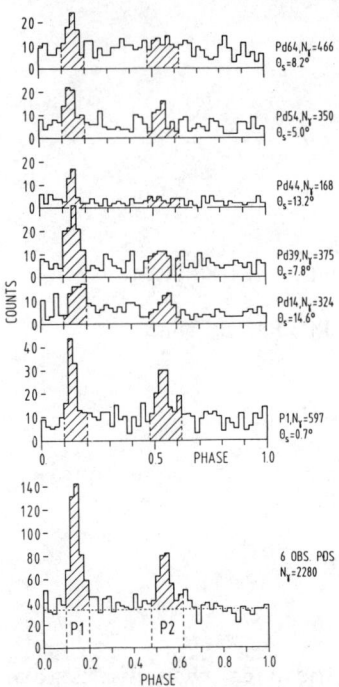

Fig. 1: Crab pulsar light curve for 50 < E < 5000 MeV γ-rays in 6 COS-B observation periods. Phase boundaries for main (P1), secondary (P2) pulses and background (BG) level estimation and the corresponding event numbers are given. For period folding parameters, see WEA. For Pd 64, parameters given by Lyne (1983) are used. (The Pd 64 light curve is a preliminary release of COS-B data, where a relative instrumental efficiency of .40 is applied.)

has been investigated by Wills et al. (WEA, 1982) and an inconclusive result with some credit to variability ($\lesssim$ 1% chance effect) was reported. The same problem is chosen as an attempt for the application of bootstrap method in γ-ray astronomy. This exercise was suitable in several respects: It would be good to compare the bootstrap estimations with WEA's "standard" calculations and also some new γ-ray data not previously used was now available for a better estimation of the variability parameter R.

2. SELECTION OF γ-RAYS FOR THE ANALYSIS

Since γ-ray telescopes have rather crude angular resolution ($\sim 1°$ at 100 MeV), one has to define γ-ray acceptance regions around the source position. When the source under investigation is embedded in a spatially structured background emission, selection of those γ-rays which are probable to increase the observed signal is not a trivial task. Full treatment of such considerations in event acceptance has been treated elsewhere (Özel and Mayer-Haßelwander, 1983). For the Crab Pulsar, which is strong and also somewhat away from the structured galactic plane regions, the event acceptance cone angle θ as a function of γ-ray energy E_γ as suggested by Buccheri et al. (1983) as $\theta \leq 12.5\, E_\gamma^{0.16}$ (MeV) is found good enough for present purposes.

With this criterium, γ-rays from 6 COS-B observations have been selected and the phase value of each γ-ray in the corresponding light curve, ϕ_{ik} (phase of the k-th γ-ray in the i-th COS-B period), are evaluated by the usual barycentrezation and folding procedures to yield the light curves of Fig. 1. For the folding procedure and pulsar parameters see WEA.

Phases for the primary (P1) and secondary (P2) peaks and baseline ("background", BG) level in γ-ray light curves are defined, in line with WEA, as

$$\Delta\phi(P1) \equiv \Delta\phi_1 = 0.10 - 0.20$$
$$\Delta\phi(P2) \equiv \Delta\phi_2 = 0.48 - 0.62 \qquad (2)$$
$$\Delta\phi(BG) \equiv \Delta\phi_3 = 0.62 - 1.00 .$$

3. BOOTSTRAP ESTIMATION OF ACCURACY OF R-VALUES

For each COS-B observation period we created "new" γ-ray light curves of the Crab pulsar by random sampling of the phase values ϕ_{ik} of the observed γ-rays by replacement. This way, we can "obtain" a high number of Crab light curves and for each we can estimate its pulse strength ratio R. As a best estimate of the error in R, we calculate its variance as

$$\sigma_{R_i}^2 \cong \frac{1}{J} \sum_{\ell=1}^{J} (R_{i\ell} - R_i)^2 , \qquad (3)$$

where J is the number of bootstrap samples created for the calculations. (Bootstrap does not give us a new estimate for R, but only the error in its measurement.) For every observed COS-B light curve, we produced J = 1000 bootstrap samples.

Fig. 2 shows the normalized frequency distributions $f_i(R)$ of R-values obtained by bootstrapping, from different COS-B observations. $g_i(R)$'s are the distributions obtained from the rest of 5 of the 6 COS-B observations combined, again by bootstrapping. Overlaps between $f_i(R)$ and $g_i(R)$ can be taken as a measure of consistency of the two distributions. For example, for periods 1 and 39 the overlaps are at a minimum. It is highly probable that

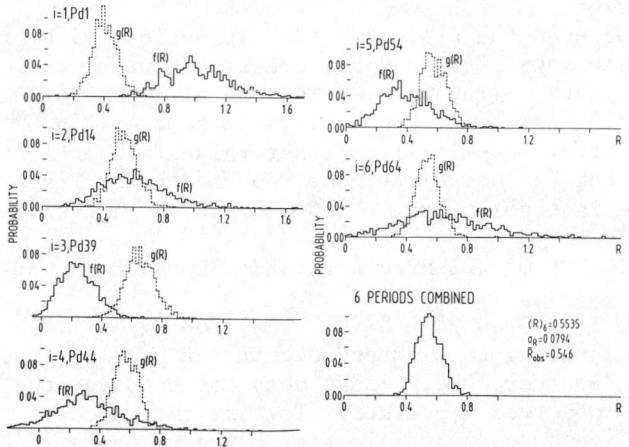

Fig. 2: Normalized frequency distributions $f_i(R)$, $g_i(R)$ for the light curves of Fig. 1, obtained by bootstrapping (a-f). Also given (g) is the result of bootstrapping for 2280 γ-rays from all 6 observations combined.

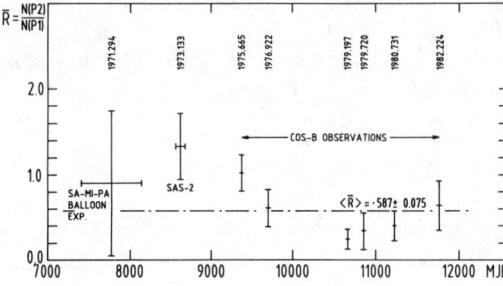

Fig. 3: Pulse strength ratio values R_i as a function of observation epoch. 2 observations prior to COS-B are also drawn. SA-MI-PA is from the balloon experiment of Saclay-Milan-Palermo collaboration (Parlier et al., 1973). SAS-2 point is for $E_\gamma > 35$ MeV and is estimated by WEA.

Table 1: Comparison of present bootstrap estimations of pulsed strength ratio value with Wills et al.(1982). Observations prior to COS-B and used in the analysis are also noted.

Observation	Epoch	$R \mp \sigma_R$ (Bootstrap)	$R' \mp \Delta R$ (WEA) (classical error propagation)	Notes
SA-MI-PA	1971.294		0.90 ∓ 0.85	Parlier et al. (1973)
SAS-2	1973.133		1.33 ∓ 0.39	Thompson et al. (1977)
COS-B Ob.Pd.1	1975.665	1.02 ∓ 0.22	1.04 ∓ 0.20	Bootstrap/WEA
Ob.Pd.14	1976.922	0.61 ∓ 0.22	0.49 ∓ 0.20	" "
Ob.Pd.39	1979.197	0.24 ∓ 0.12	0.36 ∓ 0.12	" "
Ob.Pd.44	1979.720	0.34 ∓ 0.22	0.48 ∓ 0.31	" "
Ob.Pd.54	1980.731	0.40 ∓ 0.18	0.37 ∓ 0.17	" "
Ob.Pd.64	1982.224	0.64 ∓ 0.29	—	Bootstrap

these observations have R-values that are significantly different from the rest. To quantify the variability of R-values further, we look at the graph of R_i-values as a function of epoch (Fig. 3) and test the resultant scatter of values against the hypothesis that observations are consistent with a constant <R> value (i.e. weighted mean of observations) and that the observed scatter can be explained as statistical fluctuations.

In Table I are presented the present $R \mp \sigma_R$ estimations with that of WEA values. Both sets of values are consistent with each other within 1σ error limits. (The γ-ray selection criteria for WEA are different from ours.) Fig. 3 is essentially the same as Fig. 2 of WEA, except for the first (SA-MI-PA) and last values. The first value is the result of the balloon experiment reported by Parlier et al. (1973), while the last point from the COS-B Pd 64 value is new.

4. STATISTICAL SIGNIFICANCE OF OBSERVED R-VALUES' SEQUENCE

χ^2-test applied to Fig. 3 as $\chi^2_\nu = \Sigma (R_i - <R>)^2/\sigma^2_{Ri}$ (i = 1,8) yields $\chi^2_7 = 18.38$ corresponding to a chance occurrence probability of $P(\chi^2 \geq \chi^2_7) = .0099$. However, the χ^2-test does not consider the ordering present in the observed R_i-values. To take into account this, we, further, applied a run test. In Fig. 3 the sequence (+ + + + - - - +) has k = 3 runs with respect to the mean. Its probability of chance occurrence is, as given by Fisz (1977, p. 415), $P_1 = .0078$. To be on the conservative side, we also estimated the probability of observing $k \leq 3$ runs in 8 consecutive measurements, under all possible configurations. It is $P_2(k \leq 3) = .0351$. Results of χ^2-test and run test can be combined since they are independent tests, to yield an overall significance level α_0 (Eadie et al., 1971) given by $\alpha_0 = \alpha_1 \alpha_2 (1 - \ln \alpha_1 \alpha_2)$. Here $\alpha_1 = P$ and α_2 is P_1 or P_2. For the 2 values of α_2 we have

$$\alpha_0 = .00081 \quad \text{and} \quad \alpha_0' = .0031 \tag{3}$$

respectively. These 2 values can be taken as the best and worst cases against the time variability hypothesis. A mean value of $\bar{\alpha}_0 = (1/2)(\alpha_0 + \alpha_0')$ giving

$$\overline{\alpha_0} = .002 \mp .001 \tag{4}$$

can be used for the approximate measure of odds against the time variability. (4) is a factor ~ 5 less than the WEA estimate giving stronger support to the variability of pulse strength ratio in time.

A more conservative estimation of α_0 can be obtained by using only the 6 COS-B observations, considering that combination of data from different experiments need to be avoided. In that case, one has, for $\chi_5^2 = 13.40$, $P(\chi^2 \geq \chi_5^2) = .020 = \alpha_1$ and $P_1 = .030 = \alpha_2$, which give a combined effect of

$$\alpha_0(\text{COS-B}) = .0052 \tag{5}$$

which is worse by a factor 2.5. However, it still supports strongly that the ratio of pulse strengths is variable in time. A possible explanation for this variation might be a precessional motion intrinsic to the pulsar.

5. OTHER TRENDS IN THE DATA

Bootstrap R-values for different energy bands from combined 6 COS-B observations are plotted in Fig. 4. They are consistent with a constant value for all γ-ray energy bands. It implies that P1 and P2 have similar spectra. Indeed, a preliminary analysis of

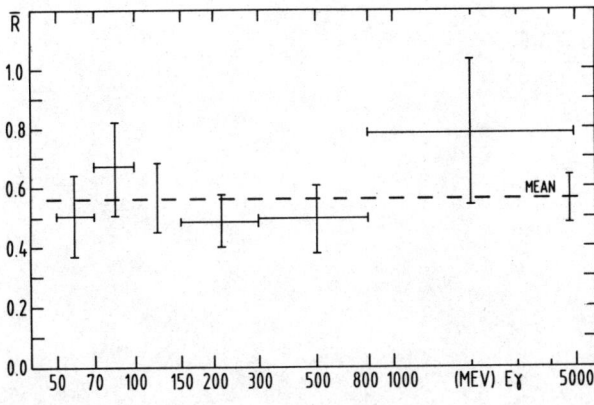

Fig. 4: Pulse strength ratio values in different energy bands from all 6 COS-B observations combined. Values are consistent with a single $\bar{R}$-value of $.56 \mp .08$.

combined 6 COS-B periods yielded, for each pulse, a power-law form E^{-s}, with index s, $s_{P1} = s_{P2} = 2.10 \mp .10$. This is consistent with Lichti et al. (1980) estimations from the first two COS-B observations. This way, we can exclude significant changes in spectra in time. Variability in pulse strength ratios, then, seems to be independent of γ-ray production mechanisms. Implications of this can be quite important for the theoretical analyses.

6. CONCLUSIONS AND DISCUSSION

The present exercise indicates that the discrete nature of high energy γ-ray data makes the bootstrap method especially attractive for γ-ray astronomy. Present analysis shows that the ratio of pulse strengths is variable with a 99.8% confidence. The nature of this variability remains uncertain and requires further work. Data is also found to be consistent with the lack of variability in spectral shapes from P1 to P2, and also, in time, within 6 1/2 yrs of COS-B observations.

We thank the Caravane Collaboration allowing us to use COS-B Pd 64 data prior to its publication.

REFERENCES

Buccheri, R. et al., 1983, Astron. Astrophys. 128:245
Diaconis, P., and Efron, B., 1983, Scien. American, Sept., p. 96
Efron, B., 1982, "The Jacknife, the Bootstrap and other re-sampling plans", Soc. for Industrial and Applied Mathematics, Philadelphia, Pennsylvania
Eadie et al., 1971, "Statistical Methods in Experimental Physics", North-Holland
Fisz, A., 1977, "Probability Theory and Mathematical Statistics", John Wiley, 3rd edition
Lichti et al., 1980, "Non-solar γ-rays", COSPAR meeting proceedings, Cowsik and Wills, eds., Pergamon Press, p. 49
Lyne, A., 1983, private communication
Özel, M.E., and Mayer-Haßelwander, H.A., 1983, Astron. Astrophys. 125:130
Parlier et al., 1973, Nature (Phys. Sci.) 242:117
Thompson et al., 1977, Astrophys. J. 213:252
Wills et al., 1982, Nature 296:723

BINARY SLICE FIT ELLIPTICITY ANALYSIS

Luigia Rusconi and Giorgio Sedmak

Universita' di Trieste, Istituto di Astronomia
Via G.B.Tiepolo 11, 34131 Trieste, Italia

INTRODUCTION

Many astronomical objects show intensity sections that approximate an elliptical shape. Optimal estimates of the parameters of the ellipse best-fit of the various sections are needed in the quantitative astrophysics of galaxies, globular star clusters, and other extended objects. The methods used come from the morphological characterization of the objects. The ellipse best-fit of isophotes is used for objects, like galaxies, that approximate the linear combination of an extended continuum to a quasi-normal noise (1). Current methods for objects, like globular star clusters, that show a substantial amount of non-normal noise intrinsic to their discrete-like morphology are based on interactive techniques involving the human operator, like star counting, and visual, photographic, or isodensitometric ellipse contour fit (2,3,4,5). Of course, star counting is limited by spatial resolution and visual-supported contour fit affected by personal bias, while being in any case highly time consuming. More effective impersonal methods can be defined using numerical image processing. One simple implementation based on the iterative fit-reject technique is described in the following. It was realized for the automated analysis of globular star clusters and proved to yield results satisfactorily similar to the visual

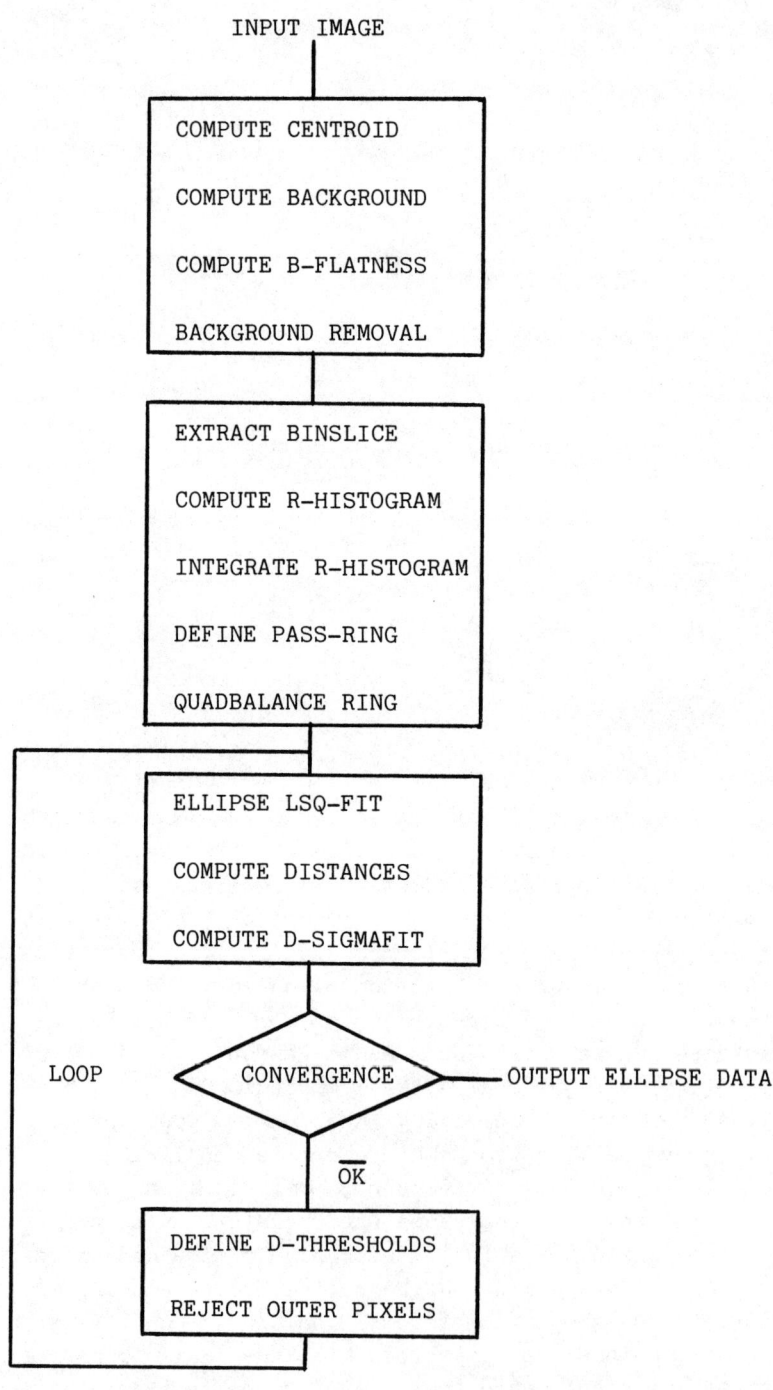

Fig. 1. Flow-chart of the binary slice fit ellipticity analyser.

estimates, when comparable, while being able to manage without relevant problems cases of low signal to noise ratio.

NUMERICAL BINARY SLICE FIT ELLIPTICITY ANALYSIS

The proposed method is described in the flow-chart of Fig.1. The image of the object is read-in as a two dimensional digital data array of (n x n) format. The approximated centroid of the object is computed by the half integrals of the (x) and (y) projections of the image. The marginal background and the background flatness are computed from a marginal square corona of width equal to 10% of image size. Objects with background flatness exceeding a given threshold (10%) are flagged for optional background removal or skipped. Binary slices of the image are generated by passing all pixels of intensity within the range defined by the assigned lower and upper slice thresholds, and giving them a constant value. The polar histogram centred on the image centroid is then computed. Thresholding the integral of the polar histogram allows to define the inner and outer radii of the circular corona containing the assigned fraction of available pixels. Threholds at 0.01 and 0.90 allow to reject most of the morphological noise with an 'a priori' loss of 11% of data. The numbers of pixels in the ring quadrants are then equalized in order to show relative weights equal within a factor 2. The passed pixels are then least-squares fitted to an ellipse and the distance of each pixel from the fitted ellipse is computed in order to define the standard deviation of the fit. The iterative fit-reject loop is implemented by the definition of an inner and outer distance threshold, followed by rejection of all pixels beyond thresholds before entering the next iteration. The thresholds are defined according to this recipe: first iteration (-3.0,+0.0) times sigmafit, further iterations (-1.0,+1.0) times sigmafit. The minus sign is given to inner pixels distances. The convergence of the loop is monitored through the current standard deviation and stopped when the sigmafit varied by less than 1% and in any case after 4 iterations.

CONCLUDING REMARKS AND APPLICATIONS

The method was tested by means of numerical simulations. It works properly for image sizes larger than about 40 times the seeing figure. Smaller images can be processed by limiting the number of iterations. One sample application to the globular star

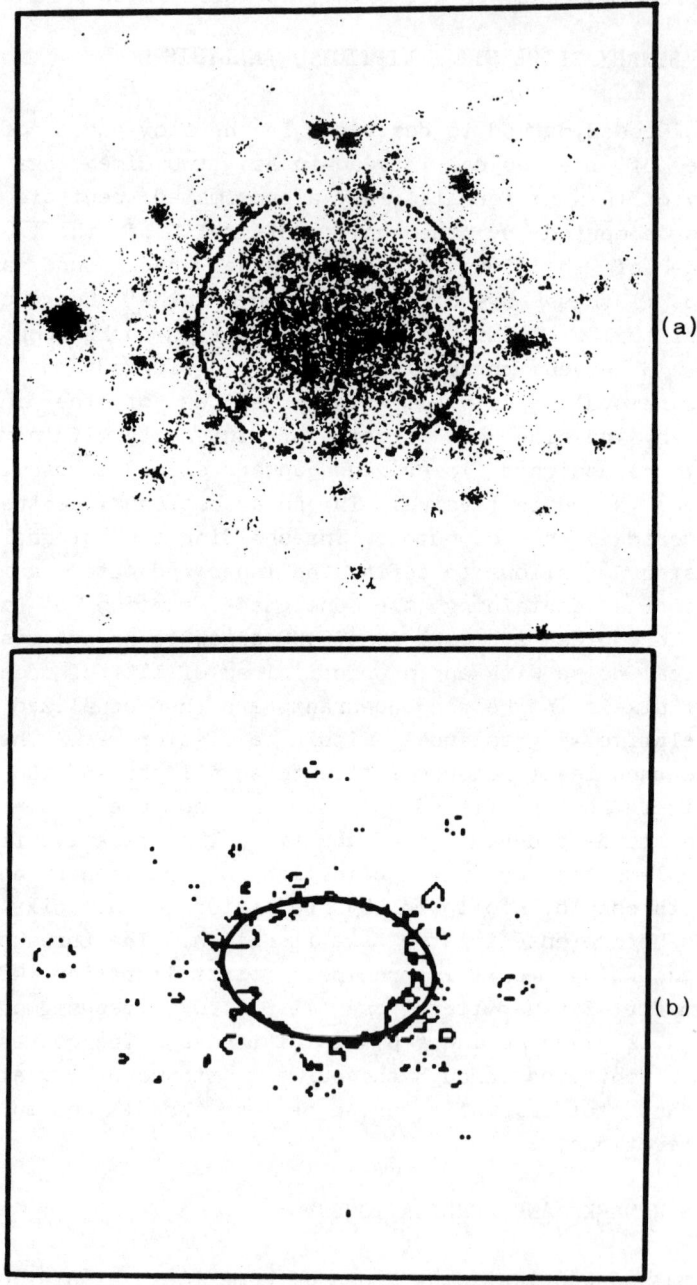

Fig. 2. (a) Object NGC121 with the ringpass circles for slice (2.0,2.2) of density range (0.82,4.28) of the object. (b) Binary slice (2.0,2.2) and final fitted ellipse.

cluster NGC121 is reported here. The object was digitized from the original plate to a (400 x 400) pixels image by means of a PDS1010A microdensitometer using a square aperture of 20 microns size and a sampling step of 10 microns. The result for the slice (2.0,2.2) of the density range (0.82,4.28) of the object, as shown in Fig.2, is quite satisfactory. The approximate computing time on a DEC VAX11/750 computer was 40 seconds for pre-loop processing plus 10 seconds/iteration for the slice of Fig.2, that contains 500 pixels. The time per iteration increases linearly with the number of pixels. The numerical binary slice fit ellipticity analyser reported here shows a reasonable efficiency and accuracy for automated processing of large volumes of data. It is now proposed to the astronomical community for constructive criticism and further implementation.

ACKNOWLEDGEMENTS

This work was carried out at Astronet Trieste center. The Starlink software was used for data presentation. The plate used for the tests on NGC121 was given by E.Kontizas and M.Kontizas, who also suggested us to try solving the problem of numerical ellipse fit of digital images of globular star clusters. We thank F.Pasian, M.Pucillo, and P.Santin for cooperation during the work, and E.Kontizas for discussion on this paper.

REFERENCES

1. Barbon,R.,Benacchio,L.,Capaccioli,M.,   Astron.Astrophys. 51(1975)25.
2. Khogolov,P.N.,   Publ.Astr.Sternberg Inst. 23(1953)250.
3. Frenk.,C.S.,Fall,S.M.,   M.N.R.A.S. 199(1982)565
4. Geyer,E.H.,Richter,T.   (1981) Astrophysical Parameters for Globular Clusters,   Eds. A.G.D.Philips,D.S.Hayes(Schenectady, L.Davis Press) p.239.
5. Kontizas,E.,Dialetis,D.,Prokakis,Th.,Kontizas,M.   (1983) Astronomy with Schmidt Type Telescopes,   Ed. M.Capaccioli, IAU Colloquium No.78, p.363.

AN AUTOMATED METHOD FOR VELOCITY FIELD ANALYSIS

Gustaaf van Moorsel

National Radio Astronomy Observatory
Edgemont Road
Charlottesville, VA 22901, U.S.A.

INTRODUCTION

This contribution concerns velocity fields of galaxies and other objects in which the kinematics can be described by a rotating disk with radial symmetry around the center. It should be emphasized, however, that the method presented here is probably also applicable in a number of different astronomical situations, but it was primarily developed for disk-like spiral galaxies.

Our information about a randomly oriented object in space is severely restricted by projection effects. We are only able to see the projections of true spatial distances onto the plane of the sky, and we can only measure the line of sight components of the real velocities. One way to overcome these restrictions is to adopt a model for the true spatial distribution of the object under study. One such model, the symmetrically rotating disk, has been proven to be very successful to model spiral galaxies. In this model, each point is in a circular orbit around the center, and all points at a given radius have the same tangential velocity. This means that its kinematics can be described by a function $V(R)$, which gives the rotation velocity as a function of the radius. The function $V(R)$ is commonly referred to as the rotation curve of the galaxy.

Apart from whether the galaxy is viewed from above or from below, the orientation in space of such a disk is fully described by 4 parameters: the position of the center (x and y), the position angle of the major axis (p.a.), and the inclination of the plane (i). Given an observed point of such a disk, those four parameters suffice to convert from observed distance to the center to true distance, and, if the overall systemic velocity of the galaxy is known, from

line of sight velocity to true rotation velocity. Thus, assuming
such a model, and given an observed velocity field, 'deprojection' in
space as well as in velocity, requires the knowledge of 5 parameters,
four of which describe its spatial orientation, and one its location
on the velocity axis. Apart from these 5 orientation parameters,
there are a number of other parameters, describing the rotation curve
of a galaxy. The number of these parameters can range from one
(namely the rotation velocity, in cases where the rotation curve is
flat throughout), to as many as one wishes (e.g. in the case of
elaborate polynomes in R).

Spectral line studies of galaxies usually result in lists of
position − radial velocity pairs, which, when shown as a map, are
referred to as velocity fields. A variety of ways to construct such
a velocity field exists; the actual construction, however, is not
the subject of this contribution, and we will assume that we have a
velocity field which truthfully represents the correct line of sight
velocities at all locations. The next task is then to find a best
fit of the model velocity field to the observed one, and to determine
the values of the orientation parameters at which the best fit
occurs. The parameters describing the rotation curve may be
determined in the same process, or can be determined afterwards.
Here first a short discussion is given of the hitherto used methods,
and their advantages and disadvantages, followed by a description of
a new method.

EXISTING METHODS

The first observations made to study the kinematics within a
galaxy can be characterized by their small number of data points,
either covering only a fraction of the optical image of the galaxy,
or observed at a resolution larger than the angular size of the
galaxy. Examples are optical spectroscopy, where the slit of the
spectrograph was made to coincide with the major axis of the galaxy,
and single dish observations of galaxies in HI, resulting in global
velocity profiles. These methods have in common that the kinematical
data contain little or no information on the orientation parameters,
which had to be determined using optical information, such as the
apparent major and minor axes.

The development of radio interferometry techniques in the 70's
made it possible to map a galaxy in velocity. Initially, when the
beam sizes were large compared to the HI diameter, similar methods as
described in the previous paragraph were adequate: cross cuts along
the major axis give the rotation curve, and the systemic velocity can
be determined very accurately applying symmetry arguments to the
whole velocity field. When resolutions became higher, and sensitivities better, the beam size/galaxy size ratio dropped steadily, and
methods were needed to include the off major axis data points in the

analysis, to make better use of the information present. A method which became popular, and has been in use until this very day, is the one described by Warner et al. (1973). In this method virtually all data points are used, and therefore the parameters of the spatial orientation are determined from the velocity field rather than to serve as input parameters. This is particularly useful in the following cases, where optical determination of the parameters proved to be difficult: 1) Optically irregular and distorted galaxies, 2) galaxies dominated by loosely wound spiral arms, 3) galaxies viewed at a low (< 30 degrees) inclination angle.

The method is based on the assumption that all material at a certain radius has the same circular velocity. For the five parameters mentioned previously reasonable initial values are chosen, and using these values all velocities in an annulus at a certain radius from the center are 'deprojected'. Projected onto the sky, such an annulus appears as two confocal ellipses. The mean circular velocity in that annulus is calculated, as well as the spread around the mean. This is repeated for all other annuli containing data points. This results in one number, the total velocity spread belonging to that particular combination of parameters. This procedure is repeated for different combinations of parameter values, in search for the combination resulting in the lowest total velocity spread.

This method has found wide applications, as well in neutral hydrogen studies as in optical mapping. In the course of the years, however, a number of disadvantages and limitations of this method have become apparent, showing the need for an improved method. The five main disadvantages are listed below: 1) The observed velocities are corrected to fit the model rather than the other way around. In particular the velocities near the minor axis must be corrected by a large amount to find the true rotation velocity from the tiny radial component. This necessitates a weighting of the corrected velocities, the weights ranging from 0 on the minor axis, to 1 on the major axis. 2) It is essentially a trial and error method, in which one try does not give an indication in which direction the parameters have to be changed to cause the total velocity spread to decrease. Only by allowing one parameter to loop through various values, keeping the other parameters fixed, an estimate of the 'best' value can be made. 3) Even if one parameter has caused a minimum total velocity spread, it is not necessarily true that this parameter still will do so for a different combination of values of the fixed parameters. This is so since the parameters are not totally independent. As an example, allowing the systemic velocity to vary, while keeping the central position fixed, the solution will be shifted from the 'true' value towards the value observed at the fixed central position. Another example of closely related parameters are the maximum rotation velocity and the orbital inclination. 4) The trial and error nature of the method does not make it clear what errors are to be assigned to the results.

A NEW METHOD

We installed in the AIPS astronomical image processing system a new method of velocity field analysis, a description of which follows here.

Each observed velocity is the result of a projection of the true circular velocity at that position. This circular velocity is determined by its distance to the center R, and by the rotation curve V(R). So, if there are five orientation paramters and N rotation curve parameters, a single observed velocity can be described as an equation with 5 + N unknowns. Typical values of N range from 1 to 5. In cases where the ratio beam size/galaxy size is very small, the total number of observed velocities can be as much as many thousands, so that a typical velocity field can be described as a system of 5000 equations with 8 unknowns. This new method consists of solving this nonlinear system of equations for the total number of unknowns by means of a least squares algorithm. The AIPS implementation uses the LMSTR algorithm from the 'MINPACK' package from Argonne National Labs.

Each unknown has to be assigned a reasonable initial value, after which the solution is usually reached in less than 10 iterations (for the example given above). The key feature of this method is that an observed line of sight velocity is compared with the value which would have been observed if the model (specified by the 5 + N parameter values) were correct. To solve the nonlinear set of equations, the partial derivatives of the model velocities with respect to the unknown parameters have to be provided. This implies that the rotation curve V(R) must be supplied in an analytical form. Examples are: 1) the one parameter constant rotation curve, valid for most flat curve galaxies beyond their central regions. 2) the one parameter linearly rising curve, describing the solid body rotation in the innermost regions of galaxies. 3) the two parameter curve which increases linearly in the central region, and reaches asymptotically a constant maximum value. 4) the three parameter 'Brandt' curve which attains a maximum velocity, and decreases Keplerian for large values of R (Brandt, 1960).

It must be emphasized that it is not necessary, even not advisable, to try to make an elaborate many-parameter fit to the rotation curve. The fitting of the orientation parameters is by far the most important task to be performed first, and supplying a functional form for the rotation curve must be seen as a means to attain that goal. In fact, it has turned out that the solution of the orientation parameters is to a high degree insensitive to the exact functional form chosen for the rotation curve. Once the orientation parameters have been established satisfactorily, the rotation curve can be determined in better (i.e. physically more meaningful) ways, e.g. integration in rings.

How does this new method perform in comparison with the one by Warner et al.? In the following, the four disadvantages of the latter method, as described in Section 2.2, are reviewed in terms of the new method. 1) The model is corrected to fit the observations. This means that the observations remain uncorrected, and no special weighting on the basis of the position in the galaxy is necessary. 2) The only 'trial' the user is involved in is the choice of the initial values of the parameters. After this, the method itself takes care of assigning new values to the parameters, leading to an as fast as possible convergence. 3) The mutual dependence of some parameters is automatically taken care of in the least squares routine. 4) The method assigns formal error estimates to the solutions.

A velocity field can consist of thousands of observed velocities, which, with only eight unknowns, forms a heavily overdetermined system. One can select subfields of such a velocity field, and apply the program on the subfields alone. Obvious examples are: solve for a given sector only (e.g. the approaching half of a galaxy), and solve for an annulus only (to study the radial dependence of position angle and inclination). To facilitate this, the program allows the user to hold any combination of parameters fixed during the fitting process. To judge the quality of the fit, a residual map (observations minus model) is made.

The fitting program has been tested extensively using Very Large Array 21 cm neutral hydrogen observations of NGC 6503, and was also applied successfully to a number of other galaxies, in which inclinations as low as 14 degrees could be determined precisely. Presently, a related program to analyze more complicated cases, in particular galaxies with warped orbital planes, giving rise to more than one velocity in the line of sight, is under development.

REFERENCES

Brandt, J.C., 1960, Astrophys. J. __131__, 293.
Warner, P.J., Wright, M.C.H., Baldwin, J.E., 1973, Mon. Not. Royal Astron. Soc. __163__, 163.

THE STAR OBSERVATION STRATEGY

FOR HIPPARCOS

S. Vaghi

ESA/ESTEC, Hipparcos Division

Noordwijk, Netherlands

SUMMARY

The Space-astrometry satellite HIPPARCOS, which is being developed by the European Space Agency, will perform systematic observations of about 100,000 preselected stars over a period of 2.5 years.

An important and complex function to be implemented on board of the satellite is the selection in real time of the stars to be observed and the allocation to them of appropriate observation times.

This function is performed by what, in the HIPPARCOS terminology, is called the Star Observation Strategy (SOS) algorithm.

The present paper describes in detail the SOS selected for HIPPARCOS.

After a short description of the main mission concept, the requirements and constraints which have been considered in the selection of the SOS are discussed. The SOS algorithm is then introduced and the general conclusions which can be drawn from the results of extensive simulations performed with realistic data are finally indicated.

1. INTRODUCTION

The main objective of the HIPPARCOS mission (Bouffard and Zeis, 1983; Perryman, 1983; Schuyer, 1983) is the accurate determination of the positions, proper motions and trigonometric parallaxes of about 100,000 preselected stars, programme stars, mostly brighter than magnitude B=10 and with a limiting magnitude of about B=13.

Target accuracies depend on the magnitudes and colours of the stars (Table 1); they are of the order of milliarcsec (rms) for positions and parallaxes and of the order of milliarcsec/year (rms) for proper motions. Star magnitudes will also be determined.

The HIPPARCOS observations consist of direct measurements of the relative angular distances between stars. Stars in the same star field and, more importantly, stars located far apart on the celestial sphere can be observed together. The latter type of observations is made possible by superposing on the focal plane of the telescope two star fields separated in the sky by a constant angle of 58°, called the "basic angle" (Fig.1). Large relative angular distances are thus effectively reduced to small apparent ones, which can be measured with high precision. The superposition of the two star fields is realized by a beam combiner mirror which combines two half-circular pupils in the entrance pupil of the HIPPARCOS all-reflective Schmidt telescope (Fig.2), thus collecting the light from two separated fields of view and projecting it on the focal plane of the telescope.

The satellite operates in a scanning mode, called "revolving scanning", which ensures the complete coverage of the celestial sphere several times during the mission. The revolving scanning can be described in terms of the movement of a properly defined telescope reference frame with respect to the ecliptic satellitocentric system.

The telescope reference frame (Fig.3) is the rectangular right-handed frame (OXYZ)
- where the origin, O, is the intersection of the two lines of sight, OP and OF, of the telescope
- the plane (X, Y) is coincident with the plane $(P_\perp F)$
- the X-axis is the bisectrix of the basic angle POF.

The revolving scanning is then defined in the following way:

the Z-axis of the telescope reference frame rotates at a constant angle $\xi = 43°$ around the Sun direction, following the Sun in its apparent motion along the ecliptic and performing $K = 6.4$ revolutions per year. At the same time the Z-axis rotates around the Z-axis in the (X, Y)-plane

Table 1

TARGET ACCURACY OF THE HIPPARCOS MISSION

Colour Index / Magnitude	B ≤ 9	B = 12
B - V = - 0.25	2.6	5.2
B - V = 0.5	2.0	4.0
B - V = 1.25	2.6	5.2

N.B. The values in the table are rms errors on star positions and parallaxes in milliarcsec and on star proper motions in milliarcsec/year.

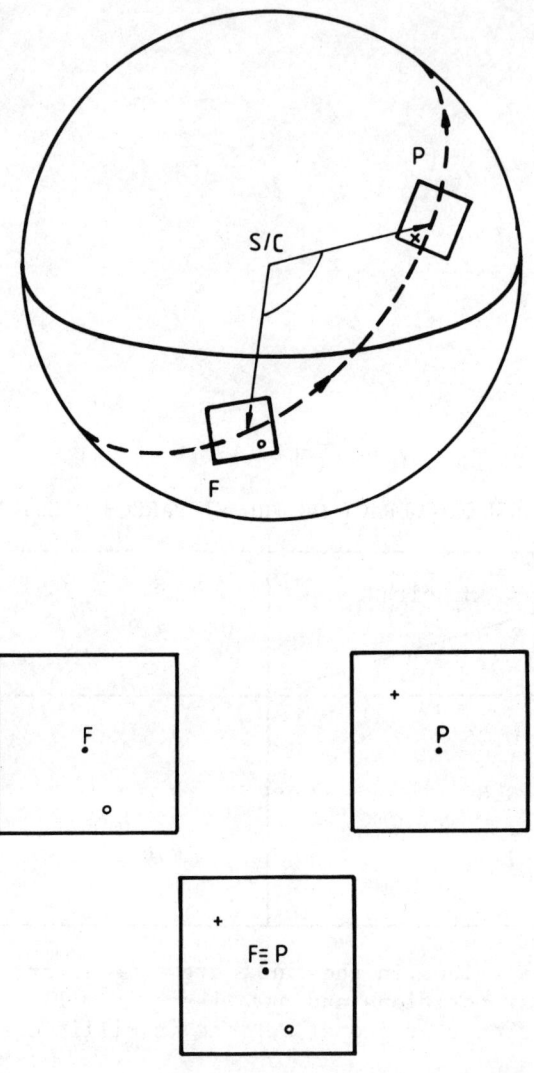

FIG.1 - PRINCIPLE OF THE HIPPARCOS OBSERVATIONS OF STARS
FAR APART ON THE CELESTIAL SPHERE

The beam combiner mirror of the telescope
projects two star fields located about
58° apart on the celestial sphere onto the focal surface.

A star (+) from the preceding viewing direction
and a star (0) from the following viewing direction appear
in the same combined star field, and their relative position
can be measured as if they were actually close together in the sky.

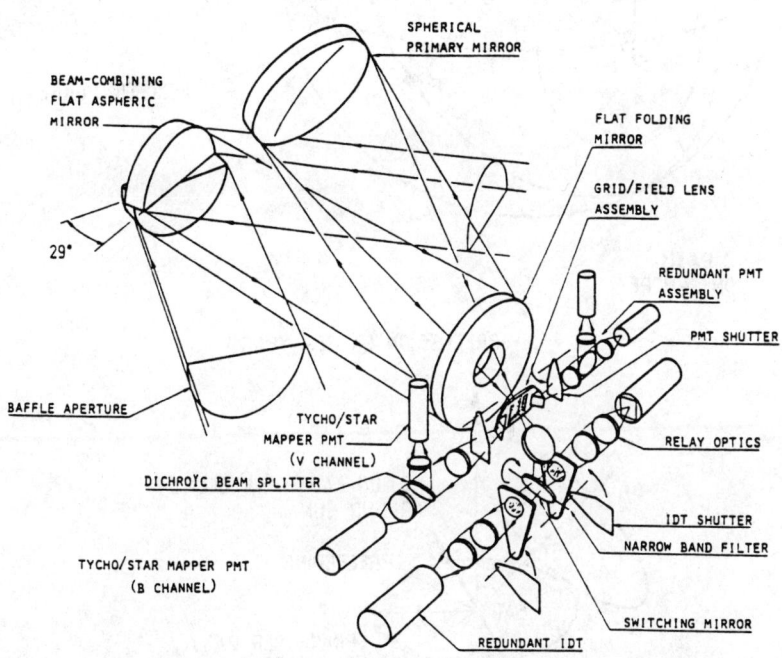

FIG.2 - EXPLODED VIEW OF THE HIPPARCOS PAYLOAD OPTICS

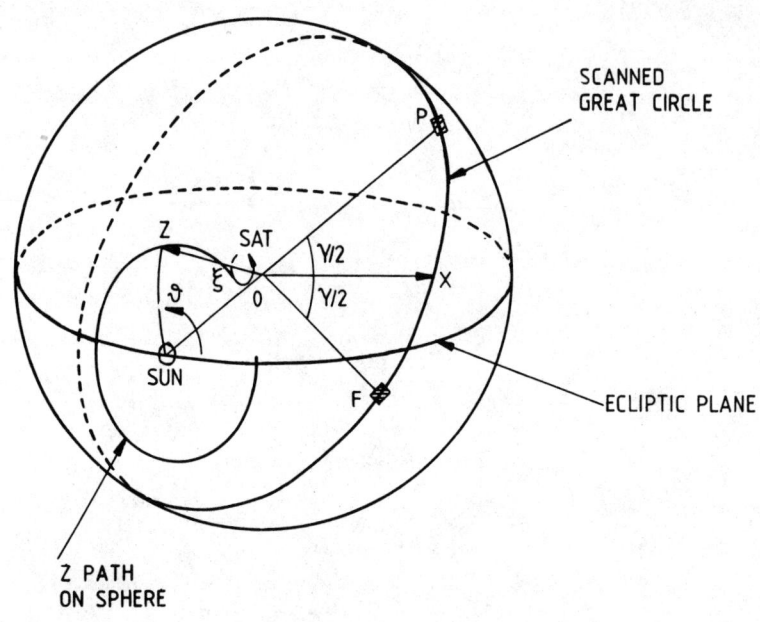
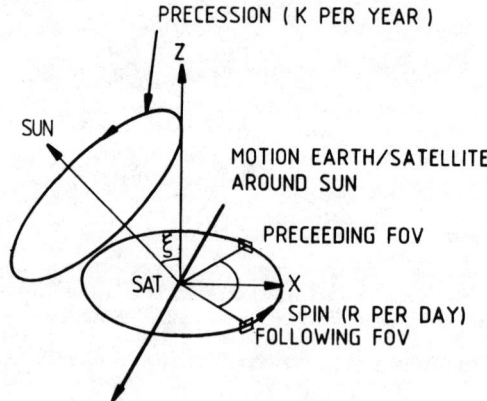

FIG.3 - TELESCOPE REFERENCE FRAME AND REVOLVING SCANNING

performing R = 11.25 revolutions per day. The motions of the axis OP and OF of the Preceding and Following fields of view which scan the celestial sphere are consequently determined.

At focal plane level the result is a continuous flow of stars from both viewing directions. Each star crosses the field of view (FOV) of the telescope in 19.2 s, and 4 programme stars, on average, are in the combined FOV at any time.

A grid covering a square field of 0.9 deg x 0.9 deg is situated at the focal surface of the telescope. It consists of 2700 parallel slits perpendicular to the scanning direction, with a period of about 1.2 arcsec. The grid modulates the light of each crossing star, which is then collected by an image dissector tube (IDT) and converted into photoelectron counts.

The sequence of photoelectron counts obtained during the transit of a star can be used to derive its phase. From the phase difference of two stars observed almost simultaneously in the FOV, their angular distance can be derived.

The instantaneous FOV (IFOV) of the IDT is circular with a diameter of 37 arcsec. The IFOV position can be directed at any point of the FOV by varying the currents applied to the deflection coils. Only one star can be followed at a time, hence strictly simultaneous observation of two stars is not possible. Quasi simultaneous observations are, in practice, achieved by switching very frequently the IFOV from one star to another, following a predefined scheme.

The scheduling policy used to reposition the IFOV this way and allocate the observation time to the various stars is, in the HIPPARCOS terminology, called the star observation strategy (SOS).

The aim of the presend paper is to describe the SOS selected for HIPPARCOS.

In Section 2 the objectives of the SOS are indicated, and the constraints discussed.

In Section 3 the SOS algorithm is introduced, and in Section 4 the results of extensive simulations performed with realistic data are briefly summarized.

2. OPTIMISATION CRITERIA AND SYSTEM CONSTRAINTS

The SOS algorithm, implemented in the on-board computer, selects at regular time intervals among the programme star which cross the FOV those which have to be observed, and allocates to them appropriate observation times.

The choice of the algorithm depends on optimisation criteria related to the performance of the mission, and on system constraints due to the hardware environment in which the SOS has to operate.

The main optimisation criteria are the following:

- Minimisation of jitter effect.
  Since observations of stars contemporaneously present in the combined FOV are not strictly simultaneous, the attitude jitter of the satellite could introduce considerable noise in the measurements.

  The SOS must be able to minimize this effect by proper interlacing of star observations.

- Even distribution of observations between the two viewing directions.
  Peculiar to the HIPPARCOS concept is the capability to measure relative angular distances between stars located far apart on the celestial sphere. It is one of the objectives of the SOS to make sure that, when several stars are present in the FOV, angular distances are preferably measured between stars coming from different star fields.

- Compatibility with the global observing programme.
  In order to achieve the target precision at the end of the mission, stars of the various classes of magnitudes are supposed to receive, in average, a certain, pre-defined, global observation time. One of the functions of the SOS is to ensure that, at local level, stars receive a time allocation compatible with their global observation time.

- Special emphasis on the observation of bright stars.
  Observations of bright stars ($B \leq 9$) are particularly valuable. Their positions can be measured with high precision, and subsequently used in the reconstitution of the attitude of the satellite during the relatively long time periods (400 s in average) between control jet firings. This "smoothing" of the attitude reconstitution makes, in turn, possible the establishment of connections between pairs of bright stars not

contemporaneously present in the combined FOV. These
additional connections can considerably improve the final
astrometric results.

The SOS must be such that bright stars are observed as long
as possible, that is as soon as they enter the FOV and until
just before they get out of it.

- Minimisation of wasted observation time.
  No observation time should be wasted. Thus, ideally, no
  star is observed only when no star is present in the FOV.
  Although this, in practice, is not always possible, the
  performance of the SOS must come very close to this
  objective.

- Minimisation of errors due to grid imperfections.
  Imperfections in the manufacture of the grid can induce
  errors in the phase extraction of a star. These errors
  can be considerably reduced by a proper choice of the
  observation scheme.

The main system constraints related to the hardware environment in which the SOS has to operate are:
- Memory and CPU time limitations of the on-board computer
- Synchronization with the telemetry format, the attitude
  and orbit control system and the cold gas thruster firing
- Constraints due to the downlink resources.

The establishment of a satisfactory SOS can thus be regarded
as a fairly complicated optimisation problem in the presence of
several constraints.

3. THE SOS ALGORITHM

The SOS algorithm is built around a rigid time hierarchy and
its operation is driven by three star-dependent parameters uplinked from ground.

The time hierarchy is based on the following definitions:

- Sampling period T_1

  T_1 = 1/1200 s is the sampling time during which photoelectron
  counts are accumulated by the IDT.

- IFOV repositioning period T_2

  $T_2 = 8\, T_1$ is the shortest time interval (or "slot") during
  which the IFOV remains pointed on a given star. Each
  star is always observed during an integer number of slots.

- Interlacing period T_3

    $T_3 = 20\ T_2$ is the period of time during which a group of up to 10 stars are observed.

- Frame period T_4

    $T_4 = 16\ T_3$ is the period of time during which essentially the same group of stars are observed in the same order and with a given observation time allocation (exceptions are discussed below).

- Transit time T_5

    $T_5 = 9\ T_4$ is the time taken by a star to cross the FOV of the telescope.

The star-dependent parameters uplinked from ground are

- The selection index b.
  b is the parameter used to calculate the priority with which a PS must be observed with respect to the other programme stars contemporaneously present in the FOV.

- The minimum observation time x.
  x is the minimum number of slots of T_2 which, at frame level, must be allocated to a star in order to achieve a sufficient precision in phase extraction.
  x is a function of the magnitude of the star.

- The target observation time y.
  y is the observation time which, at frame level, must be allocated to a given star in order to achieve, at the end of the mission, the global observation time associated with that star.

The values of the star-dependent parameters may vary with time, to take into account the past and projected observational history of the star during the mission, but they are fixed during each crossing of the FOV by the star.

Other parameters included in the SOS algorithm are the time of entrance of the star in the FOV, its magnitude, and a flag indicating whether the star is observed in the preceding or following viewing direction.

As mentioned in the introduction, the programme stars are pre-selected stars contained in an Input Catalogue prepared by a scientific consortium of astronomical Institutes and delivered to ESA well before the satellite is launched.

During the satellite operations the ESA Operations Control Centre elaborates the star data derived from the Input Catalogue and the observational history of the programme stars to produce a so-called Program Star File (PSF) containing the information to be uplinked to the satellite.

The PSF essentially contains star identifications, magnitudes, FOV entering times, viewing direction flags and the current values of the three star-dependent parameters discussed above. A scheme of the data processing involved in the generation of the PSF is shown in Fig.4.

Before describing the SOS algorithm, a few more definitions have to be introduced.

Programme stars which are found in the combined FOV at a given time are called "current stars".

When a frame of T_4 seconds is considered, current stars which remain in the combined FOV during the entire frame are called fully observable stars (FOS); those which remain in the combined FOV only for a part of the frame, but not less than 3 interlacing periods T_3, are called partialle observable stars (POS), and can be stars which either leave or enter the combined FOV during the frame period.

One way to visualize the flow of programme stars in and out the combined FOV is by means of time-diagram such as the one shown in Fig.5.

A star S_i enters the FOV at time $t_o^{(i)}$ and leaved it at time $t_o^{(i)} + T_5$.

The "transit line"

$$f_i(t) = \frac{t - t_o^{(i)}}{T_5}$$

represents the fraction of transit time spent by the star in the FOV

$$0 \leq f_i(t) \leq 1$$

One can easily identify the stars which are present in the FOV at a given time t' by drawing the line $t = t'$: they are those whose transit lines are intercepted by $t = t'$.

By marking on the t-axis the time intervals corresponding to subsequent frames of T_4 seconds, one can also identify the stars present in a given frame. For example, in frame N° 10 of Fig.5 there are three current stars, of which two are FOS and one is a POS, leaving the FOV before the end of the frame period.

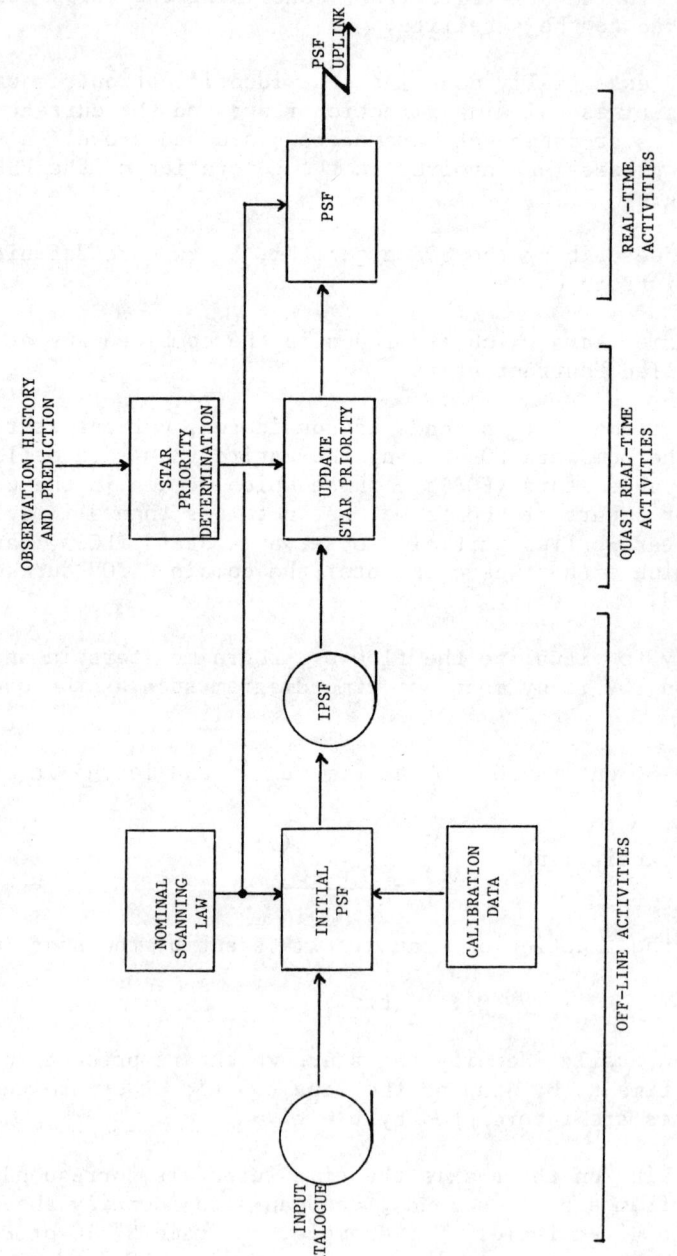

FIG.4 - PROGRAMME STAR FILE GENERATION PROCESS

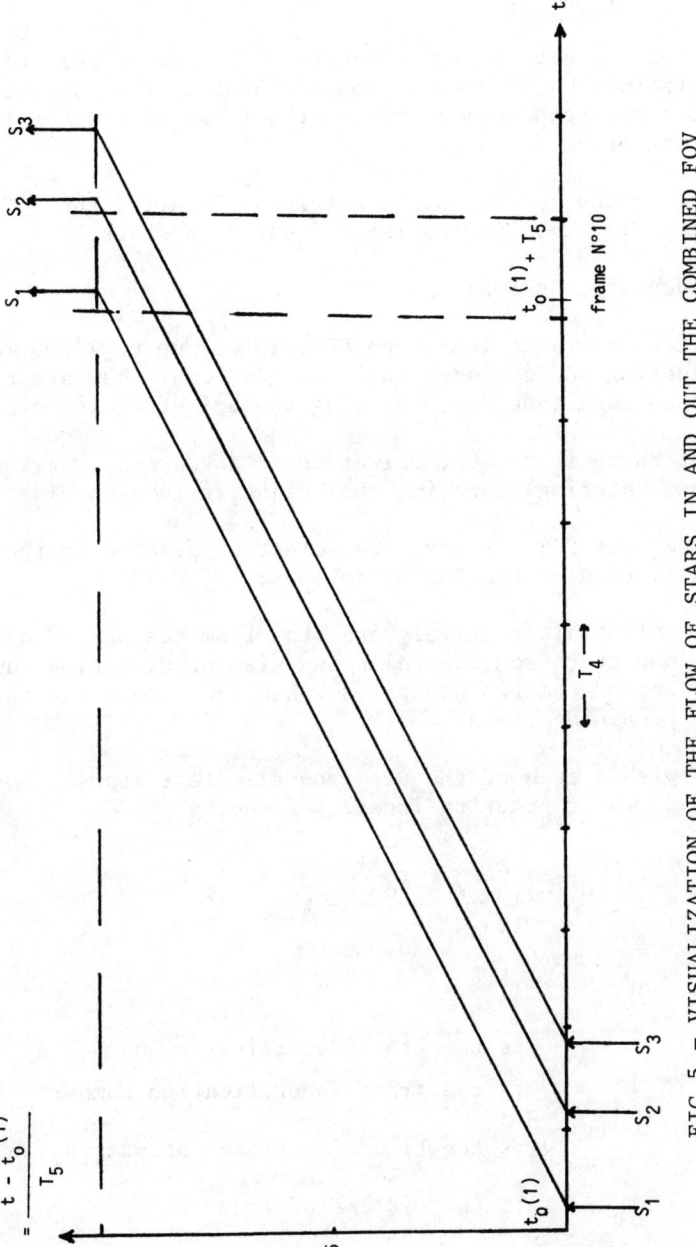

FIG.5 - VISUALIZATION OF THE FLOW OF STARS IN AND OUT THE COMBINED FOV

The SOS algorithm provides two functions, star selection and observation time allocation, according to the following prescriptions:

(a) Star selection.

1. From the data contained in the PSF and the real time attitude calculated on board, identify the current stars for the next frame; retain the first 15 of them for further selection;

2. Among the current stars retained, identify the POS and classify them in leaving and entering stars;

3. Identify the FOS;

4. If there is at least one FOS, select up to 2 POS (one leaving and one entering) provided that they are brighter than magnitude B=9, choosing the brightest in each class;

5. If there is no FOS, select up to 2 POS (one leaving and one entering) choosing the brightest in each class;

6. Once the POS, if any, are selected, proceed to the selection of the FOS as follows:

   - select alternatively one star from the preceding and one star from the following viewing direction, until one of the two groups is exhausted; then add the remaining stars;

   - within each of the two groups, select stars according to their priority index, defined by

$$P_i = (-1)^k \frac{t_k - t_o^{(i)}}{T_5} b_i$$

where

i	is the star identification number of star S_i
k	is the frame identification number
$t_o^{(i)}$	is the time of entrance of star S_i in the FOV
t_k	is the midtime of frame k
b_i	is the selection index
T_5	is the transit time.

Give priority to stars with the higher values of P_i.
(It should be noted that, due to the factor $(-1)^k$, priority
indexes change of sign from one frame to the next, in such
a way that priorities are alternating: the star with the
highest priority in frame k is, as a rule, the star with
the lowest priority in frame k + 1);

7. The complete list of stars selected consists of the POS, if
   any, plus a number of FOS, if any, such that the total
   number of current stars selected does not exceed 10.

(b) Observation time allocation.

(Time is allocated over an interlacing period of T_3 seconds.
The allocation is identically reproduced for the 16 interlacing periods contained in a frame, except for the change
described at point 5 below).

1. If there is no FOS in the FOV, allocate

   - 10 slots of T_2 seconds to each POS, when two POS are selected,
   - 20 slots of T_2 seconds to the one POS, if only one POS is selected;

2. If there is at least one FOS in the FOV, allocate 2 slots
   to each POS selected.

Once the observation time allocation to POS is completed,
allocate the remaining slots available to the selected
FOS as follows:

3. Allocate, in sequence, to each FOS its minimum observation
   time x_i, until either the list of stars is exhausted or
   the number of slots available (20) in the interlacing
   period is exceeded. In the latter case drop the remaining
   FOS from the observation list and no longer consider them.

   In absence of POS, at least two FOS, if present, shall
   always be observed, by allocating, if necessary, an
   observation time shorter than the minimum observation
   time (for the faintest stars the nominal x_i is larger
   than 10 slots);

4. Allocate the remaining slots, if any, one by one to the
   FOS actually observed, on the basis of their so-called
   "performance index" z_i, defined as

   $$z_i = \frac{n_i}{y_i}$$

where n_i is the actual number of slots already allocated to the star in the interlacing period, and y_i is target time. Allocate the first available slot to the star with the lowest performance index, and so on;

5. If POS are actually observed, allocate the 2 slots which are free when these stars are not present in the FOV to the FOS with the lowest performance coefficient.

Once these two functions are completed, the actual observation sequence is executed in the following order:

- in each interlacing period the FOS are observed in order of entrance in the FOV;
- all slots devoted to the same stars are contiguous;
- during the interlacing periods in which POS are present, the observation slots of each POS are contiguous with the ones of its associated FOS;
- the entering POS is observed after its associated FOS;
- the leaving POS is observed before its associated FOS;
- if there is no FOS, the leaving POS is observed before the entering POS.

The above description contains only the main features of the SOS algorithm. Full details are found in the documentation of the SOS computer program.

4. SIMULATIONS AND RESULTS

A simulation program has been developed to test the performance of the SOS algorithm. The general functional diagram of the program is shown in Fig.6.

The program can accept, as input, stars from a simulated input catalogue (SIC), randomly generated stars based on a given star distribution mode, or simpler lists of stars prepared by the user.

Output from the program is the description, frame by frame, of the resulting sequence of star observations, and general summary statistics on the performance achieved in terms of number of stars observed and time allocated.

Extensive simulations have been performed with realsistic data taken from the SIC, over time intervals covering up to one great circle, i.e. a complete revolution of the viewing directions about the Z-axis (2.133 hrs.). Particular attention has been given to the cases of high star density, when the telescope is scanning in proximity of the galactic plane.

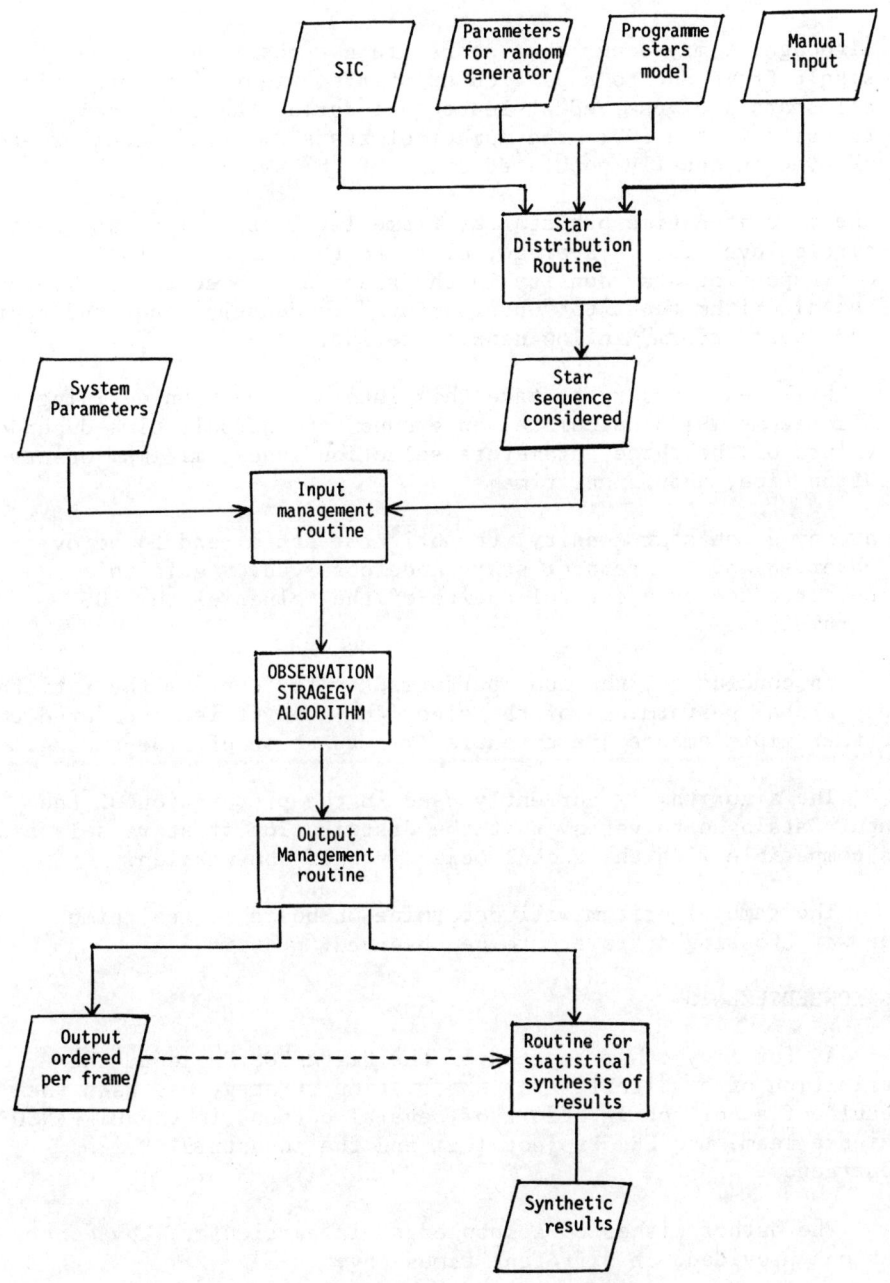

FIG.6 - FUNCTIONAL DIAGRAM
OF THE SOS SIMULATION PROGRAM

Some general conclusions can be drawn from the simulations performed:

- although it may occur that stars are not observed in some single frame due to a particular star configuration, essentially all stars are observed at least once during their two complete transits in the FOV; the number of stars "missed" never exceeds 1% even in densily populated areas of the sky;

- the allocated time per star at frame level as well as at great circle level is, in average, close to that expected for the corresponding star density in the great circle scanned. Typically the algorithm tends to "underperform" in densely populated regions, and "overperform" in low density region.

  It will be necessary to take this into account when defining the algorithms which calculate, on ground, the actual, time dependent, values of the three parameters selection index, minimum observation time, and target time;

- at any given star density, the brighter stars tend to be over-observed and the fainter stars underobserved. Again this can be corrected by a careful choice of the values of the SOS parameters.

In conclusion, the runs performed so far confirm the satisfactory global performance of the algorithm, and it is considered that further improvements are essentially a question of fine-tuning.

The algorithm is currently used in the preparation of the Input Catalogue to verify that the distribution of stars selected is compatible with the global objectives of the mission.

The same algorithm will determine on-board in real time the way crossing stars are to be observed.

ACKNOWLEDGEMENTS

As for many other aspects of the HIPPARCOS Project, the definition of a suitable star observation strategy has been the result of a collective effort of several persons in the HIPPARCOS Science Team, the ESA Project Team and the industrial Prime Contractor.

The author wishes to acknowledge, in particular, key contributions provided, at different times, by:

- M. Crézé, who introduced the concept of target observation time;

- J. Kovalevsky, who stressed the special importance of bright stars observations and suggested the introduction of the alternating priority scheme;

- L. Lindegren, who first introduced the concept of a rigid time hierarchy and proposed a preliminary SOS algorithm;

- J.J. Tiggelaar, who has been responsible for the development of the SOS simulation program, and contributed to the clarification of many fine details of the algorithm;

- E. Zeis, who managed, with much skill, to reconcile scientific creativity with technical reality and propose a star observation strategy both feasible and scientifically acceptable.

The present paper should hence be regarded as the summary of a collective effort, and by no means as a report limited to the author's own contribution.

REFERENCES

Bouffard, M. and Zeis, E. (1983), The Hipparcos Satellite, in Bernacca, P.L. (ed.), The F.A.S.T. Thinkshop: Processing of scientific data from the ESA Astrometry Satellite HIPPARCOS;

Perryman, M.A.C. (1983), The HIPPARCOS Space Astrometry Mission, Adv. Space Res., 2, 51-58;

Schuyer, M. (1983), HIPPARCOS - A European Astrometry Satellite, in Napolitano, L.G. (ed.), Space 2000.

PANEL DISCUSSION ON

DATA ANALYSIS TRENDS IN X-RAY AND γ-RAY ASTRONOMY

30/5/84, $11^{00} - 12^{00}$

        Participants: M.E. Özel   (MPIfR, Bonn), Chairman
                            L. Scarsi   (IFCAI/CNR, Palermo)
                            R. Buccheri   (IFCAI/CNR, Palermo)
                            J. Friedman   (Stanford University, USA)
                            G. Simpson   (Univ. of New Hampshire, USA)
                            P. Mussio   (Universita' di Milano)
                            V. di Gesu'   (Universita' di Palermo)

[*The text of the panel has been edited by Dr. Özel (with indispensable help from Gabi Breuer, secretary of MPIfR) from a tape recording. The words not completely understandable are noted by (?), while various inclusions for the continuity of the text are indicated by [ ]. The slides and viewgraphs presented in the panel are added as Figures and Tables.*]

ÖZEL: Before addressing the problems of data analysis in X- and γ-ray astronomy, I will try to make a very brief introduction to these two sister-fields, relevant to our discussion, remembering that many in the audience are not the experts of these fields. In these energy bands, particle nature of electromagnetic radiation dominates completely. In the spirit of the previous talks in the workshop, we can represent any recorded celestial X- or γ-ray as a point in a multi-dimensional space with space (α,δ), photon energy (E), polarization (in short, $\vec{P}$) and time dimensions: $\gamma(\alpha,\delta,E,\vec{P},t)$. Table I summarizes various statistical and observational properties of X- and γ-ray photons and also gives a rough idea about the observational status for these two fields.

Table 1: X- and γ-ray Astronomies (A quick, comparative look)

Property		X-ray Astronomy	γ-ray Astronomy
Photon energy E		100 eV > E > 100 keV	E > 100 keV Most explored part: 30-2000 MeV
State of art values for	Statistics (counting rate)	good ($\gtrsim 10^2 - 10^3$ s^{-1})	poor (long integration time is necessary) ($\gtrsim 1$ hr^{-1})
	Angular resolution Δθ	∼ 1"	∼ 1°
	Energy resolution ΔE/E	$\lesssim$ 10%	∼ 25%
	Polarization measurements	routinely achieved	not yet available
Satellite experiments with significant new results (date)		Uhuru (SAS-1) (1971) SAS-3 (1975) Einstein (1979)	OSO-III (1968) SAS-2 (1972) COS-B (1975-1982)
Present:		EXOSAT	
Future:		ROSAT (†) (1987) AXAF (††) (1990's)	γ-1 (*) (1985) GRO (**) (1990)

(†) <u>R</u>ontgen <u>Sat</u>ellite of W. Germany, x3 improvement over Einstein

(††) <u>A</u>dvanced <u>X</u>-ray <u>A</u>strophysics <u>F</u>acility with x5 improvement over ROSAT, it will operate $\gtrsim$ 10 years with astronauts visiting it for service

(*) French-Soviet experiment, E > 20 MeV, Δθ ∼ 2° per single photon (100 MeV), x4 SAS-2 sensitivity, coordinated X-γ observations (has a 2 - 25 keV X-ray detector as well)

(**) <u>Gamma-Ray</u> Observatory: US-European experiment, E > 100 keV, 4 γ-ray experiments to cover up to 30 GeV with energy resolution ∼ 15% at 100 MeV, 2 years duration, x20 SAS-2 sensitivity, retrival by Space Shuttle.

Now I come to the question, what to analyze or how to analyze the data obtained. These photons basically carry the following information about the emission regions/objects:

1. Temporal structure (time variability, periodicities, etc.)
2. Spatial structure (localized sources, intensity variations, spatial features, etc.)
3. Spectral structure (distribution of photons in energy, emission mechanisms, models, etc.)
4. Polarimetric structure (polarized intensity, polarization direction, relevant parameters, etc.).

The following graphs [Fig. 1] summarize the improvements achieved by consecutive γ-ray experiments, in angular and energy resolution and sensitivity, as well as in the total volume of the data base in the history of γ-ray astronomy. Lack of any levelling-off in the graphs is representative of any young, rapidly developing branch of science.

Now, we will try to address in our panel, the relevant problems in turn. I would like to start with another historical remark and question: The COS-B γ-ray collaboration was realized as an international effort including 6 institutions in four countries. And for seven years the same standards of data analysis methods and

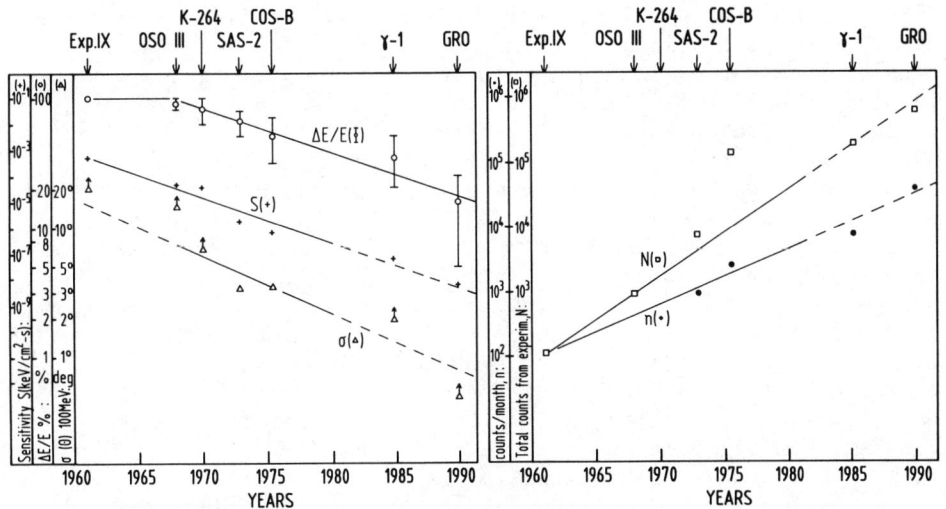

Fig. 1: Improvements achieved by consecutive γ-ray experiments since 1960's. In the lefthand graph are drawn the improvements in experiment sensitivity (minimum detectable flux level) S (keV/cm²-s), in angular resolution σ (°) and in energy resolution ΔE/E. The righthand side shows the increase in the volume of γ-ray data in unit observation interval (1 month) and in the total experiment duration. The off-line position of total volume of COS-B experiment is due to its 6 1/2 years life time.

various data reduction procedures have been kept in track. Therefore I would like to ask, first, Prof. Scarsi the following: How was this achieved in the planning phase and did you really expect that COS-B would be so much successful in terms of previous experiments in gamma-ray astronomy?

SCARSI: Well, I was involved with COS-B from the beginning. As we can see a proper management has played an essential role for the success of COS-B, for its more than 6 years of operation. This was preceded, naturally, by a planning phase for 4 or 5 years before. Well, first of all, let us look what kind of data we got: we got gamma rays detected by a spark chamber instrument. And the signature of the event is visual. It was, roughly, an inverted V, composed of, let's say, around 20 sparks. They contained most of the information that you could use to deduce energy and direction of arrival. Same sparks tell us that this is a gamma ray. It was preceded — the organisation — by a long period of instrument calibration in accelerators, by injecting, on the spark chamber, gamma rays of known energy and known direction. And the first problem to be solved was a pattern recognition problem. How, automatically, to extract the good gamma rays from the background, taking

into account the variety of forms which could appear due to varying
energy, due to statistical fluctuations etc. This was the first
problem attacked by a data reduction group (DRG), a group that was
settled by scientists from collaborating labs. DRG stayed in
operation up to now. To tell you the difficulty encountered in
pattern recognition, we should remember that one over about one
hundred recorded events was considered acceptable; so it had to be
a sophisticated system. The second problem to solve was that the
spark chamber telescope was degrading during these years and moreover, every two months, almost all the gas in the spark chambers
has to be changed, and then you have a picture with a changing spark
noise, i.e. the spark efficiency is changing during the observations.
So the pattern recognition problem has to be adaptive and in a way
easily controllable. Other colleagues in the panel will also talk
on that. We come now to another technical point to be solved, the
problem of management and to have a data bank, a centralized data
bank. This is the problem that we face now: at the end of the
operational part of COS-B, to release the data to the scientific
community. Release of the data collected over more than 6 years,
∿ 150000 photons that we are talking about. Every one of them, i.e.
identified, reduced, catalogued. This will be the official COS-B
data base. The problem that we will face with [the coming γ-ray
experiments] Gamma 1 and GRO I think will be simpler. The only
change that we can anticipate is, possibly, a more stable operation,
also possibly a higher statistics. The main problem is, more or
less, identical: identification of pictures photon by photon
against a background, possibly varying with time.

ÖZEL: Thank you. Now, let's come to the problem of temporal analysis of gamma-ray data. If you are aware, there is a list of
"official" COS-B γ-ray "point" sources. Here I repeat [Fig. 2]
their space distribution in galactic coordinates and the unhatched
regions are observed by COS-B and, more or less, the parts of sky
which we know best. And these circles and points are the "official"
sources. Yesterday our colleague Buccheri gave us a comprehensive
summary of how time analysis has been achieved in gamma-ray
astronomy and radio astronomy. I would like to ask him, again,
whether all possibilities for temporal analysis have been exhausted.

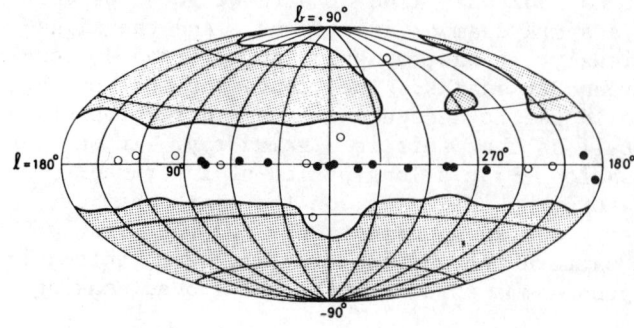

Fig. 2: γ-ray sources of 2CG catalogue in galactic coordinates. Sources stronger than $1.3 \cdot 10^{-6}$ ph/cm²-s are denoted by filled circles. Shaded areas are the sky regions not searched by the COS-B experiment.

Is it possible to look at the time variability of these sources at a different scale?

BUCCHERI: Concerning the periodicities, there is not much to add to what has been said yesterday. I want to stress here that future experiments of high energy gamma-ray astronomy should concentrate on the task of improving the signal-to-noise ratio, increasing the sensitive area and achieving a better angular resolution. In this way, also, unbiased periodicity searches will be possible and we could undertake pulsar searches independently of the radio observations which are essential nowadays. Pulsars with gamma-ray beam directed toward the earth and radio beam directed elsewhere will therefore become detectable (if they exist). In the X-ray range, where shorter integration time is requested and better counting statistics are available, this is already possible.

Now I want to rise a question related with a problem that has not been discussed much yet, but which has been very important at the beginning of the COS-B mission. I believe it is important for any experiment in gamma-ray astronomy, at least until when the kind of detector we use is the spark chamber. It is the problem of the discrimination of the good gamma rays against the "noise": the pattern recognition problem [Prof. Scarsi noted], and it's good that we have here several experts on this field in order to make some further points. The discrimination of good events against the noise has been done in COS-B by using a pattern recognition computer program which was based on the knowledge of the physics involved in the detection, the main ingredient being the scattering of the electron-positron pair by the tungsten plates and the following measurement of the position of the spark between two plates. Now, the shape of the observed tracks depends on the thickness of the plates and their distances and is contaminated somewhat by the spurious sparks arising from the inhomogeneities of the gas filling the spark chamber. For COS-B the pattern recognition program has been settled at the beginning of the experiment and has been used without changes all along the life of the experiment which lasted almost 7 years. So, the problem is this: This longevity could have caused a degradation of the sensitivity of the experiment, which in fact at the end of the mission was reduced to the half with respect to the beginning. Since we used the same pattern recognition program, it could be that the shape of the tracks changed during the life of the experiment due to poisoning of the gas. The gas was refilled from time to time [$\sim$ every 2 months] and therefore the physics of the system could have changed and some kind of inhomogeneity could have arisen. So, if the shape of the tracks changes with time, then one should, in principle, change the programs for the discrimination. This can be done only learning from the transmitted photons. Now I wanted to ask Dr. Pierro Mussio, who was in COS-B at the beginning of the experiment and

worked very much on this problem, what is his opinion about this
problem and whether he shares some of my concerns?

ÖZEL: Please, Dr. Mussio. I hope you would like to address to the
problem of automatic pattern recognition which, I think, will also
be very useful for the next generation spark chamber experiments.

MUSSIO: Well, first I shall speak about a more general problem.
Because the problem with the COS-B pattern recognition program was
tied to the larger problem of the necessity of a cultural exchange
among computer scientists and physicists to arrive at a real interdisciplinary collaboration, and this is the main point for the
success of a scientific data analysis program. Actually the tools
used are deeply tied with the logic of the experiment. Therefore,
one must know the experiment and understand what the computation is.
At present, generally, it seems to me impossible to combine in one
person the two professionalities. Therefore a collaboration is
necessary. As to the specific problem pointed out by Lino Buccheri
we proposed a structural pattern recognition program which worked
with good results on real data. It was a table-driven program, because we thought that it was necessary to adapt in time the program
to the new data. And we proposed a Bayesian approach to the evaluation of the physical parameters. Because we thought that it would
have been possible to adapt such a scheme in time. But 'a posteriori' I must emphasize two aspects of my experience. The first is
that the data reduction group, at least at the beginning, until I
was in the group, was much more concerned with technicalities which
are essential to the good results from the experiment, but are not
essential to the good understanding of the experiment. I mean, we
lost a long time, many many months, in defining the exact format of
the data to be recorded on the tape, and this was obviously
essential for the exchange of the data among the different centers.
But in my opinion, it was not exactly the kind of task for a
physicist. I mean it is not necessary for the physicists to go
into the tape formats etc. On the contrary, there was very little
effort, in my opinion, in understanding what a computational tool
is, in understanding what was the impact of the software tools on
the data which were now treated in a different way from the one in
which physicists were used to. The data were digitally worked with
digital machines, while people — at least at the time I was there
— thought in a continuum way, and the metaphor underlying the
thought of a physicist was the real line. This is the point which,
in my opinion, our design was weak at the time, and this point had
two main consequences. The first is that the software tools were
never considered as real devices, were never considered something
which has to be tuned, tested and maintained, and tested again
against the data which are changing in the life of the experiment.
As a consequence, sometimes there was no real control of the
physicists on their computation and the underestimation of the
problem of maintenance of the software tools. I think that the

physicists have to be in a situation in which they control both the data and the result and to control the computation performed by their tools. For example, we had a lot of samples of data and only controlled programs on samples from input to output, and never got a real insight into what happened inside the tool.

ÖZEL: Thank you. To address these problems we have various methods which are, many of them, common to many fields. Some are like cross-correlation analysis has been really developed and tested for gamma-ray source identification, but yesterday you have heard about a new method in gamma-ray astronomy: this is the bootstrap. At this point, I would like to ask Dr. Friedman, as a pure statistician let me say, how do you feel about this method? Is it really so revolutionary as our colleague Dr. Simpson described it yesterday?

FRIEDMAN: Yes, it is. I think the bootstrap is the most important development to come out of theoretical statistics probably for the last 20 or 30 years. But you have to keep in mind exactly what it is doing. It is intended to allow you to assess the variability to the sampling fluctuations. Because they are not the only kinds of variabilities you have in astronomy or any other kind of science. A way to think about it is to consider, instead of having a finite sample, say you take data, imagine that in the time interval that you were taking your data you can take an infinite amount of data. Statisticians call that the parent population or the universe population. You take an infinite amount of data. Now, of course, you can't, you have to take a finite amount of data, and that then reduces variability because the sampling is usually random and so the finite random data set may not be wholly representative of the infinite data set that hopefully you could have taken had you had enough time and money. Now, the bootstrap is intended to assess that variability, it does not assess variability due to the fact that your instruments were drifting out of calibration, or something shaking, or something wasn't tightened down, or anything like that. It only addresses variability due to sampling fluctuations because you couldn't take an infinite amount of data, and the finite data set that you took was taken at random which meant that it has a chance of not being representative of the infinite data set. So that's the only kind of variability it is intended to deal with, but of course that's the only kind of variability that the whole field of statistics deals with. So, that's why it is so important. But in actual science you have many other sources of variability than sampling fluctuations. One has to be very careful when using the bootstrap for hypothesis testing, which means, that's the statistician's way of saying, asking the question: I see an effect in my data, is it real or is it due to random chance, just bad or good luck depending on your point of view. And one has to be very careful when using the bootstrap for that, the reasons are slightly overlooked (?), but I'll try to explain them on the chance that maybe one person or two might understand. The idea is that if you assume that the underlying process does not contain the effect, and the effect was only due to random chance, then the underlying nature gives you a distribution that does not have the effect. What you would like to do is know how many times if you do a sample of data from that distribution that you would see the effect. If you only see it once in a million then you would say the fact that I saw it is significant, it probably is real. If I see it one time in five or one time in two, then I would say, well it's a good chance it was just a random fluctuation and wasn't really due to a systematic effect. Well, what the bootstrap does is it says that the best guess for this parent population is the data sample itself. But that's dangerous because the data sample itself has the effect which you are asking the question: what's the chance of seeing the effect if it wasn't there? It is in the data, so when you use the data, when you use the bootstrap to assess that, you will be asking: how many times would I see the effect given it's actually as strong as I see it in my data, not from a parent population which does not have the effect, and therein lies the danger. Another important restriction on the bootstrap is that your data must be independent and, as we say in statistics, identically distributed. That means that if you take a piece of data and that has one set of values, that set of values does not affect the set of values you will see when you take the next piece of data. That your observations were in fact independent. That's a big assumption, if that's not true the bootstrap will give you misleading answers. But again, almost all statistical techniques that you would apply do assume independence and usually identical distributions, so that any other technique that you are likely to use to assess sampling variability will have the same danger as that. The dangers that the bootstrap has are the same dangers that all other statistical techniques have, but it has many fewer of the other dangers that most of the classical statistical techniques have. So, and as I said, the conclude is: it is the most useful and probably most important single development in mathematical statistics in a long time.

That's all I'll say on the bootstrap. I have some other comments on other aspects. I will speak about them maybe later.

ÖZEL: Thank you. Actually I would also ask Dr. Simpson about bootstrapping. Basically, bootstrapping is sampling with replacement, but in another sense we never observe the same experiment or same event a second time. I mean, is there something wrong with this counter-idea? And what other limitations does the bootstrap method have?

SIMPSON: Well, I will answer your last question first. I am in the process of learning about its limitations by talking to Dr. Friedman and other colleagues here. At first I thought that it applied to all statistical parameters. In these conversations I learned that it is not a proper tool for cluster analysis — or any statistic in which a sample point appearing twice violates the underlying principle. But most statistical analyses don't have this property. The concept of the bootstrap troubles the mind a bit at first. Clearly one can never observe the same event twice. But nature often gives us, from the universe of possible events, more than one event which produces the same signature in our instruments. The paradigm that Dr. Friedman reiterated just now of making a model of the parent population by copying our sample many times, acknowledging the fact that the model contains all the information you have about the parent population, and sampling from it to find a statistical distribution of your parameter does seem logical; after you think about it a while.

ÖZEL: At this point we can return to another aspect of presentation of gamma-ray data in a most conceivable shape, that's in terms of two-dimensional sky maps. In this direction there has been a considerable effort by various groups of [COS-B] collaboration, and also there have been several presentations of the same data. Let me note historically something I find interesting. (Sorry, there is no one over there for the slide projector to show the slides. Probably we will see them later.) The problem is, in the era of SAS-2, the previous gamma-ray experiment, really we were not so sure about what we really see. There was no previous comparable experiment. Establishment of the strongest unidentified [yet] gamma-ray source Geminga shows this problem for the first time. For some time it was discussed in the collaborating groups of SAS-2 [GSFC-Maryland, Middle East Tech. Univ., Ankara] whether this is a separate source or is it something spilling from nearby Crab. Because, this was the first time we met something entirely new: a gamma-ray source. There was nothing very peculiar in this region of sky, and it took some time to gain confidence that it was a separate source [see Fig. 3]. Now, after COS-B experience, we have a better feeling of what we have, with a better control of our data and therefore also with a better feeling of the methods we apply.

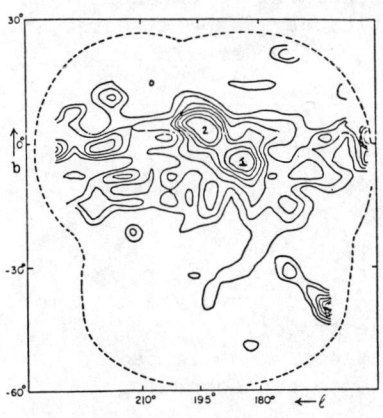

Fig. 3: γ-ray intensity contours in the galactic anticenter direction as seen by SAS-2 (Fichtel et al. 1975, Ap.J. 198, 163). (1) denotes the position of the Crab Pulsar (the first identified γ-ray source) and (2) denotes the first unidentified γ-ray source γ195+5 (Geminga).

In this respect I would like to address Dr. di Gesu on the problem
of a spatial representation of gamma-ray data in general, because
his institute has put some effort into this, at various occasions.

di GESU: One of the main problems of the digital pattern recognition is to restore data convolved by a detector response function
and distorted by the structured background and by the internal
detector noise. The restoration, also called deconvolution, is
preliminary to further digital image processing as: edge detector,
skeleton computation, boundary analysis, classification and scene
description. The problem could be stated in a more precise way as
follows: "given a set of X of data, determine their probability
distribution function T starting from the probability distribution
function M of the measurements X' and from the response function,
R, of the detector". In the context of this talk $T(X)$ is referred
as 'True image' while $M(X')$ as 'Measured image'. From the mathematical point of view the deconvolution operation is equivalent to
solve the following integral Fredholm equation:

$$M(X') = \int T(X) * R(X,X') dX + N(X') \qquad (1)$$

where $N(X')$ is the instrumental noise. The noise component is,
usually, not known and is not a translational invariant; therefore
it is not easy to solve equation (1) exactly. Approximate direct
methods have been proposed to solve it:

-- Maximum entropy (R.H. Bryan and J. Skilling, 1978),
-- Iterative solution algorithm (Van Cittert, 1931),
-- Bayesian deconvolution (Richardson, 1972).

Many applications of the previous methods to optical astronomical
data have been performed (Perriman Maximum Entropy, V. di Gesu and
M.C. Maccarone Bayesian technique to the Jet M87) in the past and
good results have been achieved. In fact, whenever the background
and the point spread function (PSF) are well defined and the
statistics of the data is such that the error on each pixel count
is well defined, the deconvolution may be helpful to enhance signal
against noise, and the statistical significance of the effects may
be calculated. Unfortunately, the nature of the data in X- and
gamma-ray astronomy is very complex:

-- the background is very structured and not well known,
-- the PSF of the instruments are varying with the energy of the
   detected photons and its analytical form is approximated,
-- the number of counts per pixel is very small.

Therefore the following open questions arise:

-- How to define the probability measure for deconvolved data in
   order to decide which effects are true or not?
-- How to define the statistical significance for extended
   sources? Does it also depend from the object morphology?

At the end one consideration on the computational aspects. The computer time necessary to perform deconvolution is:

$$T_{cpu} = 0 \ (N^2 * M^2) \ ,$$

where N is the size of the input image and M is the size of the deconvolution window used. The last consideration suggests the need for the use of advanced computer architecture (e.g. array of processors) in such kind of problems.

I can't guarantee that these methods could solve the problem uniquely and completely.

Let us look at this Figure [Fig. 4]. Here you see the problem. This is a part of the γ-ray sky map, containing the Crab and Geminga sources. These are prepared from the same original data. The first is a representation of the data after suppressing the fluctuations smaller than instrument PSF. We made [in the Bayesian map] a smoothing with a window that is smaller than the COS-B point spread function. There is also a map from maximum entropy results. Can you see the problem? We don't see much difference between the maps. Therefore I think we must work more and more on this line

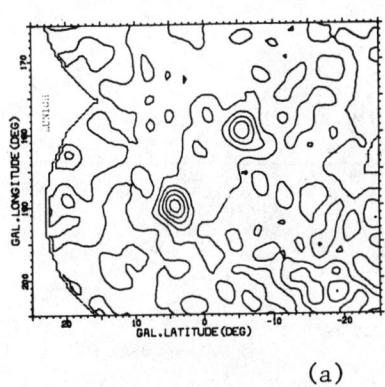

(a)

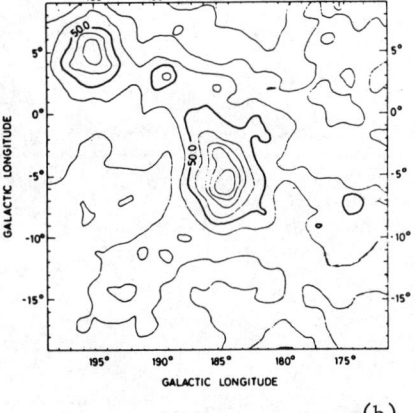

(b)

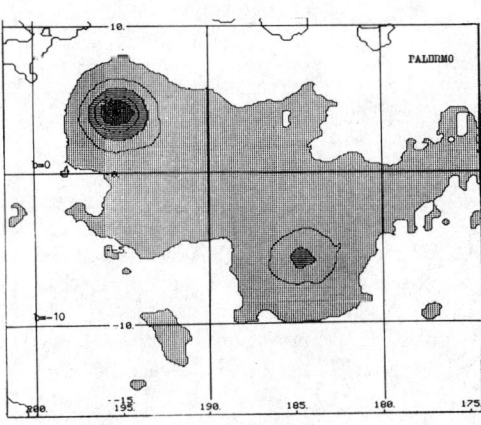

(c)

Fig. 4: Different representations of galactic anticenter γ-ray sky. (a) Smoothed, equal intensity contours as in Mayer-Haßelwander et al. (1982, A&A 105, 164). (b) Maximum entropy contours by Skilling et al. (1979, MNRAS 187, 145). (c) Bayesian deconvolution contours by Palermo group.

before to have a quantitative result and not just to see better pictures.

ÖZEL: Thank you. Actually while this graph [Fig. 4] is here, let me try to make one more point. Fig. (a) is originally what COS-B collaboration published some time ago in Astronomy and Astrophysics, and it is basically the presentation of recorded data after some smoothing. On the other hand, deconvolution methods Bayesian and maximum entropy work in order to derive the real photon distribution that has produced the maps. However, as our colleague mentioned, there is still not much difference between the maps and we need much effort to understand, to improve our quality in the structures of real gamma-ray sky.... I think now we have our projector operators there. I can show you two slides [Fig. 5] I was intending to show a while ago. The first one is a never-published SAS-2 all-sky gamma-ray map. We were not confident about the features we see and

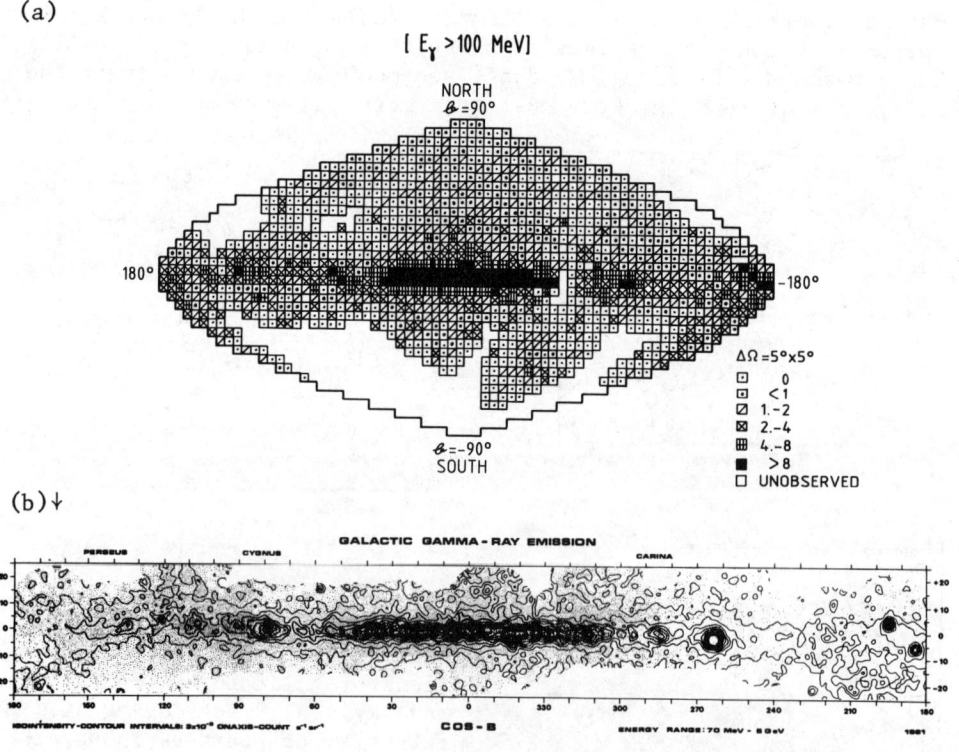

Fig. 5: γ-ray sky as observed by SAS-2 (a) and COS-B (b). This never published SAS-2 map indicates the total γ-ray counts divided by the exposure, in 5° x 5° bins. Especially the region around $\ell = 300°$, $b \cong -50°$ has contamination due to high elevation of the earth's atmosphere in this direction from the satellite orbit. The COS-B map is from Mayer-Haßelwander et al. (1982).

statistics were too bad, especially at high latitudes. Also there were problems in accounting several instrumental effects. Now this figure is almost obsolete by the COS-B statistics and sensitivity level which is the second slide [Fig. 5b]. You are all familiar with this one I believe. Highly improved statistics reveal many features which were never anticipated before.

At this point I would like to ask Dr. Simpson again, on a problem which has not really been addressed in high-energy (SAS-2, COS-B) gamma-ray astronomy. That is the measurement of γ-ray's polarization. This information was (is) also present in the SAS-2 and COS-B data. However, in reality it has never been extracted from the data, considering the high level of background noise and also the difficulties in the availability of the relevant satellite housekeeping information at accuracies necessary. For the first time now there is a planned effort for this: in GRO, by COMPTEL telescope. GRO consists of 4 different gamma-ray telescopes, and COMPTEL is the one which aims to detect gamma rays by double Compton scattering, and the polarization of gamma rays up to 30 MeV will be measured for the first time. I think he would like to say something about COMPTEL and its relevance to our discussion.

SIMPSON: Oh, yes, I am very interested in speaking about COMPTEL, being part of that team. It's an interesting sociology we have here too, because up to this point in time gamma-ray astronomy has been COS-B and SAS-2: both pair production type experiments. One of the 4 GRO experiments — EGRET — will continue observations of this kind. But in COMPTEL we have something totally new. This is the first time a Compton telescope will be in space, and the problems are quite different. COMPTEL will make measurements in the 1 - 30 MeV range where the flux is a factor of 1000 higher. We will be collecting some 10^9 photons [in 2 years] as opposed to a few hundred thousand detected up to now at the higher energy range. We have qualitatively different problems in developing the software systems to manage this information, but more importantly, we will make qualitatively different observations, such as nuclear lines and polarization. Polarization is a secondary goal of COMPTEL but perhaps one of the most interesting ones physically. In the telescope we have single Compton scattering in an upper detector of liquid scintillator followed by absorption in NaI detectors a meter and a half below. We locate each event in both the upper and lower detectors, so we know the direction along which the scattered photon travelled. From the energy deposits in the upper and lower detectors we calculate the angle that the photons scattered by. The missing information is the azimuth of the incident photon about the scatter direction. This is one of the different things about the COMPTEL: instead of having ℓ and b or RA and DEC for photon coordinates, we have a direction and an angle defining a circle on the sky. It's an unusual kind of imaging, having something in common with coded apertures in which the location of each photon is partially indeterminate but the total dataset contains an image which can be extracted. The polarization signal arises because the azimuthal distribution of polarized photons is unsymmetric, polarized gamma-rays tend to scatter more to the left and the right than to up and down. COMPTEL should be able to detect polarization of photons from a strong source.

ÖZEL: Thank you. I would like to hear anyone from the audience to express his/her ideas on the methodologies in data analysis in X- and gamma-ray astronomy, the data in general or X- and gamma-ray astronomy results in general. This way we can have a really interactive panel.

SPIZZICHINO, A. (Istituto Tesre/CNR, Bologna, Italy): Is it possible to apply the bootstrap method to the data from a coded-mask telescope?

ÖZEL: Would you like to answer this question, Dr. Simpson?

SIMPSON: The limitations of coded aperture systems are not statistical, they are systematic. Systematic changes in the background which can alter the detector rates in unknown ways by far dominate the uncertainties. So the bootstrap would not reveal the relevant errors. However, if one can succeed in completely controlling the rate drifts or else somehow compensate for them, then the bootstrap could be applied to determine the statistical error. On the other hand with a Compton telescope, where you do have the record of individual events, it is certain that the bootstrap is directly applicable.

ÖZEL: Any other questions? Or remarks? ... OK, I'll ask Prof. Scarsi a question I am sure many here would also like to know about. When and how will this COS-B data base be available to interested scientific community so that non-gamma-ray astronomers, statisticians, everybody can enjoy their own satisfaction or whatever the proper word is, say, their own experience.

SCARSI: Well, what will be available to the scientific community is a completed data set, and it will be released by the data reduction group of COS-B [in a year or so]. It will consist of these 150000 gamma rays tagged with energy, quality — because this is again one of the things that you have to attach to every single photon, which has been extracted from such a high background — and timing: each event is timed in absolute time to a quarter of a millisecond and it will have the form of a photon-by-photon record — this can be strange for optical astronomers — and the instantaneous efficiency of the telescope, at the moment in which it has been detected because the efficiency is changing in time. Well, that's all what will be released.

ÖZEL: For those who really want the data base, where should they apply at the time it is available?

SCARSI: This is to ESA, European Space Agency Headquarters in Paris.

ÖZEL: Thanks for the information. You have a further comment, Dr. Mussio?

MUSSIO: I have a general question which I address to Dr. Vito di Gesu, but I think can be addressed to all the panelists and perhaps to all the participants of the workshop. The question is: Each time we have to match with inference problems, we find that the probability is undoubtedly an unsatisfactory tool. We heard from Vito di Gesu that he has some reservations about the Bayesian approach. Now, why not to leave probability and try new tools? In other fields new tools have been proposed: from fuzzy sets to multivalued logic which are used to make inference when imprecise and uncertain data and procedures are present. And this approach seems tuned to the youth of the discipline of gamma-ray astronomy and it may be a good case to test these techniques.

ÖZEL: OK. Anyone to make a further comment, or to contribute on this point?

di GESU: Yes — I think that you addressed the question to me — I agree with you that the sort of data we have in gamma astronomy, for example the shape description of the sources, would be better explained if you use fuzzy sets. What I am really not clear is how also the fuzzy set will change first of all and before the deconvolution, because you must have some deconvolution, some operation on the data. Therefore it is just the concept. I agree that there are some phenomena that cannot be described by probability. You cannot say the probability to be tall or blond and so. But the possibility of the fuzzy function that you use is also changing with some operators.

ÖZEL: Dr. Friedman?

FRIEDMAN: I know almost nothing about fuzzy sets, I really can't help you. I guess the only comment I would make would be that in problems where a probabilistic approach is logically inconsistent, nevertheless if you do construct estimators and various statistical quantities, they often show not to be quite useful. And in natural things that you would use anyway, even when the notion of a probability distribution and identical sampling and all of that, it doesn't make any sense. That's often the case in many, many applications. When you look at, say, data from the countries of Europe or what is the infinite population, I mean, can you think of an infinite number of worlds with an infinite number of countries? I mean, it makes no sense to talk about probability distribution. Yet, techniques that had derived from those concepts of a probability distribution work just fine in talking about statistics, economic statistics and things like that. So often even though the notion of a probability distribution and probability in identical sampling and all of that doesn't make much sense. Often the constructs that come out of such thoughts are very useful in contexts where the basic ideas don't make sense.

ÖZEL: I think we have to close our panel discussion, if there are no other comments from the audience. I will just say a few closing words: In our panel, we have mostly discussed the problems of γ-ray astronomy. We see that the young field of gamma-ray astronomy is also reaching to a sophistication level that data will be really addressing to more intricate questions in astrophysics. Also in terms of data analysis and methodologies we have now our own experiences which go back to, like 10 to 15 years, I think what we can hope from next generation of gamma-ray experiments is that they will come nearer to X-ray astronomy results and contribute significantly to our understanding of the universe. Thank you.

PANEL DISCUSSION ON TRENDS IN OPTICAL AND RADIO DATA ANALYSIS

Chairman: R. Albrecht, Space Telescope European Coordinating Facility

Participants: D. Cesarsky, Meudon
Ph. Crane, ESO
G. Sedmak, Obs. di Trieste
P. Wallace, STARLINK
D. Wells, NRAO

Albrecht: What I want to do is to give a brief five-minute introduction to the subject, justifying the title which puts optical and radio astronomy in one and the same category, which I believe it is, as far as data analysis is concerned, and then I will ask the panel members to give us two-minute statements of their opinions on the subject and then I would like to ask the audience to fire questions at us.

Up here on the screen I have a list of current activities, and a list of issues. Let me begin with what I think are current trends, current activities. There are certain requirements that are being introduced, simply by the data that we deal with. Both in optical and radio we are faced with big multi-dimensional frames or data cubes, or even data bodies of a dimensionality higher than three, which of course then leads us into structures, for people do not want to deal with these bodies of data just as arrays or cubes, they also want to have history information, processing information, intermediate results going into tables; there needs to be a system to do that. Of course, this list is not complete.

The user also introduces certain requirements. The user wants to have a high level interface to interact with all of that. Sophisticated help, and by help I mean an on-line facility, there should be an index function, a cook-book-type manual on line.

There is a requirement to do data transport, because most users work on more than one system. Solutions have been drawn up. Software needs to be transported, again, in response to user demands. The user is concerned about through-put because it takes a long time to process big frames. There are system level requirements, how to produce an interactive system. "Macro" means the possibility to execute batch procedures, batch-like procedures, but in an interactive way, and of course there is batch automatized processing, as we have heard in the first paper today. This is very important for the big frames because you don't want to sit around and wait until an eight thousand by eight thousand raster is processed. Remote access of course becomes more and more important, as is parallel processing, there will be a session on that this afternoon.

There is also application software development and it is important, it is where the astronomy lives, but as far as this discussion is concerned, we will probably spend more of our time on the system software development, in particular, and that is very important, because it concerns decisions that are being made right now, about high level command languages, the users interface, graphics, image display, a question of GKS and what do you do for imaging, should you use a GKS type approach, or should you use something more hardware specific, which is faster. Data base interface of course, and the question of data base management, which we do not have an awful lot of experience with in astronomy, but it is becoming more and more important. I wrote down a number of keywords, buzzwords to give everybody a chance to find his favourite buzzwords and fire questions.

General issues are transportability, how far can we go in terms of transportability. Of course it costs money, it costs time, is it worth it, to what degree should we push it? Documentation and the related subject of configuration control. Again, how important is it, how much effort should we spend on it? Computer links, networks, distributed processing, user support. Mike Disney in yesterday's introduction said: Well, I want to have a chap who talks my language and knows a lot more about computers. Is it really justified to ask somebody to spend his time learning all about computers so Mike Disney can do his science better? I mean, this guy is making a sacrifice.

Software maintenance, software exchange and library, all those are issues that we have to deal with. The application software, the question of languages comes up, if we use preprocessors, which ones, and what should be the target language. There are utility libraries, should we support them, which ones, graphics and image display libraries, system software, which high level command language, which host operating system, UNIX versus VMS, or both, or something completely different, and how should we standardize our interfaces.

Again, this is not a complete list. I would now like to ask a volunteer out of the panel to either complete the list, or to come up with a completely different interpretation of the subject, not more than two minutes, because we have a total time of one hour, that is why I try to go fast. Pat, for instance.

<u>Wallace</u>: That is a very complete list and the only addition which I can make from the STARLINK experience is the observation that the general purpose computer is playing an increased role in the work of all astronomers, radio, optical, X-ray, and they now require as part of their office a general purpose computing capability quite apart from the data processing, and that, to some people, is more important than much of the data processing itself. The network is important and that is part of that general purpose system. So far, our experience has been that the different classes of astronomers do use more or less separate packages although there are some central packages which can do work in all the various areas and these are used to some extent. But everything that you have listed is important and there are a lot of tradeoffs in there which we can talk about ad nauseam.

<u>Albrecht</u>: You mentioned the subject of work stations. I guess this is what you meant by interactive computers in everybody's office.

<u>Wallace</u>: Oh no, I simply meant that so many background jobs, odd calculations, word processing, communicating with other astronomers, involve a computer now, and a few years ago that was not true. Yes, the computer has certainly overtaken the role of the telephone and the pocket calculator as well.

<u>Albrecht</u>: Maybe Phil should continue.

<u>Crane</u>: I had a somewhat different interpretation of what we ought to be talking about, so maybe what I am thinking is perhaps not appropriate this time. I thought we ought to be talking more about what kind of operations we are doing on the data, what kind of algorithms we should be developing, where can we interact on development of particular algorithms like object identification, various fitting routines, questions of what pieces of software make up, what actual algorithms are there needed to make a data processing system in radio astronomy and optical astronomy. What kind of filters are necessary, what kind of imaging, rubber sheeting, and similar kinds of things are required? Do we need a table filing system, or data base system at the hands of the individual astronomer and so forth. Those are the questions which I would have thought we should address in this panel and I am not sure that they

are addressed in another panel in this conference, so ... Maybe we can touch on those subjects also.

<u>Albrecht</u>: OK, we shall bring it up in the discussion.

<u>Cesarsky</u>: Actually, looking at the list ... To begin with, most of the people here are responsible for developing software. Well, I do develop software sometimes, but now I am here, I think as a responsible of a computer centre in Paris, so my main concern is both software and users. Looking at the list of Rudi, and listening to Phil, I don't see the users anywhere in those lists. First, is there such a thing as a typical user? Because everybody knows that users come in an enormous range of abilities. There are some people you have to tell, you have to explain why you have to push the return key, and there are other people who try break into the computer code to see whether they will find the system manager password.

<u>Albrecht</u>: The user, I guess, is hidden under the high level command language, HLCL, which means the whole issue of user introduced requirements in my list, with help functions, index features, and all the associated problems.

<u>Cesarsky</u>: Yes, but my question is, is there such a person as an average user, or should a piece of software be flexible enough to cater to a big range of users.

<u>Albrecht</u>: I agree. There is an issue: Who is our ultimate customer? Is it the astronomer who comes in cold from the street in a visitor oriented institution, or is it a sophisticated user who wants to push the data analysis system to its limits? Maybe Don as the person responsible for a large-scale system, oriented towards visiting astronomers can comment on the issue?

<u>Wells</u>: Rudi is referring to the software packages that are dealt with by the group that I work with at NRAO, that's AIPS. In the case of AIPS, I think the vast majority of processing is made by canned algorithms, so most of the users are looking at a product which they did not develop, although they influenced its development by complaining about what they did not like in it, and over a period of years their complaints were mostly satisfied, or at least they got tired of complaining and quit complaining. In that sense the users get what they want and do influence software development. There is also the situation that many of the users do create their own algorithms so the software evolves by that means as well. I don't know if that answers your question.

I had several points I wanted to make further than what you had said. I thought your list was quite complete but there are a couple of things I want to add. First, I do not think I see in your list the discussion of data interchange between different systems.

Albrecht: In my first list, which I admit I went through rather fast, there is the user introduced data transport requirement. The reason why it is not on this list is that I think that it has been solved rather nicely with the Flexible Image Transport System.

Wells: Agreed, except that I want to say something about that. There is an attitude of complacency in the community, a feeling that we have solved that problem and we do not have to worry about it any more. I want to stress that this is like preserving one's liberty. You cannot be complacent about the matter. These problems will come back to haunt us if we do not continue to assure the interchange between different systems and that requires testing, certification, collaboration, cooperation between the different software groups worldwide. Also, no data interchange standard can ever remain static for a long time because user needs change with time, so there has to be a continuing effort to maintain and extend interchange standards.

I have two more issues I want to add. One you have sort of covered. It is the question of automation. You called it batch processing or macros or things like that. I want to stress this issue because the data volume in astronomy is rising very rapidly, in fact it is rising more rapidly than the number of astronomers. I see major problems in the future in that the astronomers will be swamped with data. It follows that the algorithms must become more automated and must be more intelligent in selecting data to bring to the attention of the users and I think this is one of our great frontier efforts in data analysis. We must have better classification algorithms, better automated data management and interesting signal detection. I want to say that's got to be one of our future efforts because of the increasing data volume.

The final item I want to stress will be discussed somewhat later this morning and that is the question of vector and parallel processing. It is an item, which is not on your list, it is another frontier area for the future and it goes along with this question of data volume. We must learn how to use new kinds of computers which do pipeline processing and multiple unit parallel type processing. I and some of my colleagues that in NRAO have been studying in this area closely in the last year, and I can tell you that it is a new world. It requires very careful thinking and we don't have much experience in this area.

Sedmak: I want to make a few comments more oriented towards the astronomical side rather than the data processing side, talking essentially about experience made in the three years of Astronet work. Of Astronet, as you know, the Astro is working, the net is not yet working, at least until our telephone company will open the access to the packet switch network.

A few years ago, we knew that the most demanding request from the astronomers was to have a flexible system and a transportable system, that is a system more or less exhaustive and capable of facing all problems the typical astronomer could find in the astronomical work of data processing. Now, after a few years of experience at least, I repeat, within the Astronet environment, my opinion has a little changed. I suppose that we must separate the two categories of users: a very high level or very qualified user, that is an astronomer who does astronomy but is perfectly able to manage a modern complex data processing system, like, for example, a system such as the ESA computing center, or the systems which are going to be built at the Space Telescope Institute, or a typical Astronet pole for example; and the average astronomer, who normally is not capable to exploit all the capabilities of such advanced data processing systems. This ratio between the two categories is extremely important because if the number of people, of astronomers, being able to use modern advanced data processing systems, is after all smaller than the other category, then what we need is no more, in first priority, the superflexible, the superadvanced system and so on, but a system more oriented towards what the Starlink experience has shown to be very useful to astronomers, that is packages dedicated to some well defined astronomical problem. For example, one comprehensive package for IUE data reduction, one comprehensive package for, say, radio astronomical image cleaning and storing and so on. Of course, this approach does not fill all the applications and so we have also to provide for the future astronomical problems, and in particular we have to provide some room for new problems, and in this case, of course, a flexible system is mandatory, because otherwise you cannot build simply the procedures which after a certain time of development will become a consolidated product, able to be used by anybody. So, I suppose that this reflects into hardware requests for an approach like work stations, but completed over software simply useable by a non-expert astronomer, for non-expert I intend non-expert in advanced data processing. This is one side of the problem. On the other side, we need somewhere, for example in large institutions, in large computing centres in large institutions, very advanced systems to develop the new data processing for future astronomy, and I suppose that some compromise should be made between these two sides.

Albrecht: I guess at this point we should give the audience a chance to point out to us where we make mistakes, where the gigantic blunders are, the traps that we are falling into.

Fosbury: It seems to me that we have made several retrograde steps in image processing, data analysis in the past five or so years. I think that instead of talking about trends in data analysis, we should talk about trends in computers, and see how we can perhaps retrace our steps a little bit and get back to a more ideal situation.

Certainly five or seven years ago, the few people who were doing data analysis, at least to my experience, were doing it on rather small, single user mini computers. They got very quick response times, they wrote their own software, the software was not portable between computers. We have moved out of that regime, into a regime where we now have bigger computers, VAX'es, superminis, and we sacrificed speed for the goal of software compatibility, software sharing. Of course we gained some advantages, the data analysis facilities are now available to many more astronomers, but most of the people are frustrated by a number of characteristics of these data processing facilities.

I think there are two items I feel are important: one is that we think of data processing as being quite separate from the observing. Now I haven't heard anybody yet in this meeting trying to move the process of data processing closer to the telescope. Don Wells, just a moment ago, was telling us about the vast quantities of data coming from the telescopes. I think one solution is to design more intelligent observing programs, which move more of the data analysis very close to the telescope.

<u>Crane</u>: Can I respond to this? I agree with what you say, of course, and the trend is exactly what you say, but the problem is not that we have made retrograde steps, in fact we have increased the number of people who have access to computing facilities perhaps one order of magnitude, and only can increase the computing facilities by a factor of two, so naturally things would run, if the ratio is right, only one fifth as fast. So really what we should do is get five times as many computers or ten times as many.

<u>Fosbury</u>: Well, that is an important question: should we get five times as many computers, or should we not really ask the question how to work in a different way? I am proposing we should indeed work in a different way. Two proposals: one is to do more at the telescope. I think that is important, I won't discuss that any further now. But if we did more at the telescope and came back with a rather clean set of data, perhaps we would be thinking more about data analysis than about data reduction. Then we could regain some of the experience that people have on single user systems, by using personal computers, or microprocessor based work stations.

<u>Wells</u>: Well. I think that the supermicro work stations and things like that are bound to happen, they are happening already and it has got to be a major trend. I must also emphasize that large computers are becoming much more powerful and that will also increase the computing power available. I must warn the people who still want to buy VAX'es that in my opinion, just looking at things, I think that the VAX era is coming to an end and I see IBM

and the Japanese as the major vendors of the future. I realize that
is hard to believe in 1984 when we have 50 VAX'es worldwide to do
computing in astronomy, but that is my calculation. Finally, I
wanted to remark about something you said about bringing more
intelligence to the telescope, data analysis software to the
telescope. I think you are right. I agree with you. I would like to
see the data analysis software closely associated with the data
acquisition software, I would like to see servo loops built around
the telescope for optimizing observing parameters, based on results
from data analysis routines running in real time. This is a
heretical view, I have not been able to sell it in any observatory
in North America. I might add that I see major management problems
in implementing it, because all observatories that I have looked at
or have talked to people about that concept, cannot consider it,
mainly because they have separate software development groups
building data acquisition software and data analysis software and
those groups do not talk to each other. It is a major problem in
the observatories I have seen, in fact it leads to data interchange
problems between data acquisition software and data analysis
software. Data analysis people do not get data that is properly
documented in general because they don't have any control over the
people who create the data. This also leads to what operating
systems do you run on the data acquisition machines, because that
controls what data analysis software you can run, what computer do
you choose, the data acquisition people have different ideas of
what kind of computer to buy and what operating system to run and
how to run it, how to manage it. These are major problems for the
concept of bringing data analysis closer to the telescope, but I
would like to see it happen.

<u>Crane</u>: Maybe I should point out that I think most people who
have used the ESO facilities realize that we have made at least a
little bit of progress. In fact, we do have the same data analysis
system running in the computer that does the data acquisition and I
hope that we can continue to do this, but we face exactly the
problems that you point out; the requirements are different, and it
takes a major effort from the user to make sure that this does not
go astray, because the problems that you mentioned clearly exist.

<u>Wells</u>: It is a management problem.

<u>Crane</u>: Yes, it is a management problem, but management has
not recognized it yet.

<u>Wallace</u>: In what way can the observing be optimized by on-
line data processing? I mean, I am aware it can be, I just want Don
to elaborate.

<u>Wells</u>: Yeah, I think it is obvious to people who have been in
both sides of it. Often times there are adjustable parameters on

the instruments, alignment parameters, optimization parameters for the observing, and one gets these parameters by analyzing data recorded with the instrument and so I think it is obvious that the process of adjusting instrumentation would frequently be improved by having more sophisticated software closer to the telescope. I can tell you that, for instance, in aperture synthesis instruments it is now possible to do real time phase corrections on the antennas, self-calibration algorithms, you could in principle do self-calibration in real time to improve the quality of data, better phase-tracking in the telescope. I don't know if the improvement would be worth the effort, but it is certainly possible.

Albrecht: Giorgio Sedmak has a comment on the subject?

Sedmak: Yes. I want to make another comment regarding how, at least in our community, the astronomers see the problem of data processing. I will separate clearly two stages of data processing. The first stage is the removal of the instrumental signature. My feeling, I believe that all that has been told up to now by the colleagues participating in this panel, applies perfectly to the first phase of instrumental signature removal, but I suppose that astronomers may be wrong in this considered this very technical part of the data processing chain and at least hope, maybe they do not say clearly, but they hope in the heart that this instrumental signature removal is made by the engineers, or at least by the astronomers having the technical task of building the instrument and maintaining the instruments operating, so that the data which arrive to the astronomer or the scientist are completely clean from the instrumental signature. Now, the second part of the data processing chain is the scientific analysis of data, that is the data analysis which as an output has the science astronomy and astrophysics, and I suppose that we should give more importance in this discussion to the second phase, because otherwise we open a very detailed technical discussion which may be very interesting, but not so much for the astronomer as for science operators.

Albrecht: But was not the point exactly that large parts of the second stage should be pulled into the first stage because of the quantity of the data, because the throughput requirements, and is there a concern, especially a user concern, that many of the astronomical scientific decisions could be pre-empted in that first stage?

Sedmak: Exactly, but the point is that while in the first stage more or less an astronomer is involved of course, but as a user, in a very passive sense, in the second stage the astronomer is involved in a very active manner and according to our experience, in the second stage, that is the scientific data analysis, we are able to identify a few things which are very requested by astronomers, for example to have standard graphics, to have a standard data base organisation, to have a certain standard in, for example, say, transportation of data, FITS, just to say one thing. This is a matter pertaining to the second stage, not to the first one. Think for example to a matter like the Space Telescope. The standard user will get a tape of data and will not, in general, see what is first from the Space Telescope to this tape. He will know of course, but he will have to from the tape on. So, I suppose the main interest of the astronomer is in the second part and I propose to discuss this one.

Albrecht: But let me ask a user. Are you comfortable with being handed a tape by an engineer who says: That's good data? Is there a user out there with problems?

Fosbury: I think that is a very important point. We have to make the user confident that what we have done to the data beforehand is correct. I believe there are many instruments around now, producing data which the user thinks of as raw data, but in fact have been quite highly processed.

Albrecht: Piero, do you want to follow up on that?

Benvenuti: Yes, again on this point I think that the difference between software which is used immediately after the observation, and I am thinking now of satellite data or Space Telescope data, and the software which is used later on to analyze the data is that the calibration, or the software which takes out the signature of the instrument, is thought to be well tested and proved. Of course sometimes that is not true. We have the example of the IUE that for a long time had a wrong calibration. But that was then, later on, found out and measures to correct that mistake were taken. What I do not see, however, is in the development of analysis software, which later on is applied, one has not a similar testing procedure. I mean, as a user I would like to have the same confidence, if you like, on the analysis software that I have on the calibration software. I think that as a user I would like to see more documentation and more specific validation, both of the coding and of the scientific application of the analysis software, especially if it is a sophisticated one. Because I think that anybody can pretend to see how the calibration is made for an instrument and in any big instrument that you see, a specific team will look after the calibration, while I do not see the same thing for sophisticated analysis software, especially if this software is exported and transported from a system to another.

Albrecht: Pat, do you want to comment on this?

Wallace: Yes, I just wanted to make a comment. There are a number of things that good software requires. One is testing, thorough commissioning. But what if the price of doing that commissioning is fewer application programs or fewer astronomers? It is quite easy to lay down requirements without considering the costs, and what we have found is that the costs of doing the job properly are not available. The astronomers invariably would prefer a share in a new satellite or another telescope to doing the job properly.

Albrecht: Massimo Capaccioli is another very critical user, and he had his hand up a while ago.

Capaccioli: Well, I must say I am a little bit disconcerted about this discussion, because it seems to me that we go again in the same direction, towards the one we took several years ago, and which led us to this point, and the point is the following: We meet together in an ideal meeting where you have astronomers ready to learn and computer people ready to teach. But in fact, one is not willing to hear, and the other one is not willing to be understood. The language computer people speak is incomprehensible to most of us. They look like physicians who try to terrify you instead of trying to comfort you. This is the first point. So, clarification in language! Don't try to scare people with these changes in the hardware. Remember, somebody said that some users don't even know that you have to press the return key to use the terminal. OK, let's kill these users and keep the others. But still, we do not want to see a new terminal every 25 seconds. We do not want to see a new computer, a new operating system, a new language. I spent my life trying to learn a system and after that, I discovered it was obsolete. And that is a real pity, because I could have done something a little bit more interesting, like go sight-touring here around in Erice, for instance. Now, I think that the main point is to try to understand that the collaboration between astronomers and computer people can in fact work provided we are willing to collaborate. So far, it seems to me that we are not willing to collaborate, and I can demonstrate why. Look at the papers that are presented. Astronomers speak of astronomy. They don't care of the fact that the audience has some computer people that want to know something about astronomy in order to provide tools. On the other hand, computer people speak about something that is incomprehensible to most of the astronomers, so it is useless. It can be useful for the other computer people in the room, but not for the astronomers. Until we do not find this kind of link between the two categories, well, I think it is a hopeless process.

Albrecht: I saw Mike Disney nodding his head in violent agreement.

Disney: I was just going to mention the question of validating the analysis software. It is very important, and one of the things that may help in that respect is that when somebody writes a package, which is for general use, is that it contains a list of

calibrating sources which are ideal for that particular package so
that every time an astronomer gets data using that package, he also
gets a very clear and good image of one of these analyses' sources,
he can then examine it and cross-correlate it with previous work
that has been done, so that he is very certain that the data that
he has got, or fairly certain, the data that he has got are sound.
So I think it is up to the people making the packages to some
extent to consult with astronomers and try and for each package to
come up with a list of sources for which there are already
published data, or data are within the package.

Albrecht: The people who I know who have experience with that
kind of problem are the FIPS group working for ESTEC in the area of
FOC calibration software, and mayby either Wim or Jim would want to
comment on that? The problem of testing versus producing new
application software.

Kegel: Perhaps I should explain. I am a computer man. I used
to be a nuclear physicist before, but I am working on computers all
the time now. Because of that I have always felt that testing was
essential, because we are a company, we have to deliver a product,
we have to give a guarantee on it, so we do test everything. One of
my major problems is making sure that I have a correct test
procedure, here and that is why I support the statement, I do not
know who made that remark. There is a price to pay for testing in
that it takes about half of the time of actual program development.
If you develop a program, you spend about a quarter of the time
thinking of how you will do it, you spend a quarter of the time
actually implementing it, and you spend half of the time documenting and testing. I think that especially if you are going to talk
about portable software and about enormous amounts of data, you
cannot afford to ignore the issue of testing, because if you do
that you will end up with a lot of work you have done and you will
find after five years there were some minor errors somewhere in the
beginning of all your procedures, and you will have to re-do all
your articles.

Cesarsky: My own experience is that computer people do
perform reasonable tests whereas users do tests that somehow manage
to bomb a system. So my suggestion for testing is to have a test
site or test users. If everything is done all right it is not
wasted time because at the same time as someone is testing a
product, he is doing some science with it.

Albrecht: I know that AIPS has test sites and selected test
groups. Do you care to comment on it?

Wells: Well, one of our major development projects right now
is a UNIX port, which I will speak about when I speak, but yes, we
have some test sites currently, but that is not the same as a

rigorous testing of algorithms. I know that certain algorithms in our code have been tested by writing programs that create test data sets with known properties and then feeding them through the software. I think that is a generally good procedure to spend time creating simulation programs that create simulated data sets. I have seen data sets of star fields, for instance, created to mimic CCD images and I have seen such simulations that were so good you could not tell them from real data. This is very useful in testing photometric analysis programs.

Kegel: Maybe one more comment. Selecting test users is a good idea I think, but mainly for practical things like user interfaces. You want to have those to have comments on the practical aspects of a system, but I don't think this is sufficient to detect all the numerical problems you might have in the internals of a program. I think users only will spot the obvious flaws in your program and not the sophisticated tiny mistakes which only occur for very special combinations of input data, or things like that.

Albrecht: Let me change the subject to something more controversial, since we have only about twenty more minutes. High level command languages. I have heard comments like: sophisticated user interfaces are necessary, but I have also heard comments like 'I am getting sick and tired of all these command languages'. Could I again invite a user to give his or her impressions?

Rots: Actually, I think that in a sense this goes back to what was said a little while ago, the communication gap between the computer person and the astronomer, myself being simultaneously on both sides of the fence. My feeling is that you can't really blame it on the computer people only. I think one of the basic problems there has been that the astronomer on the whole has been very easily satisfied. You give him a product, you give him a system and he is quite happy. He goes off and uses it without ever critically asking whether it is a good system. So if people spend their lifetime learning, and that ties in with the command language, learning how to use a system, that is ridiculous. The astronomer should require from the computer people that his system has such a command language, such a user interface, that he does not have to learn, well he may have to learn some things, but that it does not take him more than a day, to get familiar with the system and be very proficient in it. Again, let me tie in with something that Bob Fosbury said before. He said, we made some retrograde steps, Phil Crane denies that. I still think that Bob is right. One example: as Phil said, our user community expanded more than the computer capacity. That is partly true. One of the other things that we have done in the meantime is embraced transporability which is a good thing in itself, but unfortunately has brought down the efficiency. Our systems today are far less efficient than they were five - ten years ago. However, another retrograde step I think has happened is that people have turned more and more to, let's say conversational, no, to keyword, certain keyword-type command languages, whereas sophisticated menu-driven user interfaces are much easier to master for a novice user. I think there are certain things that should be considered in that respect, too, which way we want to go in the future.

Albrecht: Could I ask a representative of one of the big systems to comment on that?

Crane: I think I was agreeing with Bob Fosbury. I think, nevertheless, I don't believe our efficiency has gone down. I think per operation we are probably more efficient, it's just we are doing a hell of a lot more. Well, I have seen it certainly in comparisons between the systems that we have in ESO. In fact, the system we have on the VAX is more efficient, but if we put five times as many people on it, it is not more efficient. That is really I think what is going on. I think we are trying to do a lot more than we did five years ago, or ten years ago, and not that we are doing it in an inefficient way. I am not at all sure the transportability, I have not seen the evidence at least, that transportability is having a severe effect on the efficiency.

Rots: Maybe I am prejudiced by the AIPS example in that case, I don't know.

Kegel: I'd like to make two remarks to the two previous speakers. The first is that there is an inherent conflict between the sophisticated type of user interaction systems he is talking about, and the menu driven systems. Although in principle menu driven systems are very simple to implement, there is a conflict between a menu driven system and the kind of pipeline processing I suppose all image processing people have to do every now and then. That is, if you have worked with a system which is menu driven, you will have to convert it one way or another to do the same job in batch mode, so I am not much in favour of this kind of thing, although it is quite easy to provide as an extra in any system, I think.

The second thing I would like to remark is about the conflict which was suggested between the complexity of the system and the fact that an astronomer has to be able to use it one day after he started working. I think both is possible. What you are actually asking, what we should ask ourselves is: What possibility should there be for people who care to go into the details of the system and explore the ultimate possibilities of such a system? How much of these should be present? Because, and I think I can say I have some experience in that, I think it is always possible, even in the most complicated system, to provide a mode of operation where an inexperienced user will be able to productive work within a day.

The question is really how much facility should be present above that?

<u>Rots</u>: May I make a comment? I am not advocating just simple menu driven systems. It should be possible to build some hybrid type things to satisfy both. What I am warning about in the ease of use of systems, regardless of what they look like is: yes, it should be possible, a system should be complex, or at least it should have various degrees of sophistication as far as the user is concerned. You should be able to get deep down into it and do very powerful things. On the other hand, you should be able to stay in the upper layer, so to speak, and do simple things. The point is, though, that often, when the user gets in this upper layer, there are some hidden things that he is not yet aware of, which are deeper down, which he has to be aware of or he might fall into one of these traps. So the thing is that if one designs the system, the upper layer should be designed such that there are no hidden things which may trap the user.

<u>Albrecht</u>: Massimo Capaccioli had his hand up.

<u>Capaccioli</u>: Yes, I think we might try instead of looking at each byte, trying to look at a sequence of a scientific paper from the beginning to the end. The beginning is an idea. Suppose we have the idea that stars are cubes instead of spheres so we decide to go to the telescope and check. And this is the first part of the pain, because you go to the telescope and you discover that the telescope requires a number of operations that you do not know, so you search for a manual, and you find that the manual is incomplete, you find that the list of filters is not in the right order, and so on. After a while, the sky covers, and therefore, it is the end of the story. But, in case you are lucky enough, and you come home with a number of tapes, then that is the second part of the pain, because you have first of all to be able to read the tapes. That is trivial for most of you, but that is a drama for a lot of astronomers. We discover that our tapes cannot be read anywhere. There is no one place in the neighbourhood, where we can read our tapes. But then we find that somewhere we can read the tapes, and there is the algorithm that could be used to reduce the data, but we find that the algorithm is not implemented, or it is not documented. We find a colleague who says, "Well, I've used it but it is no good", and so on, because there is no general control on algorithms. What to me seems fundamental is not the fact that I can press the key and that my measurements were taken. That is very useful, but I understand that. In other words, I can understand not the fancy structure of the program which does it in milliseconds but I can understand the principle. What I hardly understand sometimes is the principle of a very sophisticated algorithm that is used for my science and that is not documented. There is not even a write-up in a book where you find algorithms doing this and that: OK, start

from here, you have to do this, you have to do that and so on, exercise for stupid people, and so on. Nothing of that is available. So, if we continue this way, we will witness what we are witnessing now, that most of the astronomical output comes from people working with HP45's on a desk, a piece of paper and a pencil. That is the fact. Or, I must add, the big amount of the output will come from people providing you two billion stellar magnitudes that cannot be published, cannot be handled, cannot be analysed. Therefore, computer people should provide us a way also to be able to use the huge amount, the huge volume of data that we have, but this is another aspect.

I would insist that in this long process going from the idea to, eventually, the publication, we have a lot of stops which are caused by the fact that there is not enough effort to provide the tool, but rather to show how clever people are in doing algorithms. That is the difference. We should remember that what we want is the tool.

Albrecht: Giorgio, you wanted to comment.

Sedmak: I want to stress something which I believe is hidden below the discussion regarding the command language. It is that point that for example the approach of the command language, or the control language, versus the approach of single programs, or packages, after all is a matter which is time evolving together with the technology available for making informatics. As new computers, new terminals, new work stations, new graphics and pictorial devices are available on the market then new control languages come on. And people after all have to deal with. So I suppose that all the discussion can be summarized in this concept, which is a request I suppose many astronomers are, I hope, willing to share with me. That is, astronomers absolutely need a standard for the informatic side of astronomical data processing. For standard, please understand me, I do no want simply to say: Oh, let us use IBM computers instead of UIVAC computers. I intend the data processing environment oriented towards astronomy which is once for all selected and used not because it is the best possible, it will never be the best possible, but simply because we choose it, like we selected FITS for example to transport the data in astronomy. Now, we have some holes in standardization in the data processing side of astronomy. I stress one problem for graphics, one problem for pictorial hardware, and one problem for the environment. There is a lot of discussion we know on using or not using UNIX operating system instead of the native system of the various computers. Certainly, this is one decision we will have to take as a community, because evidently we cannot continue to change our software environments as the technology changes. In this sense, I agree completely with Massimo Capaccioli. We cannot think to a normal astronomer spending months of time simply to follow, to track, the evolution in data technology. This is the first thing. The second thing, and also in this sense I agree completely with Massimo Capaccioli, is the fact that when we talk of data processing software for astronomy we must distinguish between what they call the instrumental signature removal from the real scientific data processing. Certainly, there is a cross-talk between the two but there is a large part of really scientific data processing, which normally is completely not documented for the very fact that normal journals, say APJ, Astronomy and Astrophysics, and so on, do not publish this. Simply, they do not publish this. So, you find a paper telling that these observations have been reduced in such a place and you have the results. You absolutely do not know how the results were obtained. So, I suppose that we have to also generate such a new idea, new in the sense to activate, not new in the sense that we do not know this problem, that is to have the standard to document the data processing we make because otherwise the system will not converge, whichever your command language, how your environment and standard will be. So we need a standard in the sense of standard environment for performing of data processing in astronomy. First point. Second point, a standard in that we take the committment as astronomers when we perform new data processing tasks, to document this in a scientific sense. I hope that also Rudi will give some comments on this point which is a very old point in astronomy, that is that we do not have a common environment where to publish this kind of things.

<u>Albrecht</u>: Pat Wallace was nodding his head.

<u>Wallace</u>: I want to go back to another topic. I just wanted to go back to the point that we are told that the astronomers want the right software, that it must be reliable, it must be fully tested, it must be thoroughly documented and it has got to be supported long-term. Now it is widely known in the industry that the software costs greatly exceed the hardware costs. It is also widely known that the cost of support far exceeds the costs of originally developing a given piece of software. It is also widely known that you need good computer people, professionals. Are the astronomers willing to pay? Are they willing to have the right numbers of people? Are they willing to let them document things rather than to go on to new things? Are they willing to pay them what industry pays? If so, you can have good software but I suspect that none of those costs are acceptable. You will always prefer a new telescope.

<u>Borsenberger</u>: There is something which is against good documentation. That is, astronomers are always in a hurry, and if you want good documentation, tested software, you need to be quiet for a moment, because to document software, to test it fully, it takes twice the time to get it to work. I would like to get back some comments on Bob Forbury's suggestion, to increase real time software. What you do when you increase real time software, is you

have got a more complicated instrument. The problem is that the lifetime of a telescope is between 20 or 50 years. The lifetime of a computer is 10 or more rightly 5 years. The lifetime of a software version is around six months in good cases. How can you be sure of your data after that?

Unknown: It seems to me we have been discussing models rather than trends in data analysis. Another element that has been missing is what this is all about, rather than how this gets accomplished. So, is there going to be a single data analysis environment all over the place in astronomy? If so, then one does not need to worry about algorithms, presumably all application software could be part of that environment. It seems to me that if you have more than one system you need to talk about what it should be able to do rather than how it should be able to do that.

Crane: I think that is a topic for the panel discussion on Saturday. I hope that you will join us.

Benvenuti: I think that we all agree that documentation and testing of the software costs a lot, and the astronomer must be aware of that and must be ready to pay for that. A solution to that of course is that if you reduce the number of environments or systems that are running the same software, you can easily reduce the cost. It is obvious that if you diverge and each centre is developing its own software and has the same need for documentation and testing, then the costs will rise enormously. So I think that one should really stress that we should aim to a much better coordination of effort and in that way you can reduce the costs.

Disney: I just want to raise a question that Rudi addressed at the beginning of this summary, and that is the question of transportability. For somebody like myself it sounds a fine word and it might be a solution to many of our problems, especially if we are going to have new computer systems forced on us. But I'd really like to know what transportability means, and whether or not the experts agree that transportable software is going to be feasible.

Albrecht: I'd like to ask Don Wells to comment on that, 30 seconds or less.

Wells: In general I think what it means basically is a layering in your software, so that the interfaces to the hardware and the operating system and to the various modules of software are through well-defined layers, so what one is controlling is the interfaces between software modules and that is such a fundamental concept, it is motherhood and apple pie for software design these days, and has been for years. So there is nothing very dramatic about asking for software transportability, it is just the application of principles that are generally agreed to and the question you posed is "Is it feasible, is it practical and what does it cost?". The question of feasibility has been answered by a number of projects over the years and the answer is yes, of course it is feasible. What does it cost? Well, we were faced in NRAO several years ago with answering that question, and we did a study of it, and we estimated the costs as about 30% extra. That is not a very accurate number, of course, you understand there is an RMS on that, but that was the rough estimate. Performance? In fact, the argument is that we probably gained on performance, because in the act of enforcing transportability we designed to a higher quality standard.

Rots: Another aspect of both transportability and also keeping the costs of software down. I know it is heretical in this sort of environment but nevertheless. One of the issues on your list, Rudi, was graphics software, and that is one of the classical headache areas for transportability because nobody has the same peripheral. Why don't we seriously consider buying commercial packages for doing the graphics. Not for image display, but for graphics. The classical answer always has been: It is far too expensive and our user community cannot pay for it. There are two answers to it.. One is that we do expect our users to buy a computer and to pay a license for a compiler, we do not provide compilers, so why not use the same philosophy for, say, graphics packages? The other answer is that at least one of the more successful software houses for specifically graphics packages in the US does give a 70%, seven zero, discount to educational institutions, so I don't think that really the price of commercial graphics software licences is prohibitive for most of our users.

Albrecht: I guess most of the bigger systems do use commercial packages like arithmetic libraries, statistics packages and graphics packages so I think we are in complete agreement there. Wim, you wanted to comment?

Kegel: I would like to make one comment on the price of portability. I think that the only price you pay in some areas is performance. There is not a price you have to pay in terms of manpower or something like that, because transportability is essentially the same thing as doing a good design, and doing a good design always pays off in the maintenance area. So you may pay a little extra in the implementation stage, but you gain I suppose much more during the maintenance phase. So you should always do it.

Wells: I am uncomfortable at having to disagree, Arnold, but I cannot let what you have said go unchallenged. What Arnold is referring to is the possibility of adopting a commercial plot package because that package is very well developed and the particular one he is referring to is a package that conforms to the CORE standard for example, and that is an inhouse argument in NRAO and it is quite relevant in the community. Similar discussions have gone on at Space Telescope Science Institute in the past, you had troubles with the DISPLA package you considered. Now, there are problems with that. For example, that system is not yet a national or international standard, therefore its long-term future is gravely in doubt. Being tied to a single sole source means that you are vulnerable to any problems that that vendor has and any funny pricing strategies he may wish to use to get more profit out of us. I think it is much better to attempt to get national, and, better yet, international standards, that will be supported by multiple vendors. That I think needs to be our strategy in graphics area. We do need to have standards in graphics that are universal and we don't have them and it is a major problem area and we must hope that GKS will be the answer. I don't think CORE is going to be the answer. Unfortunately in terms of commercial packages you can buy today that are really practical they conform to CORE. The trouble is they don't agree with each other. That's a major technical area.

Rots: I won't elaborate on it, but I think your argument of CORE versus GKS where you buy a commercial package is irrelevant, essentially because if GKS becomes a standard, all these commercial packages have got to migrate to where it's GKS, and that is going to happen. The other thing is that what you really buy is that you have got a working package and any of your users, when he gets the package, he can just buy the drivers off the shelf that he needs, and everything will work, and you don't have to worry about it.

Albrecht: Giorgio is going to put back on his hat as the chairman of this session and tell me that I should finish. Have we left out anything really controversial or is there anything that we should cover and did not? If not, we should go and have coffee. Thank you.

SESSION

SYSTEMS FOR DATA ANALYSIS

CHAIRMEN:

M. DISNEY
M. CAPACCIOLI

INTRODUCTION TO DATA ANALYSIS SYSTEMS FOR ASTRONOMY

R.J. Allen

Kapteyn Astronomical Institute
University of Groningen
Postbus 800, 9700 Av Groningen, The Netherlands

SUMMARY

A proposal is made for a scheme within which off-line, general purpose astronomical image processing software systems may be described. This scheme may be helpful when attempting to compare the features and shortcomings of various systems which are currently available or being planned.

INTRODUCTION

A trivial but important point to make at the outset is that a collection of data reduction programs is not a system for data analysis. All of us are surely familiar with the case of the graduate student or staff member who single-handedly writes a "package" of programs to reduce data from detection system X on computer Y. Such programs are usually efficient, fast, and very clever. Unfortunately, such packages are frequently not very durable; the decay process set in very soon after the originator of the software leaves to take up a new post elsewhere, and within a year or two the package has become so rusty that it is unusable. Systems are more than just a collection of programs; they are more complicated, with many different but related parts, often produced at different times by different people who must also relate efficiently with each other. The resulting product must have a high level of durability, outlasting at least several generations of graduate students. If it is really a good system, then besides these traits it will also contain analysis programs which are efficient, fast, and very clever.

COMPARING SYSTEMS

The greater complexity of data analysis sytems over software packages means that their advantages and shortcomings are viewed differently by different people. For instance statements such as "AIPS is really good" and "AIPS is useless" are not necessarily contradictory; they may just represent different viewpoints, such as on the one hand that of the system manager responsible for installation and maintenance of the system, and on the other that of a user faced with the problem of reducing echelle spectra. People may prefer System X because of the ease with with new programs can be added, deprecate System Y because it runs in a Harris computer instead of a VAX, and praise System Z because it works under Unix. Yet many of these remarks will be recognized by system architects as not being fundamental to their systems; if AIPS is poor at reducing echelle spectra it is more a reflection of the specific astronomical interests of the AIPS user community than of the system architecture.

So how are we to compare the systems which are going to be described in the papers to follow this one? What criteria should we use? I have no thoroughly-tested answer to this, but there is a scheme which has been developed around our own image processing system in Groningen (GIPSY, described elsewhere in this volume), and I want to take this opportunity to see whether it is general enough to be useful in the present situation. I use it as a guideline for my students when, as part of a course on image processing in astronomy given in Groningen, they have to design an interactive data analysis system on their own.

INTERFACES AND IMPLEMENTATIONS

A data analysis system can be viewed from at least three different points of view: the user; the programmer; and the system manager. For each of these viewpoints different things about the system are of importance. The first two viewpoints are sort of obvious, but I think the third is just as important and has not yet explicitly been recognized in all cases. The <u>interfaces</u> which the system presents to these three points of view can be described, I think, in rather general terms (at least that is the present thesis). On the other side the system has two types of <u>implementations,</u> one in hardware and one in software. It is perhaps the implementations which are the most completely described and hotly debated when one discusses systems; I hope the scheme I have introduced will put them into a better perspective. The scheme is illustrated in Figure 1.

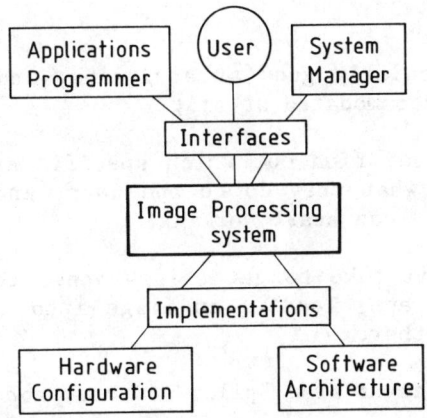

Figure 1. Scheme for describing different facets of data analysis systems.

User interface

Perhaps the best way to describe the system to a user is to present a scenario for a typical data analysis session, a sort of guided tour of what can be done and how to go about doing it. This guided tour resembles the "walk-through" analysis of program operation carried out during the later stages of a software development project. The guided tour should include some description on at least the following points:

- how does a novice user discover the way to get started; for example what are the procedures for reserving computer time, obtaining disk file space and magnetic tapes, logging on and off;

- how are queries and responses organized on the terminal screen;

- how does one tell the system what to do, i.e. keyboard command formats, "high level command language" syntax, special function keys, other input devices such as trackballs and data tablets;

- what sort of data are the presently available application

programs particularly good at analysing, and what kinds cannot currently be accommodated at all;

- how does the user find out which specific application programs are available, what they do to the data, and what they need to know before they can start working;

- how long does it take to get things done; for instance reading an image from tape, loading an image into the display, adding two images together;

- how does one recover from "pilot" errors; and

- how does the user provide feedback to other users and to the programmers concerning software (and hardware) failures, incorrect documentation, and desiderata.

Applications Programmer Interface

If your intention is to simply install one of the existing systems in your computer and not to add new software to it, then a description of the programmer's interface may be of lesser interest. However, experienced users working in research environments inevitably want to do something new with their data, something not yet available in the standard systems. Whether they write new software themselves or have someone else do it for them, it is usually desirable to incorporate those programs into an existing, proven structure. The appearance of the system to an applications programmer then becomes important. As with the user interface, a guided tour through the whole program development process is also a good way to describe the application programmer's interface. Some of the questions to be answered include:

- how does the novice programmer discover the way to get started; for example what programming languages may he use, what software conventions must be observed, and how is the documentation to be produced;

- how does one interact with other users and programmers in order to sharpen the design of the program and to inform others of progress;

- how does one go about coding and testing the program;

- what can one do when errors are detected during operation of the the program;

- what sort of refereeing will be done on the final source code and documentation; and,

- what are the procedures for bringing programs into the system library.

System Manager Interface

At a location where a system is to be imported and installed but not developed further, the system manager's tasks are confined to handling updates received from the originators and transmitting failure reports back to them. However, as in the case with the programmer's interface described above, in an active research environment new software will inevitably have to be developed in order to handle new problems not forseen by the established data processing system. To what extent an existing system may also be suitable as a <u>template</u> on which to develop new software is very much the concern of the system manager.

In software systems consisting of elements which are continually being added and modified, it is important that changes in one part of the system have as little effect as possible on the other parts. Good systems should have <u>low connectivity</u>, otherwise changes to them require excessive policing on the part of the system manager in order to maintain system integrity. On the other hand, if lots of people are going to be mucking about with the software, it is clearly very important that the system offers some assistance in <u>promoting communication</u> among them and in keeping track of the changes they make. Some aspects of systems which are of concern to the system manager then include:

- what requirements does the system impose on programmers in order to improve the quality and maintainability of the software. These requirements may include a choice of programming language and style, and peer review of program code;

- how does the system react to the addition of new software which is defective;

- how does the system keep track of software failure reporting and of updating programs; and

- what facilities does the system offer for stimulating communication amongst users and programmers.

Hardware Implementation

This is a more well-known aspect of systems and does not require much elaboration here. A description of the hardware in which a given system has been implemented usually includes a block diagram of the sort shown very schematically in Figure 2. Block M is the CPU and main memory, the horizontal line is the peripheral bus, and blocks 1, 2, 3, ... refer to the mass storage and other

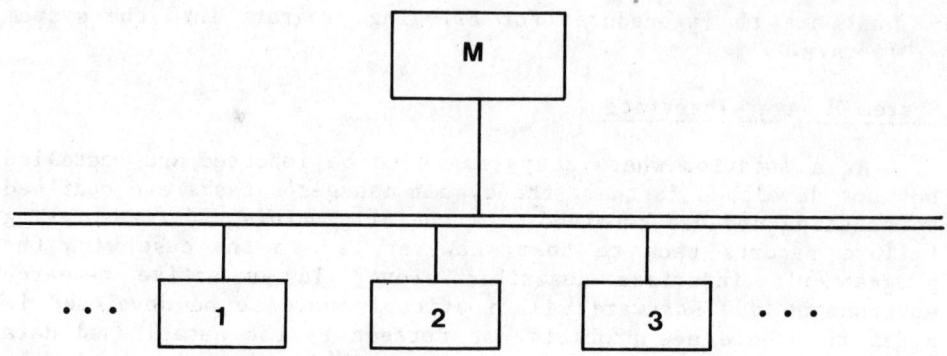

Figure 2. Scheme for describing hardware and software implementations.

peripherals. It is of course important to know if the efficient and full use of system facilities requires a particular type of computer or specific types of peripheral devices such as plotters, image display units, array processors, etc.

Software Implementation

The diagram for the hardware configuration shown in Figure 2 is organized around the computer bus. The importance of this bus depends less on the fact that it is a bunch of wires and more on the fact that it includes a set of conventions concerning such things as voltage levels and timing diagrams necessary for communication on the bus; it provides a well-defined interface to the system. The bus conventions also keep the system connectivity down by minimizing the mutual effects of peripherals upon each other; for example it is not necessary to re-wire the printer because the card reader was removed for repairs.

We recognize these features as being desirable also for the software implementation of a system, and so it should be possible to use Figure 2 for the software as well. Block M may then refer to the master control program, variously called the dispatcher, command decoder, or kernel. Blocks 1, 2, 3, ... are the application programs. The description of the software implementation then reduces to a detailed description of the interfaces through which the programs communicate amongst themselves and with the master control program. Other aspects of the software implementation which are of importance include:

- dependency on the host computer operating system;

- structure of the data base and its dependence on particular peripheral devices;

- mechanisms for defining and obtaining default values for application program parameters; and

- facilities for chaining application programs together in order to accomplish a complex data processing task.

GOOD AND BAD FEATURES

Although the speakers to follow have of course not all had the opportunity to use the scheme which I have described, it will hopefully be possible to make such a structured comparison after the proceedings of this workshop are published. In presenting this scheme I have tried to avoid value judgements, although upon reviewing the manuscript I see that my personal preferences have inevitably come through in certain places. Whether a given set of features is good or bad depends to a great extent on the context in which a particular system is to be used, and you will have to make that decision yourselves. Some general guidelines in the case of university research groups of modest size (10-20 astronomers) have been discussed by Ekers and myself elsewhere in this volume.

As you listen to the following talks, perhaps the best bit of advice I can give you is the classic one suitable for all buyers in the market place, caveat emptor!

FIPS: OBJECTIVES AND ACHIEVEMENTS

T.D. Westrup

European Space Technology Centre
Noordwijk, Netherlands

J. Gras and W. Kegel

Dynaflow Software Systems BV
Amsterdam, Netherlands

1. INTRODUCTION

The Space Telescope (ST) is one of the most ambitious projects in space astronomy. It consists of a 2.4 metres diameter telescope of approximately 14 metres length which will be placed in orbit around the Earth by the Space Shuttle in 1986, and will carry five highly sophisticated scientific instruments. A full description of the ST can be found in Ref.1.

The European Space Agency's contribution to this programme includes one of the on-board scientific instruments, the Faint Object Camera. The special features of the Faint Object Camera (FOC) make it the ideal instrument to fully exploit the angular resolution of the ST of the faintest possible sources. A full description of the FOC can be found in Ref.2. The FOC contains two complete and independent cameras, each with its own optical path and detection system. One camera is matched to the resolution performance of the ST, the other gives lower resolution but a field of view four times larger in area.

1.1 The FOC Image Processing System (FIPS)

In order to process the images from the FOC an advanced image processing system is needed, consisting of both an image processing environment and application programs to run in that environment. Although several such systems existed or were under development, ESA decided to implement a new system for the FOC. It was considered that other systems might not meet the FOC schedule constraints, were too restrictive, or did not observe the rather stringent ST Software

Standards. Accordingly, ESA decided to place a contract, which would be under its own administrative control, for the development of a FOC Image Processing System (FIPS). This contract was placed in December 1980, and FIPS was accepted in March 1983. The contract emphasised the environment aspect and it was an intended goal that FIPS should achieve a wide acceptance within the astronomical community, so that application software would become more transportable between institutes.

Many institutes had adopted the VAX computer, so this was chosen as the vehicle for FIPS. At the time, FORTRAN was still the most widely used language in astronomy and is also specified as a ST software standard, so in spite of its limitations was adopted for FIPS. To meet ST software standards the detail design is coded in PDL (Ref.5). To transfer images between institutes the FITS data format (Ref.4) is supported. The technical requirements of FIPS are summarised below:

- Means of displaying and interacting with images on a colour graphics screen
- Means to archive, retrieve, protect and delete images
- A user friendly means to process images interactively and to build more complex functions out of the basic operations
- A means to perform the basic operations asynchronously
- A facility to process repetitively a whole sequence of images
- A means to keep track of all processing operations (log file)
- A facility for comprehensive 'help' and other on-line documentation
- Support for the integration of application programs (including those written in other environments) with a minimum of effort
- A means to input images and to transfer images to other sites
- A facility to digitise pictures obtained from other sources
- A structured overall design to allow the system to be readily maintained, modified or extended to meet upcoming requirements, or ported to new generation hardware.

1.2 Hardware and Software Environment

The initial FIPS hardware (see Fig.1) consisted of a VAX-11/780 with floating point accelerator, 1.5 megabytes of memory, 134 megabytes of mass storage, two tape drive terminals, and a line printer. The raster graphics device is a Sigma ARGS 7000 with 17 megabits of memory and two video processors operable in pseudo-colour or tri-colour mode. A hard-copy device allows black and white pictures of the ARGS screen to be obtained. A TV digitiser and light table allow both prints and transparencies to be processed.

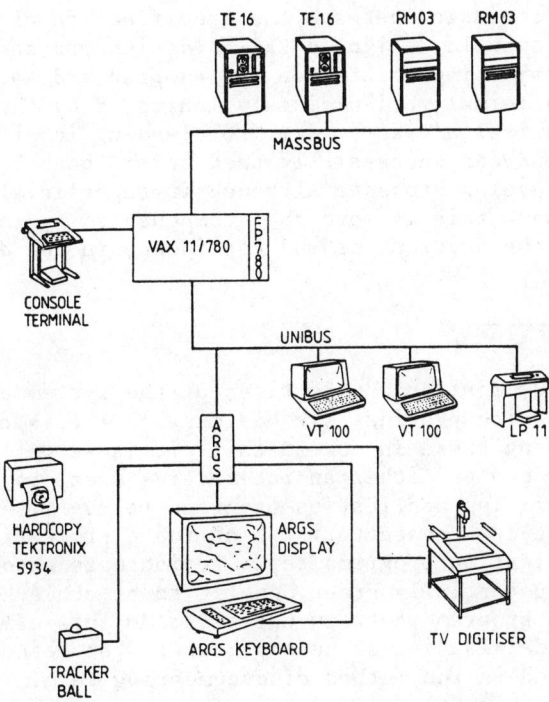

FIGURE 1: FIPS HARDWARE CONFIGURATION

Subsequent enhancements to the VAX which proved necessary included an increase to 3.5 megabytes of memory to improve performance, a further 450 megabytes of mass storage, a VT125 graphics terminal linked to a print/plot device, and a De Anza IP8400 raster display device for compatibility with other ST institutes.

The software environment of FIPS is the VAX/VMS operating system, and the implementation language is VAX FORTRAN, except for the display drivers and the ARGS 7000 order code programs. PDL is used for the detailed design documentation, together with the user documentation utility RUNOFF.

1.3 <u>Planning</u>

It is worth mentioning at this point the management techniques adopted to ensure the acceptability of FIPS to the astronomical community without sacrificing the important goal of close adherence to the planned development schedule. These two objectives often conflict, resulting either in an unacceptably rigid and restricted system, or continual slippage of the final delivery.

To avoid confusion between the specification of detailed
requirements, overall design, detailed design and coding, a method
of strictly serial implementation was adopted and each phase was
concluded by a formal review closely monitored by the ESA FOC team
and an astronomical advisory team from leading institutes. This
methodology has been successfully used before both by NASA and by
ESA for the Spacelab Project. Although it superficially appears to
create extra work this is more than compensated by increased
visibility of the development and confidence in the design.

2. SYSTEM OVERVIEW

The already mentioned objectives of the system are translated
into a design which provides six basic capabilities to the user.
Before describing these in more detail, it is as well to remember
that this user can be either an interactive user, at a terminal, or
a programmer writing application programs to run within the FIPS
environment. It is a characteristic of image processing that the
number of application programs tends to increase exponentially and
that there is potentially great benefit to be achieved from
interchange of programs between users. To do this efficiently some
degree of standardisation is needed both in the method of writing
the programs and in the method of documenting them. The environment
of FIPS forces this standardisation, for example, by defining the
form of communication with the end user, the structure of the image
data system, and the kind of documentation required with each new
application. In practice the standardisation leads to a large body
of routines that the application should call when doing I/O, e.g. to
the terminal or to the disc. This, in turn, saves effort when new
applications are written.

FIPS also provides a number of basic image processing commands,
for example, to do arithmetic, geometric transformations and data
management, but in a sense these are just 'applications' relative to
the FIPS environment. We call them 'native commands'. The six
capabilities provided by FIPS are the following.

2.1 User Interface

This consists of a single task (GATEWAY) through which all
dialogue between applications (or commands) and the terminal are
routed. GATEWAY can force therefore some uniformity in the human
interface. At the same time GATEWAY provides a true multi-tasking
capability (see Fig.2) and a powerful high level command language.
Multi-tasking is desirable because image processing tasks are often
too long for the user to await completion before executing the next
command. On the other hand, frequent user interaction with the
processing is needed so that a pure batch solution is not adequate.

In the introduction we mentioned the need to "repetitively process sequences of images with minimum operator invention". This is provided by the command language which we call FCL (FIPS Command Language) and which is described in more detail in the FIPS User's Manual (Ref.6). At this point we should mention another characteristic feature of image processing applications, which is that they often take a number of input image and transform these to a number of output image with the help of user supplied parameters. More complex processing may be achieved by "stringing together" a number of application programs in which the output of one program serves as input to the next, and so on. These capabilities are also provided by FCL. It is a fundamental property of FIPS that applications can be built from other applications (and from native commands) thus providing a processing capability of increasing complexity. And so new processing techniques can often be built up from existing ones in a fairly short time, making use of what is already available.

The user interface package also consists of a number of routines which application programs must use for communication with the terminal (via GATEWAY). The applications (and commands) are implemented as VAX/VMS subprocesses running under the GATEWAY process but the application programmer does not need to be aware of this. The GATEWAY process also arranges for the logging of all user interactions to a central file so that retracing of his actions is possible at any stage.

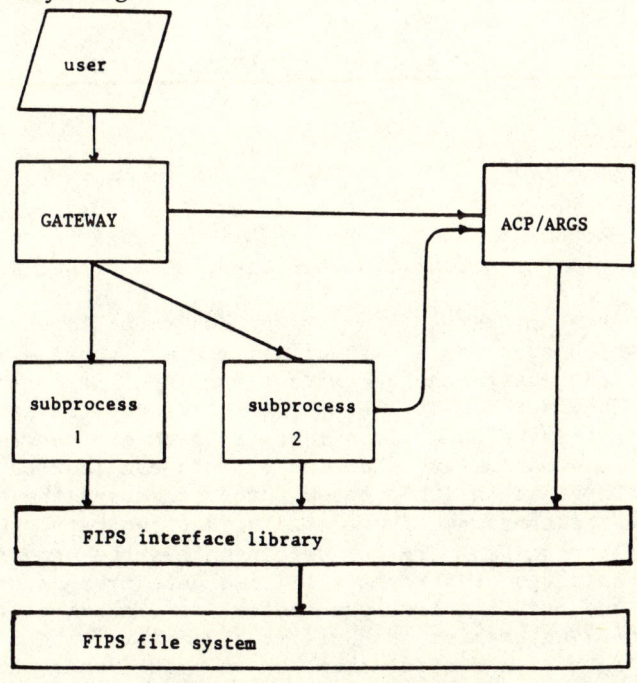

FIGURE 2: FIPS COMPONENTS

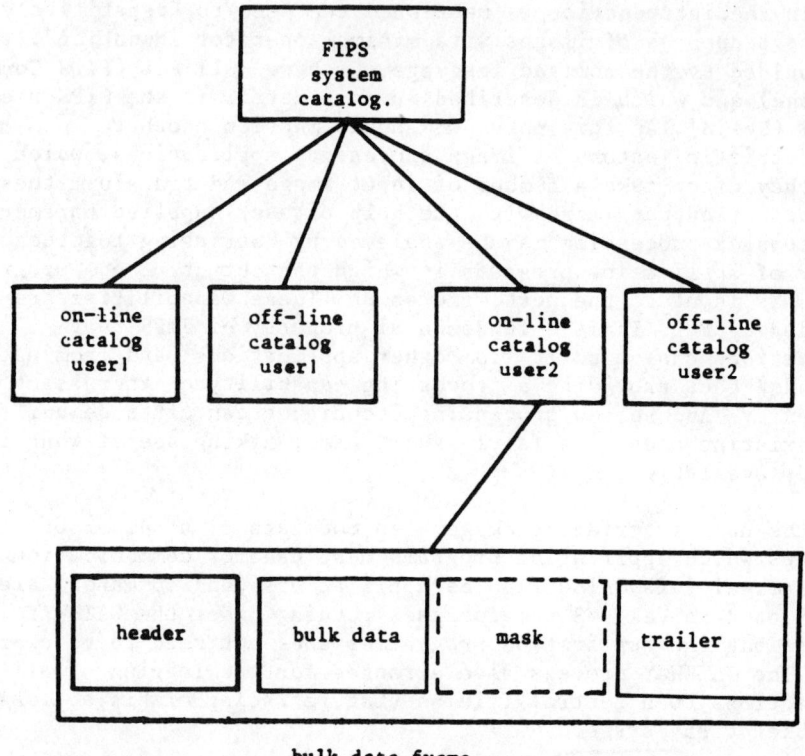

bulk data frame

FIGURE 3: FIPS FILE ORGANISATION

2.2 Data Management

This capability of FIPS is based on a rigorous organisation of files and directories that are accessible only via the software of FIPS and not via the normal VMS utilities. See further Fig.3.

At the heart of this system is the owner catalogue (one for each user) which contains a list of all frames (usually images) belonging to that user together with a number of important parameters. The user has at his disposal commands to list the catalogue and to make selections from it based on the values (or ranges) of frame parameters. Such selections can then be used, for example, as input to an FCL program which processes the entire list of frames in the same way. There are in fact two catalogues per user, an on-line and an off-line catalogue. As the name suggests, the on-line catalogue refers to all frame data present on-line (on disc) while the off-line catalogue refers to archive data which is saved on tape. The problem of building and maintaining orderly archives of data is a central theme in the data management package.

There are many commands available to the interactive user to move his data between disc and tape (in various formats including the widely accepted FITS format) to inspect and modify it, to carry out basic edit functions or to delete it. Just as important is the large suite of subroutines available to the programmer for access to data within the FIPS environment. These subroutines hide the detailed structure of the data system from the programmer, who, for example, does not have to know that there is a catalogue through which all accesses must pass.

In the above discussion we have referred to image or frame data a number of times, anticipating the definition of the basic structure in the FIPS data system called the bulk data frame. There are several types of 'bulk data' as described below. The bulk data frame is essentially the (image) data plus descriptive information contained in both a header and a trailer. In fact the header, data and trailer are implemented as three separate VMS files allowing for easy expansion of each component without disturbance to the other two. The header file contains descriptive items about the data in a FITS type format (Ref.4), for instance the dimensions, the name of the observer, creation date etc.. Allowable image dimensions are adequate to cover all the FOC formats but are in fact more general than this. The trailer contains the "history" of the image plus user supplied comments.

History has a rather special meaning in this context as it refers to the FCL command line responsible for the creation of the image, plus the actual values of any parameters specified in that line. The intention is that a user can see from this how a particular image came to be created. However, there are limits to this, for instance the creation command would probably involve other images as input, so that the user must look at the trailer contents of these to determine their history, and so on, back in time until the original "source images" are encountered (e.g. read from tape). Of course if any of the intermediate images have been deleted the "trace back" cannot continue beyond this point.

The FIPS native commands all arrange that the history is automatically included in the trailer, and so does the "creation" subroutine available to the application programmer. As suggested above, other bulk data types than "image" are implemented in the data management package. The other types are:

- graphic overlay (a 1-bit deep image)
- point transfer function (look-up-table for grey scale transformation of up to 64K words)
- control (all information necessary to restore the context of a displayed image)
- table.

There is not sufficient space here to go into the detailed meaning and use of all these types. The type "image" differs basically from the other three in that it may have an associated mask, which is essentially a 1-bit deep map telling which pixel in the image is "bad" for one reason or another. A considerable effort is made in the system design to handle "bad" pixels of images in a flexible way.

The type "table" is worth some further explanation since this data type has turned out to have a wide variety of application in practice.

The "table" data consists of a structure much in line with what the name suggests, a number of columns containing numerical values or text. The values in a column are all of the same data type (real, integer or character) but different columns may have different data types. Usually, the rows of a table will contain values intended to describe a particular object or event. As with images, there are many native commands to manipulate tables, to create them, enter data, copy data, change data and delete it. Usually these operations can be on a row or column basis, sometimes on an element basis. It is also possible to sort a table so that the elements in one particular column are in ascending (or descending) order. In section 4 some practical examples are given in the use of tables.

2.3 Image Device Interface

Currently two image processing and display devices are supported under FIPS, the ARGS 7000 (Sigma, UK) and the De Anza IP8400 (Gould, US). To enable orderly access to the image processing device, a number of routines are available to the application programmer. These routines also relieve him of having to understand the detailed workings of the devices, and in fact they should be fairly straightforward to re-program for any other device.
(At present the ARGS implementation is the most extensive).

To allow for simple expansion to a (future) ARGS configuration with multiple users, the decision was taken early that all ARGS requests be routed through a central control process (the ARGS Control Process, ACP) where requests can be properly queued and administration kept concerning what resources of the ARGS are allocated and to which users. There is also a significant amount of FIPS software resident in the ARGS memory, and a fairly elaborate protocol is involved in the interaction between ACP and this software. The resident software in the ARGS can carry out all basic functions for image processing, e.g. loading and saving of images, zoom, scroll, blinking, look-up table control and so on. The functions can be activated by application programs via standard routines. They are also available to the interactive user by means of FCL 'native' commands.

In the De Anza implementation there is presently no central
process to administrate the device and keep track of its status. This
is simply because the De Anza interface was developed under time
pressure which did not allow such a (technically desirable)
construction. A consequence is that the functionability, available
to an on-line user of the De Anza is somewhat less than that available to an ARGS user. However, this situation will be remedied in the
near future.

The De Anza is not capable of running its own local programs
so that here also the interface is less complex than for the ARGS.
On the other hand the De Anza has attractive 'firmware' features
which make the programming of its actions from the host computer
fairly straightforward.

2.4 Graphic Device Interface

Support for graphic devices in FIPS is currently somewhat
limited, although plans for a considerable expansion are being made.
The graphic functions are available as routines to the applications
programmer, again to provide orderly access to the devices and to
'hide' the details of their operation from him. The functions
supported presently are essentially orientated towards plotting.

Here again there is a central controlling task (PCP in this
case, standing for Plot Control Process) through which all requests
for access to a graphics device must go. For each new device a
separate PCP task must be started. The PCP task maintains a 'logfile'
of all the requests sent to it so that a 'playback' of this file at
a later time is capable of re-constructing the plot without re-running
the application program originally responsible for producing it. At
present the VT125 graphics terminal and LA-50 printer/plotter are
supported.

Work is presently underway to make most of the plot functions
available to an application programmer also available to an
interactive user.

2.5 A "Native" Command Set

From preceding discussion it will be clear that a number of
'native' commands are provided by FIPS to do data management,
editing, display, plotting and some basic transformations. In addition
should be mentioned the commands for multi-tasking control. In total
there are some 100 commands at present. The list will undoubtedly
expand as new commands are written and introduced to the system.

2.6 Expansion Capability

The ability to add new programs to the FIPS system, whether they be FORTRAN programs, FCL programs or a combination of the two, is fundamental to the design philosophy. In Ref.7, the exact steps involved in adding new software are defined in more detail. When these steps have been carried out, FIPS has a new "function" which is documented in a similar manner to the already existing functions, and can in turn be used to build more complex functions. New functions can even be added which imply new software in the ARGS, although for this case the rules to be followed are more complex and not recommended for beginners.

3. APPLICATION SOFTWARE

The initial delivery of FIPS included relatively little in the way of astronomical application software, apart from simple and fundamental functions such as DISPLAY, ADD, ROTATE, etc.. However, as stressed earlier, it was always recognised that FIPS would have to include a wide variety of astronomical applications to meet the needs of the astronomer, and that this software would, in many cases, originate from other environments.

The choice of applications to be incorporated was at first dictated by the development status of the FOC, and two groups of functions were, at that time, considered to be essential. Firstly, the IHAP image processing system developed by the European Southern Observatory (ESO) and running on an HP 1000 mini-computer had been used extensively as a development tool for the FOC, and it was considered essential that several IHAP applications should be added to FIPS to ensure their longer term future and continuing availability for the FOC development cycle. Secondly, the ground calibration of the FOC was supported by a suite of software developed in the VICAR environment by the Rutherford Appleton Laboratory (RAL), Chilton, U.K., (see Ref.3), and it was considered necessary that these programs should be made available under FIPS as the basis for future FOC calibration in-orbit.

3.1 IHAP and MIDAS Applications

Programs obtained from ESO were from the two image processing environments supported there, namely IHAP and MIDAS. The programs were obtained as FORTRAN sources and edited on case by case basis to conform to the requirements of FIPS. This usually turned out to be a straight forward task, taking typically 4 days per program. Preparation of test procedures with test results generally involved another 2 days per program. In total some 10 programs were integrated covering areas such as filtering, object handling, Fourier transformations and simple statistics.

The amount of effort involved in converting and testing each program varied considerably with the level of program documentation, both in the form of source code comment and user instructions. In a few cases it appeared simpler to write a new program and this was then carried out. It would certainly have speeded up the conversion effort if each program had been supplied with test procedures and results. Since these were not available, the conversion effort required much more understanding of each program to design test procedures and predict results.

Conversion of programs written under MIDAS was generally more straightforward than for those from IHAP, due to the similarity of program interfaces in MIDAS and FIPS.

3.2 FOC Calibration Applications

The RAL suite of software comprises approximately 45 separate programs running under the VICAR system, originating from the Jet Propulsion Laboratory.

The purpose of these programs is to correct FOC images for the non-linear dependence of the instrument gain on light intensity, for spatial non-uniformity over the image (a function of wavelength), for the absolute gain of the instrument (i.e. input photon flux versus observed counts), and for the transmission of selected optical elements in the FOC. A similar technique, slightly modified, is used to correct spectrographic images obtained by using the FOC spectrograph.

As each of these programs is similar insofar that every one of them interfaces with VICAR, the technique used to "add" them to FIPS was to construct an interface between VICAR and FIPS. In addition to simplifying the task of integration (the original programs requiring little or no change), this method of extending the FIPS application interface has the additional advantage that any VICAR-based application may now be readily incorporated into FIPS. However, a word of caution is in order here, several incompatabilities were discovered between the conventions of FIPS and of VICAR. To quote just one, pixel positions in FIPS start at zero, whilst in VICAR they start at one, so some manipulation is needed to ensure compatibility. Another problem relates to those RAL programs that interface directly with peripherals, e.g. plotters, where it was often found more convenient to re-write the program to interface with the FIPS PCP than to attempt to modify the original.

Testing the suite of programs also proved to be a significant part of the task. Fortunately, a version of VICAR (Portable VICAR) which runs on a VAX was available, so the comparison of the results obtained from the FIPS and VICAR suites could be largely automated. Otherwise, given differences in architecture between the IBM and the VAX, results could have only been compared by inspection, a difficult task when processing perhaps 512 x 512 separate elements in an image.

The total effort to integrate and test the 45 programs amounted to about 2 man years. Although this is by no means insignificant it does compare favourably with the effort required to write them in the first place of some 10 to 15 man years, so on balance the exercise can be considered as quite successful.

4. PRACTICAL APPLICATIONS OF FIPS

The build-up of FIPS users has been relatively slow. Like programmers, astronomers are reluctant to adopt new image processing techniques and in the same way that FORTRAN still remains a universally used programming language, astronomers seem to prefer to accept the limitations of an old familiar system rather than spending a few days learning a new one. Another factor is that astronomers, unlike programmers, are not usually subject to management control. FIPS is certainly not the first system to suffer from initial inertia, probably it will not be the last.

The first practical use of FIPS came from a user of a ground based detector built as a prototype for the FOC. This initial use concentrated mainly on the RAL calibration software and revealed the need for more testing during the integration of this software, as many small errors and misunderstandings had remained. However, after some perseverance the geometric calibration of the ground based detector was successfully realised.

Subsequently, FIPS was brought up at the Space Telescope Science Institute for evaluation, and was also used by astronomers in ESTEC for processing of IRAS data. Both these activities resulted in further improvements to the software, repair of errors and suggestions for enhancements. These initial FIPS users in fact functioned as "field testers" for FIPS, a necessary phase in its evolution since a system of such size (about 200,000 lines) cannot be perfected through testing by its developers alone. In practice it has been fairly straightforward to trace errors and implement improvements in the software, which may be considered to be an indication of a clean and consistent design.

The opportunity to use FIPS as a diagnostic tool during the calibration of the FOC at NASA Goddard Space Flight Center was also particularly welcome both as a means of demonstrating its flexibility and maturity, as well as to identify any practical limitations. Approximately 150 useful images were taken during the calibration, of which some 100 images were analysed by FIPS and the RAL developed FOC calibration software. In addition to the extensive use made of existing functions several new applications were added to FIPS. Further details of the analyses performed are summarised below.

4.1 Statistical Analysis

Selected areas of flat-field images were analysed for uniformity and noise by tabulating, for each area within each image, the mean, standard deviation and ratio of the standard deviation to Poisson noise of the counts accumulated by each pixel. For this analysis, the "Table" facility of FIPS was used extensively. Furthermore, the effect of (e.g.) wavelength on the relative efficiency of the FOC response, or the improvement of uniformity obtained by photometric correction of the image, could be similarly investigated.

4.2 Image Stability

The geometric distortion of the FOC cameras may be measured by means of a regular array of 17 x 17 chromium marks (reseau marks) deposited on the intensifier photocathode of each camera in a square grid. It was required to measure the stability of this geometric distortion as, for example, a function of time. The most graphic illustrations of this technique was provided by means of "tad-pole plots" and simple plots of reseau mark movement against time.

An example of a tad-pole plot is shown in Fig.4. The round spots show the location of the reseau marks in the reference image and the length of the "tails" the relative displacement, magnified by a user selected factor. This provides an immediate indication of the characteristics of the geometric distortion (or instability). In the case illustrated, it is obvious that the predominant effect is a rotation.

An example of the movement of a reseau mark with time is shown in Fig.5. Note that a similar plot can be obtained for the mean movement of several reseau marks, as well as for individual marks.

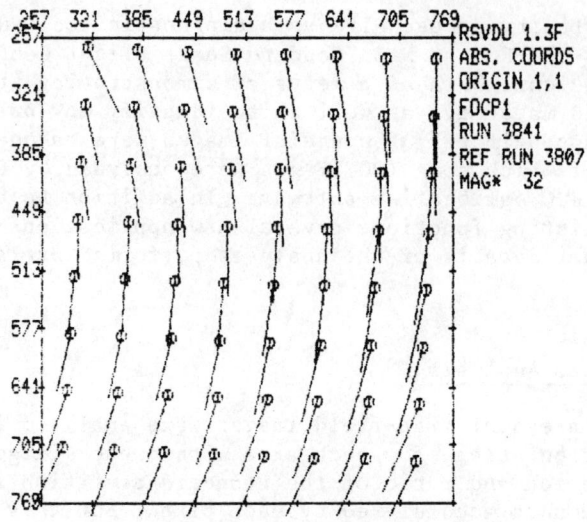

FIGURE 4: PLOT OF RESEAUX MOVEMENT

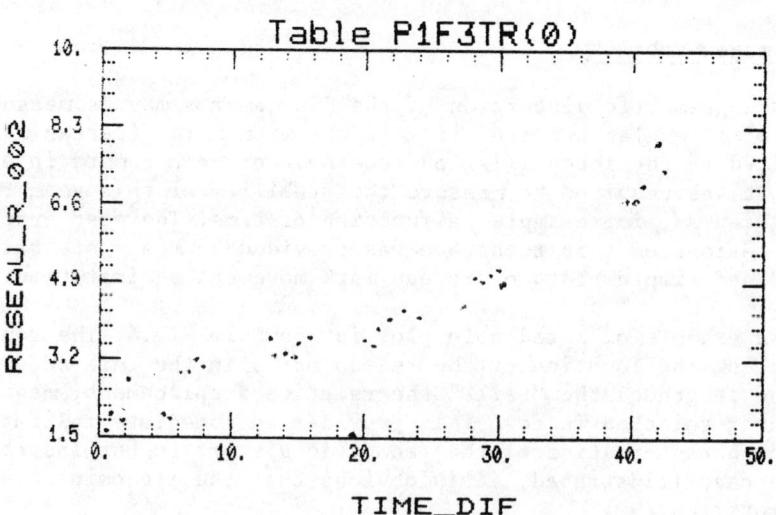

FIGURE 5: MOVEMENT OF RESEAU WITH TIME

4.3 Image Tabulation

A slightly less obvious use of FIPS during the FOC calibration was its use to tabulate selected information about the characteristics of each image. Every image taken by the FOC has an associated header which defines the image uniquely, and which also contains other information, e.g. where it was taken, when it was taken, the wavelength and intensity of the exposure, the format of the exposure, the position of the optical elements of the FOC, etc.. FIPS was used to extract this information from the headers automatically, sort it in a table and then to tabulate selected columns of this table, optionally sorted according to certain criteria. This function has proved valuable as a means to manage the calibration activity.

The calibration activity extended over a period of some 5 weeks. The lessons learned were twofold. First, a few additional incompatibilities were identified between the FIPS system and the RAL software which were readily corrected. Second, some new functions were incorporated into FIPS with relatively little effort to meet the ongoing demands, e.g. XTRA, which extracts data from image headers and stores them in a table, and several extensions to the RAL programs, proving once again the flexibility of FIPS as a vehicle for astronomical applications.

5. CONCLUSIONS

Various goals associated with the FIPS development have been mentioned throughout the paper and are summarised again below:

- Build a system under ESA control, according to ESA standards and with input from the European astronomical community

- Achieve wide acceptance amonst this community thereby facilitating the exchange of application software between institutes

- Support integration of existing astronomical software with a minimum of effort, in particular the FOC calibration and correction software so as to support FOC calibration activities.

Not all these goals have been reached with an equal degree of success. Considering each point separately, we come to the following conclusions:

- The development of FIPS under ESA control proved very successful, in particular with regard to the time schedule involved, the functionality of the final product and the adherence to prescribed standards. The formal, phased approach has led to a product which is very consistent in design philosophy and structure, this is in

contrast to many less formal developments where changing user requirements tend to drive the design into inconsistent directions. The amount of input received from the European astronomical community, channeled as it was through a number of formal review meetings, turned out to be insufficient for the full definition of such a complex and extensive product as FIPS eventually became. More continuous input from experienced users would have been preferable to sharpen the requirements definition at a detailed level.
From the functional viewpoint, the goal to provide a basic system from which new and unforeseen processing methods could be built up quickly from existing ones, was more than adequately attained. This was amply demonstrated during the FOC calibration activity.

- The acceptance of FIPS as a standard amongst different astronomical institutes has not eventuated and the problem of multiple, incompatible systems at various institutes seems today as great as when the FIPS development was started. In fact it is probably worse, since several institutes undertook the development of new systems in parallel with the FIPS development. This is a problem of co-ordination between the various European institutes and is of course not something within the control of ESA. On the positive side it may be said that the program interfaces used in FIPS for access to user supplied parameters and bulk data, after being copied from an original STARLINK proposal, have to some extent found their way into other systems. The job of transporting software between **these** systems has therefore been somewhat simplified.

- The integration of application software from a number of sources was carried out with reasonable success. Integration of software from the MIDAS environment was fairly straight forward, but this was mainly due to the similarity in the program interfaces. Integration of the RAL supplied FOC software turned out to be possible through the implementation of a "VICAR-to-FIPS" library whereafter conversion could be done in a semi-automatic way. Through this approach the whole suite of RAL programs could be integrated into FIPS with an effort considerably less than would have been involved in re-writing the programs under FIPS. The successful completion of this integration supplied and continues to supply the "on-line" facility needed for the FOC calibration activities, thereby justifying one of the major motivations for the original building of the FIPS system.

6. REFERENCES

1. ESA/ESO Workshop on Astronomical Uses of the Space Telescope, Geneva (12-14 February 1979)
   Proceedings edited by F Macchetto, F Pacini and M Tarenghi.

2. The Faint Object Camera for the Space Telescope,
   ESA SP-1028, October 1980
   Authors: F Macchetto, H C vd Hulst, S di Serego Alighieri,
           M A C Perryman.

3. IEE Proceedings of "Electronic Image Processing".
   Conference Publication No. 214, July 1982, "Processing Astronomical
   Images from the Faint Object Camera of the Space Telescope",
   P A Vaughan, Rutherford Appleton Laboratory, U.K.

4. FITS: A Flexible Image Transport System,
   Astronomy and Astrophysics Supplement Series 44 (1981) 363-370
   D C Wells, E W Greisen and R H Harten.

5. Program Design Language Reference Guide,
   Caine, Farber and Gordon Inc., Pasedena, California,
   February 1977.

6. FIPS User's Manual
   Dynaflow report FIPS-81-001, June 1984.

7. FIPS Application Programmers Reference Manual
   Dynaflow report FIPS-81-005, September 1984.

8. The Starlink End-User Environment
   Science Research Council, Rutherford and Appleton Laboratory,
   SGP 6.1, 31 July 1980.

MIDAS

Philippe Crane, Klaus Banse, Preben Grosbøl,
Charlie Ounnas and Daniel Ponz

European Southern Observatory
D-8046 Garching bei München, West Germany

INTRODUCTION

This paper describes the MIDAS data analysis system built at the European Southern Observatory in Munich. MIDAS is a general purpose data analysis tool which combines a flexible command language, powerful data analysis routines, and a set of very general data structures. It is implemented on VAX series computers and supports various image display and graphics devices.

The following paragraphs describe the hardware facilities that exist at ESO's headquarters in Garching, the MIDAS Application program interfaces and Application program structure, the command language, and the data structures. Finally, some examples of MIDAS applications are shown.

HARDWARE ENVIRONMENT AND WORK STATIONS

Figure 1 shows a block diagram of the VAX system configuration available to MIDAS Users at ESO's Munich headquarters. The major deviations from a standard VAX system are the addition of the Gould-DeAnza image displays and the Dicomed Image Recorder.

The image displays are IP8500 models with two video output channels per device and five 512 × 512 × 8 bit image memories per video output channel. Each display also contains a digital video processor which can be used to do high speed 16 bit integer arithmetic.

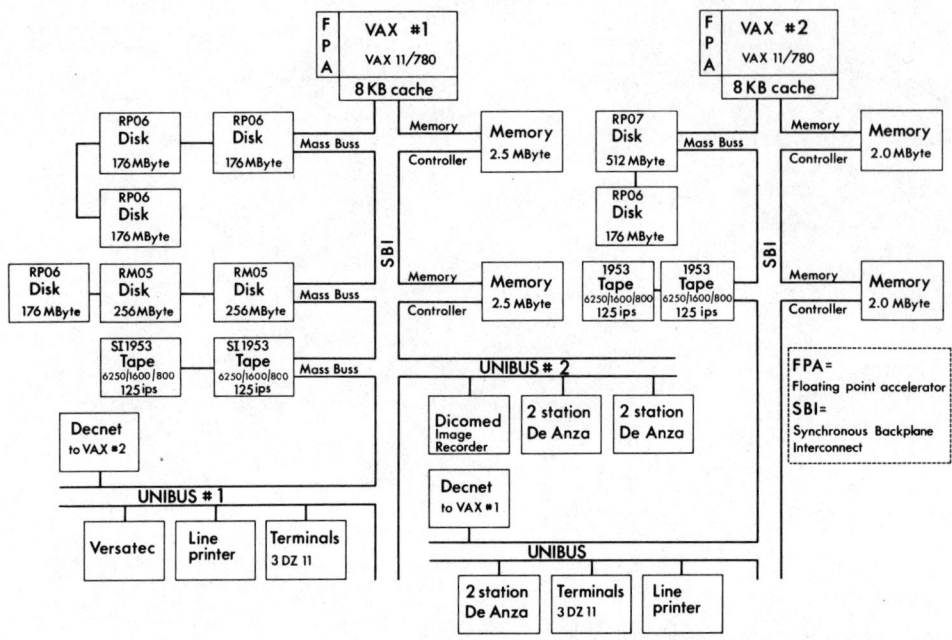

Fig. 1. Block diagram of the computer installation at ESO/Munich for data analysis using the MIDAS system.

The Dicomed D47 contains a high precision cathode ray tube which is used to write high quality images onto either black and white or color film. The device can handle images of up to 1500 × 1500 with very good resolution.

A typical MIDAS work station consists of a VT100 terminal, an HP2648 graphics display and an image display with a joystick.

DATA STRUCTURES

Since MIDAS was designed to work on astronomical data, the data structures reflect this goal. There are three basic data structures: images, tables, and keywords. However, both tables and images are data structures themselves since they both contain descriptors which provide necessary information about the data in the tables and images themselves. Keywords can be global or local named variables which are used to communicate between processes and to control the flow of processing.

Images are basically spectra, or pictures. They are limited to 6 dimensions, but not many routines handle 6 dimensional images. In fact, most routines work on 1 or 2 dimensional images. These are always stored as 32 bit VAX floating point numbers.

Descriptors are one dimensional vectors of type integer, character, real or double precision. A very convenient property of descriptors is that they can be dynamically extended so that a descriptor that was originally only 5 elements can have elements 10-14 added and the system automatically fills in elements 6-9. There is a set of standard descriptors for each image. These are required, but the user can add any number of auxiliary descriptors like a histogram, or a list of star positions.

Tables are used to store the results of various MIDAS commands as well as to serve as a general purpose structure for analysis and modelling. Each column in a table may be named and be given a value in physical units (e.g. erg s^{-1}). Operators exist to sort on columns, select on values or on relations, plot, and do ordinary mathematical manipulations.

Keywords are variables that can be accessed by any application program and by the interactive monitor. Hence, they are global variables. The main purpose of keywords is to pass information from one MIDAS application to the next, or to pass control information from the interactive monitor to an application. Keywords also serve as control variables in the command language control structures such as "DO N = 1.5". Here "N" would be a keyword. MIDAS procedures can create local variables which act like global keywords, but disappear when the procedure ends.

APPLICATION PROGRAM INTERFACE

One of the greatest strengths of the MIDAS system is the clear and simple application program interface to the data structure. Images, descriptors and keywords are accessed as follows:

```
RDDATA          to read image data
RDKEYx          to read a keyword of type "x"
RDDSCx          to read a descriptor of type "x"
```

Table interfaces are built on top of these basic routines but are quite similar. There are, of course, routines for writing data to these descriptors etc. In addition, there are higher level routines which are used to ease the work of the application programmer. For example, "RDIMAG" will provide all the basic image descriptors and the data for a specified frame.

These interfaces have proved simple enough to use that many astronomers have felt motivated to write their own astronomy specific application programs.

APPLICATION PROGRAM STRUCTURE

A typical application program has the following elements:

- Link to the MIDAS environment (INIT)
- Read required keywords (Input frame name, Output frame name, control values)
- Set up input frame
- Set up output frame
- Call application subroutine
- Update output frame history descriptor
- Disconnect from the MIDAS environment

With this structure, any application that does all its input and output through a subroutine can be quickly integrated into the MIDAS environment.

MIDAS COMMAND LANGUAGE

The command language in MIDAS was based partly on the desire to follow the tradition established with the Hewlett Packard based IHAP system and partly on the desire to stay reasonably close to the DEC DCL command language. Thus, the system is command driven as in IHAP, but uses / (slash) and spaces as delimiters. A typical command:

LOAD/IMAGE   NGC1052   3

would load the image "NGC1052" into channel 3 of the image display.

Several convenience features have been added. The system saves the last 16 commands by number. These can be repeated by typing the number or they can be edited by typing a period (.) followed by the number. Additionally, the tokens (tokens are the elements separated by a space) of the previous command can be repeated by typing a period (.) instead of whatever was there.

MIDAS provides some very powerful tools for building up procedures or combinations of commands. Firstly, there are the control structures. Secondly, there is the ability to reference values stored in the various MIDAS data structures. Thirdly, these procedures can call other procedures or even themselves. This combination of functions allows the MIDAS command language to be used as a programming language in much the same fashion as other high level languages like FORTRAN.

In addition to the normal language constructs, commands which permit assigning parameter defaults and limits are available. These are also debugging aids.

MIDAS APPLICATIONS

In this section, we show two of the more powerful application programs that exist in the MIDAS system. The first has to do with finding and analyzing images of relatively faint galaxies and stars. The second example shows the reduction of echelle spectrograph data.

Inventory is the name of the suite of programs associated with finding and analyzing images of faint objects. The commands

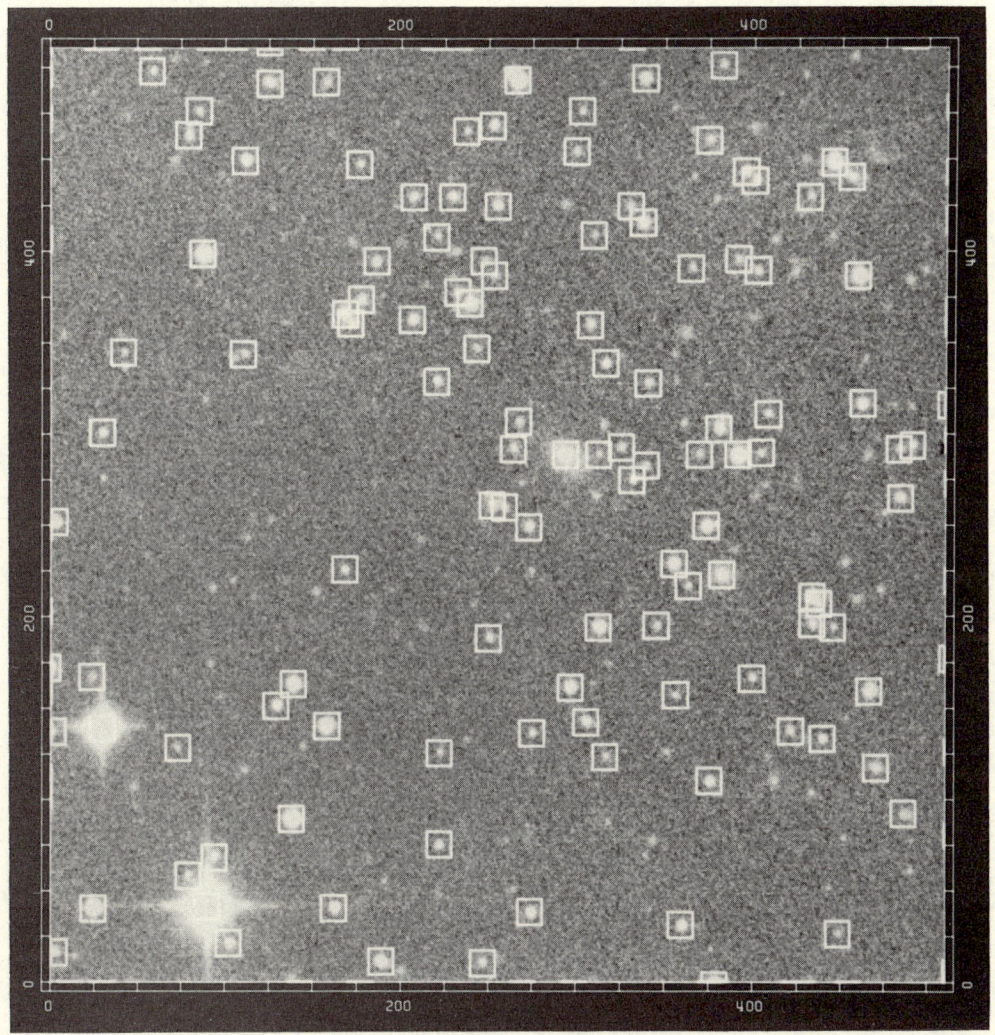

Fig. 2. Dicomed image recorder reproduction of an INVENTORY test field showing identified objects.

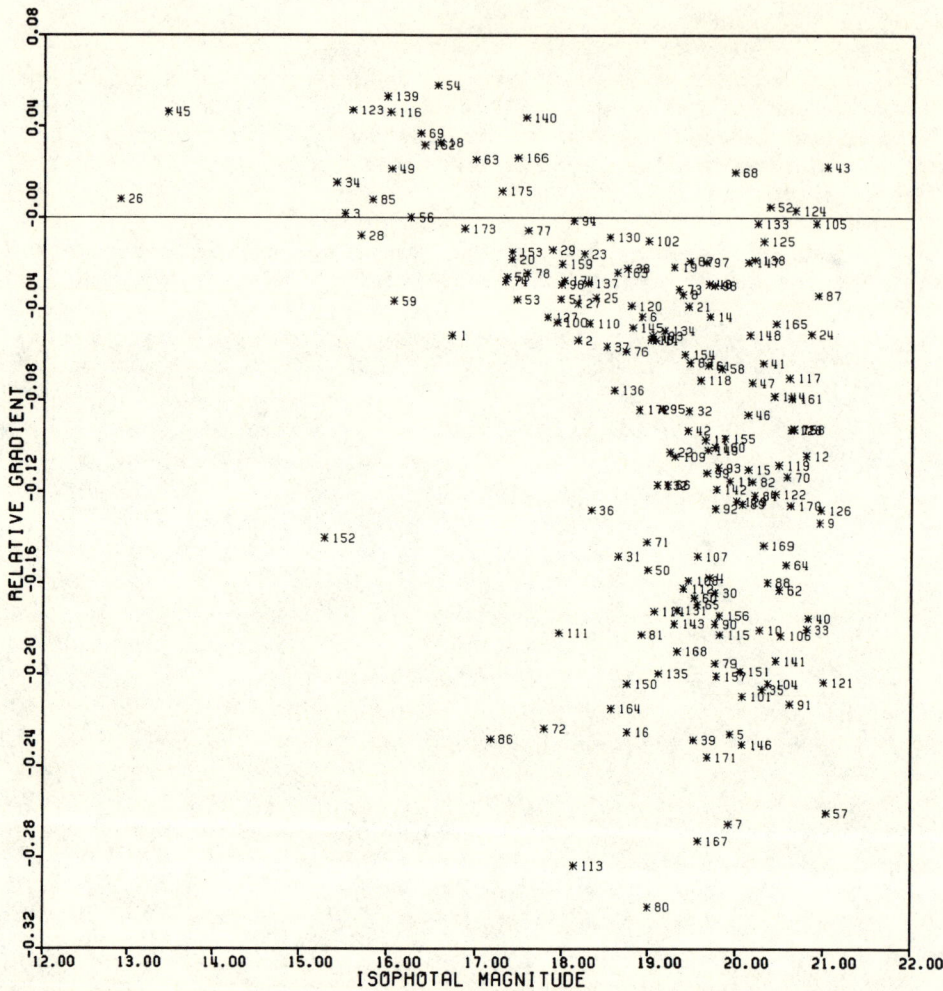

Fig. 3. Plot of relative gradient of the image profiles versus isophotal magnitude. These parameters are INVENTORY outputs plotted using the table file system.

SEARCH, ANALYZE and CLASSIFY will work in succession to find and try to classify objects into stars, galaxies or defects. Naturally, the classification scheme has some preconception of what a star or a galaxy might look like, but there is nevertheless a remarkable degree of flexibility for the user. Figure 2 shows a photograph of an image with the objects identified. Figure 3 is a plot of two of the derived parameters from the ANALYZE command which are used by the CLASSIFY command.

This set of programs has been used to perform photometry in

globular clusters and to do photometry of clusters of galaxies. The output of the search command can be used as input to any other MIDAS table driven command.

The "Echelle package" has been designed to provide wavelength and flux calibrations for echelle spectra in a semi-automatic fashion. Figure 4 gives a schematic diagram of the various steps involved.

These procedures are being used in a routine way to analyze images obtained at ESO's observatory on La Silla using the Cassegrain Echelle Spectrograph with a CCD detector. They could, of course, be used in connection with any echelle format spectrum and an appropriate wavelength table. In particular, IUE images would be appropriate.

Various tests of these procedures have shown that the wavelength and relative flux results are excellent. Figure 5 shows a plot of the peak to peak wavelength errors versus order. These errors are between 0.1 and 0.4 pixels depending on the order.

CONCLUSIONS

The MIDAS system has been designed to provide state-of-the-art data analysis facilities. This has been approached from two directions. First very powerful and versatile commands have been developed (e.g. SEARCH). Second, the command language has been

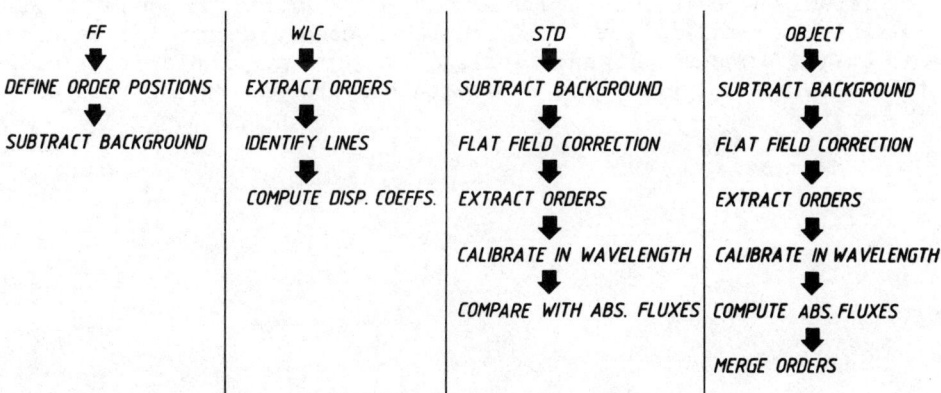

Fig. 4. Diagram of the reduction steps for reducing echelle spectra using the MIDAS echelle reduction package.

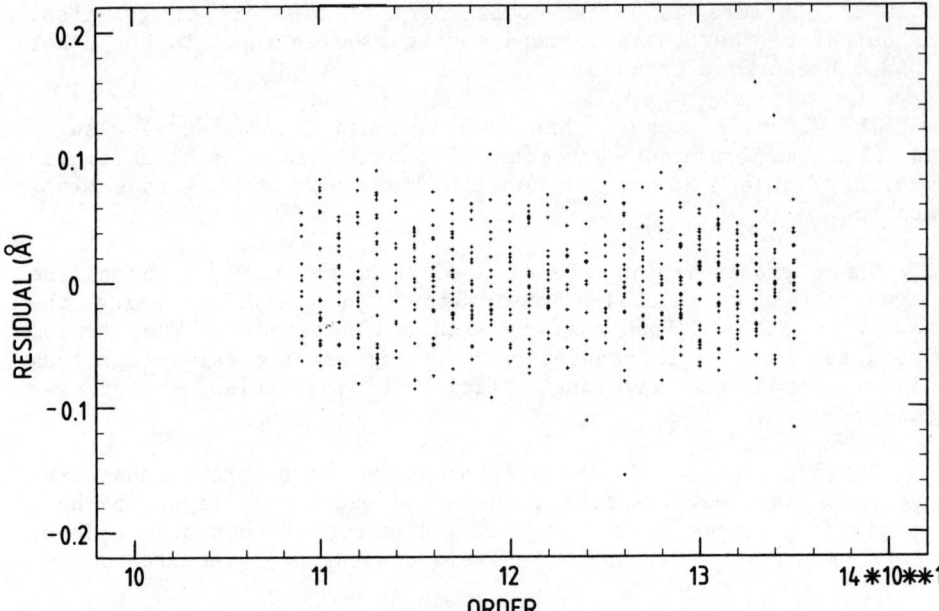

Fig. 5. Plot of the residuals of the wavelength calibration algorithm versus echelle order. The data are from the ESO CASPEC spectrograph using a CCD detector.

developed so that it approaches a programming language while retaining a set of powerful primitive commands and maintaining system performance. In addition, the clear and simple interfaces to the data structures have attracted many people to contribute to the system.

SPACE TELESCOPE SCIENCE DATA ANALYSIS

R. Albrecht*

Space Telescope European Coordinating Facility
European Southern Observatory
D-8046 Garching, West Germany

ABSTRACT

Space Telescope mission lifetime is scheduled for 15 years. Five Science Instruments and the Fine Guidance Sensors will produce a vast amount of data and put heavy requirements on the data analysis capability of the ST ScI. This paper presents the general strategy used in ST-related S/W development, and the S/W coordination between the ST ScI and other institutions.

INTRODUCTION

Space Telescope is a sophisticated astronomical observatory and will produce large amounts of complex data. It is obvious that adequate data analysis can only be carried out using state-of-the-art data analysis systems.

Concepts for the data analysis system were originally developed by NASA with input from the astronomical community, but between planning and implementation vigorous development of new image data analysis systems had taken place. NRAO and NOAO had developed AIPS and IRAF, respectively. In Europe, STARLINK and ASTRONET had come into existence. ESA had developed FIPS, an ST-specific package, and ESO had proceeded with MIDAS.

---

* On assignment from Astrophysics Division, Space Science Department, ESA.

When the Space Telescope Science Institute was founded, it was contracted to develop the Science Data Analysis System (SDAS), a minimum set of application programs to be available at launch-time, in order to do scientific analysis of the data, but also to provide operational support for ST (e.g. calibration and trend analysis). SDAS was also planned to be able to absorb application software coming from other systems.

SPACE TELESCOPE SCIENCE INSTITUTE

The ST ScI will receive ST data within 24 hours after the observations via the Nascom network and Goddard Space Flight Center (GSFC). The Ground System available to ST ScI (at GSFC and at the ST ScI) consists of a network of VAX-11/780 computers. These computers support not only data analysis but also other aspects of ST science operations (proposal management, science planning and scheduling, and real time observation support). Central element of the ground system is a commercial Data Base Management System which is mainly used for archiving, but also for the other ground system activities.

ST data analysis at the ST ScI consists of Routine Science Data Processing (RSDP) and Science Data Analysis (SDAS). RSDP is a pipeline, which will preprocess and calibrate the data with a minimum of human intervention. It will be produced within the SOGS contract by TRW.

SDAS is a package of science application S/W, tied together by a High Level Command Language. SDAS development started in 1981 and follows an industry-style system development life cycle. Currently almost all system components have been designed, about half have been coded and are being integrated and tested.

The purpose of SDAS is to provide a possibility for ST observers to adequately analyze their data. Naturally, full analysis will not be possible within the usually rather short time that observers will spend at the ST ScI. Thus, for General Observers SDAS will provide a means to prepare the data for final analysis at their home institutions. ST ScI staff, however, will carry out all analyses within SDAS.

An important aspect is the operational support of ST and the Science Instruments. Analysis of science data will provide the first indication of discrepancies. Analysis tools will also be used to trace faults and develop operational workarounds.

SDAS is normally divided in a camera section (for the Faint Object Camera and the Wide Field/Planetary Camera), a spectroscopy section (spectrographs and cameras in spectral mode), an astrometric section (for the Fine Guidance System), and a general analysis section (image arithmetic, display, etc). There is a documentation system designed to help users to understand the system in the shortest possible time, including on-line HELP and indexing facilities.

Because of the multitude of systems, because of S/W maintenance considerations, and the fact that substantial changes have to be expected both in hardware technology and in software techniques, great care has to be taken to produce a design which allows SDAS to be flexible, general, extensible, and user friendly. The concept of "limited transportability" was developed to make sure that SDAS could be ported to other institutions and/or to new technology hardware.

A layered system of interfaces was developed. These appear standard to the application programs, but they shield it from changes in the environment. The Graphics Kernel System (GKS) is being used for graphics primitives, with NCAR, a public-domain plot package developed at the National Center for Atmospheric Research, as a high level interface. A GKS type standard is being developed for image display.

The High Level Command Language was originally supposed to be produced by TRW under the SOGS (Science Operations Ground System) contract. This responsibility was recently given to the ST ScI. Based on an extensive study IRAF (Interactive Reduction and Analysis Facility, originally developed at KPNO under the UNIX operating system) was selected. A VMS version is being produced, and SDAS application routines will be integrated.

It goes without saying, that in such a complex environment, with many different organizations contributing software, configuration control and documentation is of vital importance. This has led to the implementation of a computer-resident software management system, which allows for tracking of software modules from requirements, through design and coding, to testing and operational use.

Within the available resources, efforts are made to keep the components of the ST Data Analysis System as system independent and transportable as possible. It is expected that observers will take this S/W back to their home institutions in order to continue the analysis work which they started at the Institute using SDAS.

SPACE TELESCOPE EUROPEAN COORDINATING FACILITY

Selected ST data will arrive at the ST-ECF in two ways: through European observers, who select to do their data analysis at the ST-ECF; and all data after the end of the proprietary period (nominally one year after the observations were taken).

Because of the SDAS design guidelines, it will be relatively easy to run ST ScI produced S/W at the ST-ECF. However, there are additional requirements which govern the development of the ST-ECF data analysis system. The ST-ECF is hosted by ESO, and is using ESO resources. This means that ESO and ST-ECF will use the same data analysis system. The current plans call for MIDAS, the ESO developed VAX based data analysis system, to serve as the basis of the ST-ECF system. In a carefully phased process, interfaces will be developed so application software will be executable on either system (ST-ECF and ST ScI). Since IRAF is not only a command language, but also a powerful programming language, we expect to have to support application programs written in IRAF. The current plan is to build interfaces to IRAF as soon as it becomes available. This system, plus documentation, will also be distributed throughout Europe for use at different regional centers and at the ST ScI. A software library facility will be made available to facilitate the exchange of software and to eliminate redundant developments.

NRAO'S ASTRONOMICAL IMAGE PROCESSING SYSTEM (AIPS)

D. C. Wells, for the AIPS programmers [1]

National Radio Astronomy Observatory [2]
Edgemont Road, Charlottesville, VA 22903-2475 USA

ABSTRACT

NRAO's AIPS (Astronomical Image Processing System) is a body of programs and subroutines which is the main software system for the production, deconvolution and analysis of radio images made from VLA data. The goals of the AIPS project from its inception have been to produce synthesis mapping software within the framework of a general purpose, user-friendly, portable image processing system. AIPS has evolved to include most of the necessary mapping software, to be attractive and intelligible to users and to be modular and reasonably portable (and ported) to a variety of computers, operating systems and peripherals. NRAO distributes copies of AIPS free of charge to scientific research workers and supports them in a variety of ways. Although AIPS was developed primarily to support the VLA, it was designed with considerable generality; as a result, AIPS is suitable for use with non-VLA synthesis data and even for non-radio astronomical imagery.

---

[1] DCW, who presented this paper at Erice, represented the persons listed below in Table 1. They are "virtual" co-authors.
[2] The National Radio Astronomy Observatory is operated by Associated Universities, Inc., under contract with the U.S. National Science Foundation.

PROJECT HISTORY

Scientific Goals, Design Criteria

The single most important decision in the design of a software system is the determination of who the users are likely to be, what is already available to them, and what additional requirements they will have. The NRAO, like other major data-producing centers, has accepted the responsibility to remove, at least to first order, the so-called "instrumental signature" from the data. The programs to do this are, quite reasonably, very specific to the data producing instrument and to the computer systems developed for that purpose. The user interaction with these programs should be, and usually is, quite straightforward and automatic. Therefore, the primary class of customer for our software consists of VLA observers who have reached the stage in which their data are reasonably well edited and calibrated. In other words, observers who have reached the stage at which some careful thought must be injected into the data reduction process. Most visiting observers, however, have obligations to their home institutions and cannot remain at the NRAO long enough to conduct a thoughtful final reduction of their data. And, even if they could stay, it is clear that NRAO cannot provide sufficient computer capacity to allow all users to complete the data analysis process in a convenient and timely manner. Therefore, if we are to avoid expensive duplication of software effort, our software system must be able to run not only at the NRAO, but also at the observers' home institutions. Furthermore, at current rates of funding and of change in computer technology, we cannot assume that all computers which are to run our software will be identical. Thus, a primary specification for the AIPS project was that it achieve a high degree of machine independence.

There is a second significant class of customer for our software, as well---those with data from astronomical instruments other than the VLA. This class includes, of course, VLA users and NRAO staff members who will already be using AIPS for their VLA data. It also includes scientists at other institutions who are being asked, by their VLA-oriented colleagues, to devote a portion of their joint computer resources to AIPS. A large fraction of the basic software requirements of scientists in this class are virtually identical to those of VLA users. Thus, the second primary specification for AIPS is flexibility and generality in its data base and software design. This generality also has benefits for VLA users. For example, large portions of the AIPS software, originally written for simple 2-dimensional continuum maps, worked---or nearly worked---when the VLA began to produce 3-dimensional, transposed spectral-line images.

The two design criteria mentioned above (machine independence and application independence) have been of overriding importance in the history of the AIPS project. They are also the features which have most distinguished the design of AIPS from that of other contemporary software projects in astronomy. Several other desiderata have also been important. First, AIPS must allow the user to interact with his data in order to evaluate his progress and to discover, and to point out to the software, interesting regions within them. Also, it should be "friendly" --- relatively easy and simple for beginning users and yet powerful for more advanced users. On the other hand, it must make full use of the computer's resources when performing major computations on large data bases. In particular, it must support batch-style computing as well as the interactive mode. Finally, in order to continue to develop and maintain a large, wide-spread software system, we must make it easy to add software to the system, provide extensive documentation at all levels, and maintain a uniformity in the software coding standards, conventions, service routines, and the like. The remainder of this paper will discuss how the NRAO programmers worked to meet all of these goals, plus others which will be mentioned. We begin by discussing a simple, but profoundly important, design decision.

At the beginning of the project it was expected that maps [3] would be computed with special hardware and software located at the VLA, and that AIPS would only need to be able to display the maps and extract scientific results from them. It was soon realized that the software would also need to be able to manipulate visibility data and to be able to make maps from the data. There were three reasons for this need: (1) the VLA-site map-making systems were over-burdened, (2) for an overwhelming majority of projects it is necessary to return to the visibility data in order either to edit out bad data or to revise the calibration and then to construct improved maps, and (3) it was desirable that AIPS should be able to make maps from VLBI (Very Long Baseline Interferometry) data. The iterative calibration revision procedures are called "self calibration"; the availability of these algorithms in AIPS has resulted in a spectacular improvement in the quality of VLA maps in recent years. The interactive nature of the AIPS design is a key element in the self calibration process; AIPS users have become accustomed to exploiting the flexibility which this style permits. The VLBI capabilities of AIPS have gradually increased; in 1984 NRAO decided that AIPS will be the basis for the map making software of the VLBA (Very Long Baseline Array). Thus, as a result of the early decision to support map making as well as map analysis, AIPS has evolved to

---

[3] For historical reasons radio astronomers often refer to their images as "maps". The two terms, image and map, are used interchangeably in aperture synthesis radio astronomy.

support almost the full range of needs of VLA synthesis mapping and to form the basis for the software support of the next generation of NRAO instrumentation.

The Design Phase

In 1978 two astronomer/programmers in Charlottesville spent almost a year investigating various technical options, including a review of several existing image processing systems. Although several of these systems were successful and interesting, none of them appeared to achieve -- or even intended to achieve -- the degree of generality and transportability which were, and still are, the two most important design criteria for the AIPS project. Accordingly, it was necessary to design a new system.

The initial coding experiments were made during 1979 on a Modcomp Classic CPU in Charlottesville. Several other people then joined with the first two to make the initial implementation of AIPS on the Modcomp. The very general design of the internal database structures of AIPS was heavily influenced by the FITS tape format which was being designed during the same period. In 1980, a VAX-11/780 CPU was acquired in Charlottesville and the code was ported to it; within a year it became the main development machine for AIPS. The Modcomp and the VAX computer systems in Charlottesville contained identical Floating Point AP-120B array processors and IIS Model 70E image displays. Since 1980, the portions of the AIPS code which meet the coding standards of the project have always run on both of these machines. The continuing necessity to produce machine independent, portable code which would run in these two different computers has had a profound influence on the AIPS project in many areas, such as design, implementation, policies, and philosophy.

During 1980 the mapping and CLEANing algorithms, which had been developed at the VLA for the "Pipeline" project (in array processor microcode), were adapted to the needs of AIPS. Several other mapping, deconvolution and self-calibration algorithms which use the array processors were implemented subsequently. AIPS computers do not require an array processor for the heavy computing of synthesis mapping, but an AP will improve performance by factors of 3-10. NRAO currently owns four AP120Bs for use with AIPS and about six non-NRAO AIPS sites have purchased array processors in order to exploit NRAO's microcode.

In 1981 and 1982 NRAO purchased two more VAXes with APs and IIS displays in order to utilize the AIPS software at the VLA site in New Mexico. Thus, by the end of 1981, the basic goals of the AIPS project had been met, although many details remained to be filled in. It had taken more than 3 calendar years, 10 manyears of software effort, and nearly US$2 million of hardware for AIPS to reach this stage. The filling in of details and accretion of additional features has so far consumed more than 10 additional manyears during another 3 calendar years. Note that NRAO's hardware costs have been several times higher than its software development costs (even if we use the 20 manyear figure). The integral of the continuing software maintenance costs will overtake the initial hardware investment cost within a few more years. But NRAO has two more factors which protect and multiply the effectiveness of its huge investment in image processing software. First, AIPS can be used in future NRAO computers and for future NRAO instrumentation other than the VLA. Second, AIPS runs in numerous non-NRAO computers. The first factor derives from the fundamental machine- and application-independence of AIPS. The second factor is due to the export policy of the AIPS project.

Development of the Export Policy

From the beginning of the project NRAO shipped installation kits upon request to research sites without charge. The "free of charge" policy saves on paperwork and encourages university astronomers to use AIPS to process and analyze VLA data with their own computers, thus facilitating radio astronomy research and reducing the loading on NRAO computers. At first tapes were written and shipped out as orders were received. Installation instructions were simple. As the number of sites increased it became necessary to adopt formal policies. In 1982 NRAO developed a scheme of freezing versions of the software at regular intervals for distribution. Versions of the documentation are synchronized with the freeze dates in order to properly describe the releases. Currently three types of kits are offered: VAX/VMS BACKUP, Unix "tar", and blocked card images. The installation kits for VAX/VMS systems have become highly automated and parameterized over the years. By July 1984, kits had been shipped to more than 60 sites worldwide. It is believed that about 30 of the sites are in regular production operation of AIPS. Currently about 85% of these sites utilize VAX computers under the VMS operating system.

Documentation for Users

Users depend on two kinds of documentation, online and printed. AIPS has very extensive "help" files (currently 590 files containing about 1.5 megabytes of text). The help text is closely related to the "inputs". These are the parameters (called "adverbs" in AIPS) which command language operators ("verbs" in AIPSese) and programs ("tasks") use to control their actions. For verbs and tasks, help files mostly explain the meanings of the inputs. But separate help files also exist for all adverbs. In addition, there are general help files which give brief descriptions of various classes of commands and there are also extensive writeups of many programs which are accessed with the "explain" command. In practice the online documentation of AIPS is almost completely sufficient for experienced users.

Written documentation is used mostly to inform users about changes and to introduce new users to the system. A newsletter for AIPS users called the "AIPSletter" was started in the fall of 1981 and by 1984 its mailing list had grown to contain more than 350 names. One of its main functions is to tabulate all recent changes to the software. A "Cookbook" was developed in 1981 and has become the most popular printed documentation for AIPS. The original intent of the Cookbook was to introduce new users to the system, but it also gives brief introductions to more advanced topics.

Another form of "documentation" is software problem reports. These are called "gripes" in the AIPS project, and the AIPS command language contains a verb which allows the user to enter his complaint, comment, or suggestion into a file which will be fetched by the AIPS programmers. The development of this capability and the automation of the handling of the gripes began in 1982. Currently gripes arrive in Charlottesville at a rate of about 500 per year. The AIPS programmers prepare written responses to the gripes and return them to the users. These written responses are an additional form of documentation which has the great advantage of being personalized. Copies of all gripes are kept in the computer rooms at the VLA and in Charlottesville for all users to read.

Personnel and Project Management

By 1984 NRAO has invested more than 20 man-years in AIPS, and more than 350,000 lines of text (code plus documentation) have been produced. It is very important that the fundamental system design and standards decisions have been made by one person, although extensive discussion and consensus agreement within the programming team have also been necessary. In particular, coding standards are absolutely essential for the success of any project of this size because the various programmers in the project must be able to read each other's code. Also, portability partly depends on following consistent coding practices which permit automatic translation of syntactic constructs to suit peculiarities of new machines. Note that the modular character of the system design (separate tasks, layered software interfaces) facilitates the addition of new programmers as well as new functionality.

A considerable number of people have been involved in the

project over the years; many of these people are listed below in Table 1. In the first group are current NRAO employees who devote a good portion of their time to AIPS. In the second group are NRAO employees who provide support and advice, but who are not heavily involved in the project. In the third group are people, not currently employed at the NRAO, who have made significant contributions to the project. An interesting aspect of this list is that almost all of the people are Ph.D. astronomers. There are no computer scientists, at least in the first section of the list. A number of additional programmers are expected to join in the project during 1985 to carry out the VLBA calibration development and supercomputer implementation projects.

Current Status

AIPS now contains a nearly complete set of software for VLA continuum and spectral line reductions. Most VLA users can obtain high quality maps in a reasonable time using any of the more than 30 AIPS systems running at both NRAO and non-NRAO sites throughout the world (the current estimate is that about half of VLA reductions are now done at non-NRAO sites). NRAO expects that the existing AIPS systems will be able to handle the majority of small

Table 1 - The AIPS Programmers

Name	Site	Role
John Benson	CV	VLBI applications
Tim Cornwell	VLA	Image deconvolution
Bill Cotton	CV	U-V, VLBI, array processor code
Gary Fickling	CV	VAX systems, general software
Ed Fomalont	VLA	Scientific advisor, applications
Eric Greisen	CV	Software manager; applications
Kerry Hilldrup	CV	IBM and Unix systems
Gustaaf van Moorsel	CV	Spectral-line analysis software
Fred Schwab	CV	Applied mathematics
Don Wells	CV	Management; optical applications
Al Braun	VLA	DECnet and VAX systems
David Brown	CV	VAX, Modcomp, IBM systems
Bob Burns	CV	Overall NRAO computer management
Arnold Rots	VLA	Spectral-line, image display code
Stuart Button	Toronto	Application software
Tom Cram	(CV)	Initial AIPS design
David Garrett	Texas	Unix implementation
Jerry Hudson	(CV)	POPS command language
Walter Jaffe	(CV)	Application software

and medium-sized VLA mapping problems over the next ten years, plus most of the data to be acquired with the VLBA. In order to handle the largest mapping problems NRAO needs to acquire a supercomputer. NRAO programmers believe that the portability and modularity of AIPS will permit it to utilize supercomputers effectively.

AN OVERVIEW OF AIPS AND ITS OPERATION

AIPS software is coded in a dialect which is an "extended subset" of "vanilla" Fortran (i.e., Fortran-66). Certain standard syntactic constructs of Fortran are not used (e.g., assigned-GO-TO), and certain non-standard extensions are used heavily (e.g., ENCODE/DECODE, INCLUDE, and INTEGER*2). The principal motivation behind the specifications of the AIPS Fortran dialect is to ensure the portability of the code to the maximum number of CPUs and operating systems.

One of the main programs of AIPS is actually called "AIPS". It implements the interactive command language of the system and includes a variety of utility functions ("verbs") in its command repertoire. The command language uses a typical algebraic programming language syntax, including structured control constructs, real and string variables ("adverbs"), and procedures. The user types input into program AIPS and receives most of his terminal output from it. The command language is called POPS, for People Oriented Parsing System, and was developed at NRAO in the early 70's for the analysis of single-dish data.

Time-consuming operations are done in AIPS by other programs, called "tasks", which run asynchronously as processes under program AIPS. The actions of tasks are controlled by parameters which are variables in the command language. Both string variables and real variables are utilized (e.g., file names and function codes such as '3C405' or 'ADD', or function coefficients such as BMAJ=3.56 for a beam halfwidth). One of the principal purposes of the AIPS command language is to provide convenient facilities for listing the parameters for a task (the "inputs") and modifying them until they are satisfactory. When the user is has specified the parameters for a task he issues the AIPS command "GO", which initiates the task as a subprocess, and passes the input parameters to it. When the task has begun, and has acquired the values of the parameters, it allows program AIPS to resume talking to the user. Most tasks support reasonable default values for their input parameters; it is not necessary to specify everything explicitly.

A task is free to execute in the background, often for periods of tens of minutes or even hours with current computers. The AIPS user can initiate more tasks and perform various interactive operations such as examining the results of previous computations. It is not uncommon for a single user to be interacting with the digital image display (using program AIPS) while three tasks are in progress, with all actions occurring concurrently. There is a prohibition against initiating a second copy of a task while the first copy is still executing. Two or more users may be executing copies of program AIPS and initiating tasks. Thus multiple copies of the same task may be in execution at the same time, one for each user. It is typical for a machine of the VAX-11/780 class to support two interactive AIPS users fairly well, and five tasks in action at the same time is typical.

The design of AIPS includes a heavy emphasis on high performance disk I/O. The techniques used are DMA transfers with double buffering, blocking of multiple records, and processing data directly in the buffers. Although an extensive library of subroutines supports these techniques the code complexity which results is quite visible in AIPS application code. The AIPS designers believed (and still do believe) that these design features are absolutely essential for achieving high performance in synthesis mapping applications.

A number of AIPS tasks (currently 15) are designed to utilize a Floating Point Systems AP-120B array processor when one is available. The behavior of these tasks is completely analogous to CPU-only tasks. AIPS tasks which use the array processor do so as a pipelined vector arithmetic unit; data are passed from the host to the array processor, some operation is done on the data, and the result is returned to the host. Generally there is only one AP attached to an AIPS computer. Therefore the tasks which depend on the AP are coded to time-share it with a time quantum of a few minutes. On the whole, the addition of an AP-120B increases the power of a VAX11-780 by a factor of about five for VLA mapping applications.

AIPS needed interactive capability to enhance user efficiency and to permit data analysis techniques in which user intervention is required (e.g. data editing). But AIPS also needed a batch processing capability to enhance computer efficiency for non-interactive applications. Utilities are provided in the interactive system which facilitate the creation and management of batch jobs. A special version of the command language processor processes the queue of batch command files. The command syntax used in these files is identical to that entered by the users in interactive sessions. In both interactive and batch operations AIPS users need to know very little about the host operating system.

PORTABILITY

As explained in the historical section of this paper, the AIPS strategy from its inception included an assumption that the software would need to execute in more than one hardware/software environment. There were three motivations for this policy:
(1) NRAO possessed more than one kind of computer system (Modcomp, VAX, IBM, etc.), (2) NRAO had no assurance that user sites which wanted AIPS would possess computer systems identical to any of those possessed by NRAO, and (3) NRAO wanted to be able to migrate AIPS to new computers easily in the future. A variety of examples of all three cases have occurred over the years. Portability has enhanced the value of the mapping software by allowing it to run on a variety of existing machines and it has protected the investment in the software by assuring that it will run on future machines (e.g., supercomputers).

AIPS users on a variety of CPUs under various host operating systems all use the same command syntax to invoke the same portable mapping programs in both interactive and batch modes. The portable user command interface is a vital element of the concept of a portable image processing system: it means that not only can the application programs themselves be implemented easily on a new machine/operating system combination, but also that the users will be able to utilize the programs immediately, because the command language will be exactly what they are accustomed to using. Several hundred astronomers are now trained in using AIPS, and the investment in their training is a substantial part of the whole software investment which NRAO has made in developing AIPS. AIPS has a portable command language precisely so that the skills of its users will be portable.

In 1982 NRAO formally began a project to develop a generic Unix capability. There were two reasons for this decision. First, the University of Texas at Austin had produced a prototype installation of AIPS under Unix. Second, Unix was becoming increasingly important in the computing industry; Unix is already the second most common operating system used in astronomical computing (after VMS). The goal of the AIPS Unix project is defined as: "to support AIPS under Unix as well as we support it under VMS". First distribution of (experimental) Unix installation kits will be for the 15JUL84 release of AIPS (pre-release kits were sent to four test sites in 1983 and 84).

Several of the image processing peripherals originally chosen by NRAO in 1979-80 are now either no longer manufactured or have become suboptimal choices. More and more user sites have wanted to use AIPS with devices other than the ones which NRAO chose five years ago. Both of these problems have led to a need to interface the AIPS software to a variety of new peripherals. The existence of device independent software interfaces in AIPS has made it possible to prepare, export, and support software interfaces for new devices (support for two new image displays and a laser printer/plotter were added in 1983). Other experimental device interface software exists at various non-NRAO AIPS sites and even within NRAO. Some of this code may eventually become formally supported by NRAO. The development of support for new devices will be a major activity of the AIPS project indefinitely into the future.

LAYERED "VIRTUAL" INTERFACES IN AIPS

The ability to port AIPS to a variety of different CPUs and operating systems, and to support a variety of different peripheral devices, depends on a thorough modularization of the software. The objective is that the largest possible fraction of the code of AIPS should remain invariant as AIPS is ported from system to system and device to device. Two basic strategies are used in the AIPS project (and in almost all similar projects) to accomplish this objective. First, differences between CPU hardware (such as number of bytes per word or disk sector size) are parameterized as system constants which are initialized during system configuration and are used by all application algorithms which need them. The second strategy is used for operating systems. AIPS programs do not talk directly to the VMS operating system on a VAX. Rather, they talk to the system calls of a "virtual" operating system. The layer of AIPS subroutines which implement these system calls (the AIPS programmers call these the "Z-routines" because their names begin with Z) then talks to VMS. On a VAX which is using the Unix operating system a different version of the subroutines is used which translates the functions to Unix notation. In fact, many features of Unix are similar to those of VMS, and so the Z-routines have two layers and the higher layer is actually identical for VMS and Unix!

The layered virtual interface technique is also used to implement digital television displays and graphics output devices. The subroutines for television displays are called the "Y-routines" in AIPS. Their arguments and functionality define a hypothetical image display which has a subset of the capabilities of the IIS model 70 display. A new display can be interfaced to AIPS by coding a new version of the routines which will map the virtual device defined by the Y-routines onto the hardware and functionality of the new display. NRAO currently distributes versions of the Y-routines for three different displays and it is known that other versions exist at various user sites.

Because AIPS AP tasks were developed using FPS AP-120B array processors, the way in which the AP is called uses FPS conventions. AIPS makes use of microcoded routines in the FPS standard libraries plus a set of custom microcoded and "vector function chainer" routines developed by NRAO programmers. In order to allow the use of AIPS array processor tasks on machines which do not have AP hardware, NRAO has developed the concept of a "pseudo array processor", in which a Fortran COMMON is used as the array processor memory and Fortran or assembly language routines operate on data in this COMMON. These routines have exactly the same names, arguments, and functionality as the corresponding routines in the libraries for the AP-120B. There is only one version of each of the tasks which use the AP; the choice of whether a task uses the true AP or the pseudo AP is made by link editing it with the appropriate subroutine library. The tasks do not know or care, except in the most subtle ways, if they are using a true AP or a pseudo AP. The "pseudo-AP" concept in AIPS actually amounts to defining a "virtual device interface" for vector processing. NRAO programmers believe that the pseudo-AP subroutines can form the basis for supercomputer implementations of AIPS. In such machines the Fortran code can be modified so that compilers will automatically vectorize it.

The "virtual operating system interface" and "virtual device interface" concepts have been quite successful in the AIPS project. It is very important that the universality of such interfaces be confirmed by actual ports to different systems and devices. The specifications of the interfaces must be evolved on the basis of such trials in order to maximize the portability of the design. In fact, this kind of confirmation of portability has occurred in the AIPS project frequently and the interface specifications have evolved on the basis of the accumulated experience.

DEVELOPMENT PLANS

Several areas in which development work is planned for the period 1985-86 are:

1. improved calibration and editing of visibility data (to be supplied by the VLBI effort in AIPS);
2. additional spectral line software;
3. testing and implementation of better deconvolution and self-calibration algorithms;
4. more sophisticated TV displays;
5. implementation of table data structures (analogous to spread sheets);
6. completion of the generic Unix support project;

7. conversion to GKS (Graphical Kernel System) for improved portability of graphics device interfaces.
8. development of "long-vector" versions of the pseudo-AP library in order to support AIPS implementations on supercomputers and new APs.

AIPS AND SUPERCOMPUTERS

NRAO needs to develop a supercomputer mapping capability in order to accomplish large VLA reduction tasks in a reasonable amount of time. A valuable reduction in the software development time for the supercomputer can be made if AIPS is used instead of beginning a new mapping software project. In addition, by installing AIPS on several different supercomputers NRAO will be able to make a procurement decision on the basis of benchmark trials using the intended application code. Achieving high performance in NRAO's mapping applications will require a careful study of the details of vectorization on various supercomputers. Attention can be concentrated on the vectorization of the "pseudo-AP" subroutine library in AIPS because the heavy computing load of mapping is concentrated precisely in the inner loops of these AIPS subroutines. Two important special cases are the routine that grids visibility data into maps and the routine that executes the heart of the CLEAN deconvolution algorithm. These two pieces of code accomplish a very large fraction of the heavy computing of synthesis mapping; together they amount to less than 100 lines of actual code to be optimized. A prototype installation of a part of AIPS has already occurred on a Cray-1 at the University of Minnesota. NRAO programmers expect to continue and extend this work during 1985, probably using computer time granted under the NSF's "Supercomputer Initiative" fund.

USING AIPS FOR NON-RADIO ASTRONOMY APPLICATIONS

The preferred tape format in AIPS is the FITS standard of the IAU. AIPS fully implements all keywords and functionality of the original FITS design. AIPS also fully implements the "groups" extension to FITS for the case of aperture synthesis visibility data. For the 15OCT84 release of AIPS support is included for the new "tables" extension to FITS. The full support for FITS provided in AIPS assures that data from other areas of astronomy can generally enter AIPS through FITS tapes quite easily. In fact, CCD imagery from optical observatories is read into AIPS systems quite frequently.

The image data structures used inside AIPS mimic those of the FITS tape format, although they are encoded in a machine dependent fashion on each machine for efficiency. The similarity to FITS means that not only is the disk-to-tape conversion fairly easy, but also that AIPS readily represents all images which are representable in FITS. AIPS is particularly strong in its representation of physical units and coordinate systems, especially in the cases of nonlinear projective geometries and rotated coordinates. The "ARC" geometry of Schmidt cameras and the "TAN" geometry of typical optical telescopes are fully supported in AIPS (the VLA uses the "SIN" projection).

Many of the image processing and display operations implemented in AIPS are just as useful for non-radio applications as they are for VLA data. Non-radio astronomers who use AIPS are likely to be slightly confused by the jargon of radio astronomy at first (e.g., "map" rather than "image" and "beam" rather than "seeing profile"), but they can soon find many of the tools they need in AIPS. Because "tasks" are separate programs, the additional functionality which is needed in non-radio (and non-VLA) applications can be added by just coding and installing new tasks. The easiest way to code such programs is to find an existing task which resembles the new problem and copy and modify it. In recognition of this fact the AIPS programmers have prepared a set of "paradigm" tasks which are specifically intended to be cloned. This subject is covered in Chapter 2 of the AIPS programmer's manual "Going AIPS", which is available from the address given in the next section.

INFORMATION FOR PROSPECTIVE USERS OF AIPS

The "Cookbook" and the "AIPSletter" are the best sources of information for prospective users. Requests for copies of the Cookbook or requests for addition of names to the AIPSletter mailing list should be directed to:

>   AIPS Group
>   National Radio Astronomy Observatory
>   Edgemont Road
>   Charlottesville, VA 22903-2475 USA
>
>   telephone: (804)296-0211

Written orders for AIPS installation kits are preferred, preferably using the order forms which are included in each AIPSletter, but even telephone orders will be accepted. There are absolutely no charges of any kind for scientific research sites. NRAO even supplies one magnetic tape and one plastic mailing cannister per site!

Newsletters are mailed about two weeks after each "freeze" date (currently 15 January, 15 April, 15 July, and 15 October of each year). Installation kits are mailed six to eight weeks after the freeze date in order to allow time for testing before distribution.

ASTRONET

Giorgio Sedmak

Universita' di Trieste, Istituto di Astronomia
Via G.B.Tiepolo 11, 34131 Trieste
Italia

INTRODUCTION

Astronet is the italian project of a national wide facility for astronomical data handling and processing. The geographical distribution of Astronet centers is shown in Fig.1. It was born in 1981 after the study carried out by a commission of the National Research Council (CNR) of Italy and following the guidelines selected by the Scientific Council of the National Group of Astronomy (GNA) of CNR, which took the responsibility of Astronet.

The objectives of Astronet include the identification, procurement or realization, installation, operation and maintenance, management and user interfacing of the hardware and software required in order to give italian astronomers an effective support to handle and process large volumes of high quality data. Particular concern is given to data from spaceborne telescopes and large terrestrial telescopes.

The design of Astronet is network oriented and takes into account the positive experience of the UK Starlink project. Astronet is planned to be integrated into the European astronomical environment by means of networking to the national and international data processing centers and data banks dedicated or of interest to astronomy and astrophysics.

Fig. 1. Geographical distribution of Astronet centers 1984.

The Astronet project is funded by the Italian Ministry of Education (MPI) through the University National Research Plan and the Astronomical Observatories, and by CNR through its National Space Plan (PSN) and of course the GNA. The budget for the first three years foundation plan totalized about 2 million dollars.

The foundation of Astronet was planned for completion within 1984. On June 1984 all Asronet centers are operative and the implementation of the network is expected to be at least partially operative on time.

The future activity of Astronet is now going to be defined within the next three years plan 1985-1987 expected from GNA on October 1984.

MANAGEMENT

The Astronet project is managed by a national committee consisting of the project responsible, one local coordinator for every Astronet center, and the coordinator of the astronomical application software as the representative of users.

The activity within Astronet is structured in local and national tasks carried out by working groups coordinated by one Astronet local coordinator.

The management structure is shown in Fig.2. The members of the national management committee 1981/82 to 1983/84 are listed in Table 1. The local activities are listed in Table 2. The national acitivities are listed in Table 3.

HARDWARE STANDARDS

The Astronet hardware standard is the 32 bits, hardware virtual memory DEC VAX11 computer with DEC VMS operating system. This standard is expected to be maintained for at least 10 years, with the possible replacement of VMS by UNIX operating system if the international trend in astronomical software will make it necessary.

The basic configuration of a medium sized Astronet center is

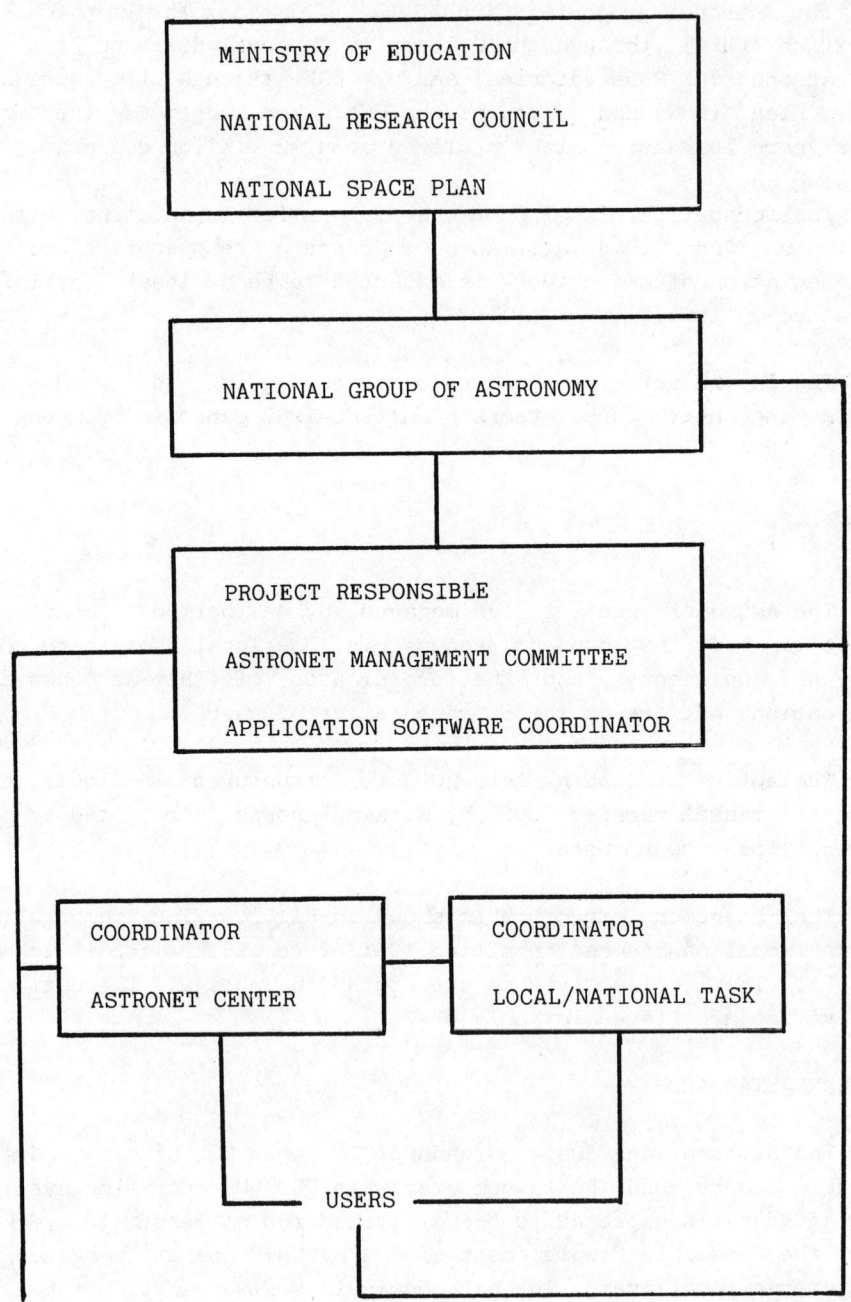

Fig. 2. Management of Astronet. Nine Astronet centers are operative on June 1984.

Table 1. Astronet management committee 1981/82 to 1983/84.

Astronet Project Responsible	G.Sedmak
Astronomical Application Software Coordinator	M.Capaccioli

Astronet center	Coordinator
Bari	C.De Marzo (*) A.Distante
Bologna	A.Ficarra
Catania	S.Catalano
Firenze	L.Fini
Napoli	G.Russo
Padova	L.Benacchio
Palermo	V.Di Gesu'
Roma	G.De Biase
Trieste	M.Pucillo

(*) Until 1982.

Table 2. Astronet local activities 1981/82 to 1983/84.

Activity	Coordinator
1. Image processing workstations using an array processor and a low level pictorial subsystem.	G.De Biase
2. Integration of VAX resident NASA JPL VICAR and procurement of IBM resident NASA JPL VICAR.	A.Distante
3. Integration in Astronet of ESA FIPS software for the Faint Object Camera of the Space Telescope.	M.Pucillo
4. Project of an Astronet oriented documentation facility and realization of a prototype documentation system.	I.Mazzitelli (*) F.Pasian
5. Networking Astronet centers.	L.Fini
6. Integration in Astronet of UK Starlink software collection.	A.Distante
7. Integration in Astronet of ESO MIDAS software system on DEC VS11 video display subsystem.	V.Di Gesu'

(*) Until 1982.

Table 3. Astronet national activities 1981/82 to 1983/84.

Task	Coordinator
1. Basic software environment and network.	L.Fini
2. Graphics and pictorial software.	M.Pucillo
3. Documentation and data base.	L.Benacchio
4. Astronomical application software.	M.Capaccioli
5. Integration of DEC VS11 to Starlink software.	G.De Biase
6. Astronet documentation service.	F.Pasian

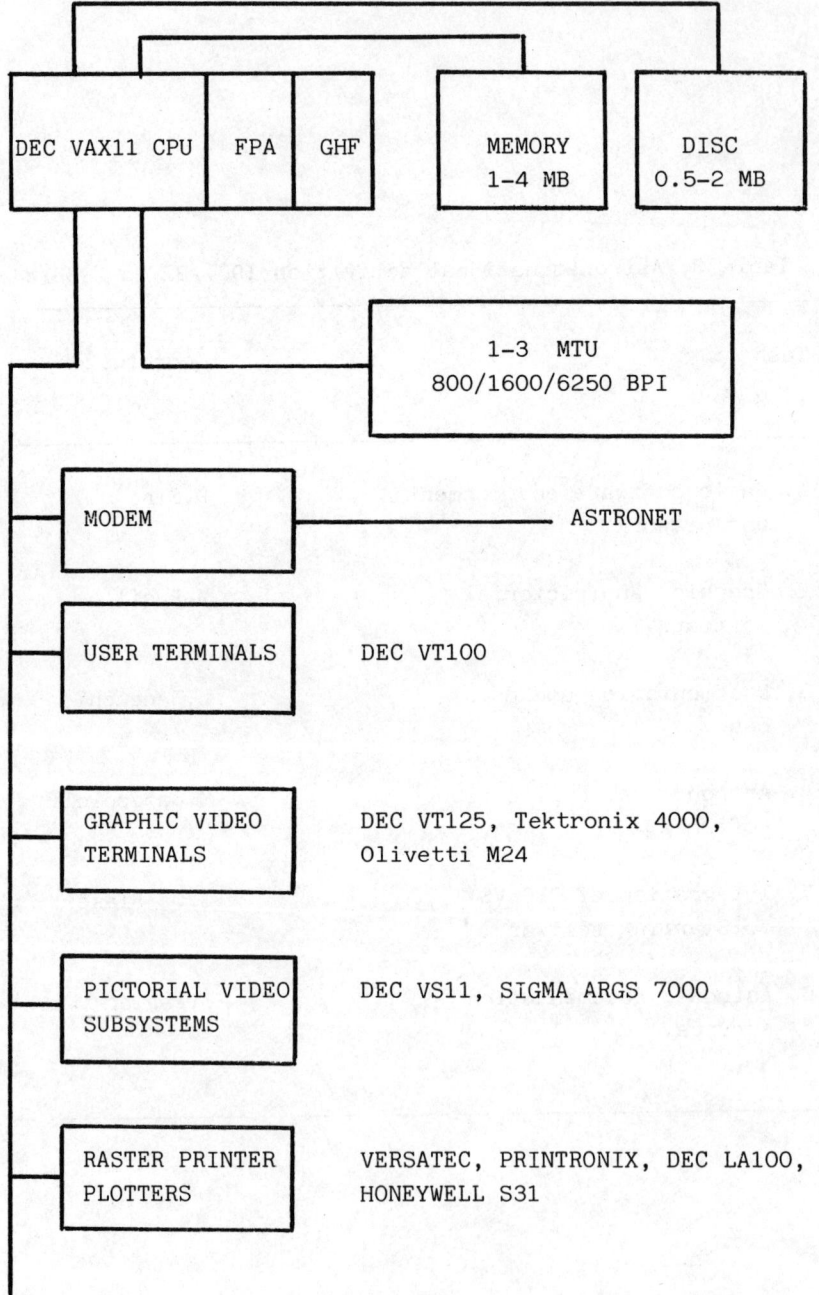

Fig. 3. Configuration of a medium size Astronet system.

shown in Fig.3. The basic configuration can be varied according to current demand of resources by the users.

All alfanumeric video terminals are ANSI standard DEC VT100 devices or equivalent units. The graphic video terminals supported in Astronet include the Tektronix 4000 series, the DEC VT125 devices, and the planned Olivetti M24 device. All graphic video devices are intended to support colour graphics.

The pictorial video subsystems include the low cost DEC VS11 device with 512x512x4/8 format and the SIGMA ARGS 7000 video display processor with 2x(1024x1024x8) format. On recommendation by Astronet and other interested users, the PSN CNR accepted to prompt the italian industry for an advanced pictorial video system with fully programmable architecture. This machine should be in any case ARGS 7000 compatible and is planned to replace the original SIGMA units at lower cost when available.

All video display devices may be coupled to optional Polaroid photographic colour hard copy units.

All hard copy printers and plotters are of raster type. They include the Versatec, Printronix, DEC LA100, and Honeywell S31 machines.

SOFTWARE STANDARDS AND DOCUMENTATION

Astronet is committed to support original Astronet software as well as the major astronomical software systems and libraries available at international level.

The international software facilities available on Astronet are the ESO MIDAS system, the UK Starlink Software Collection, the USA AIPS and NASA IBM VICAR systems, and the ESA FIPS system.

ESO MIDAS and UK Starlink software are distributed as standards all over Astronet. The planned ESA ST-ECF standard as well as the planned NASA ST-SCI IRAF-supported standard will be integrated in Astronet as soon as available, of course within the limits set by the hardware compatibility. These standards are or will be supported together with full reference to their graphics and pictorial libraries.

The original Astronet software developed up to now includes the preliminary versions of an application oriented software environment (Astronet Personal Environment : APE), two general purpose astronomy oriented graphics packages (SGL, AGL), one guided software documentation program, and the general documentation facility software. This last product is a generalization of UK Starlink documentation system and is committed to allow a distributed access and updating of the network oriented data base of the documentation service.

ASTRONOMICAL APPLICATION SOFTWARE

The astronomical application software available in Astronet includes the software available through the international standards supported by Astronet, the software classified or produced by the Astronet Group for Astronomical Application Software, and any software contributed to the Astronet Documentation Service. However, Astronet supports and maintains only the international standards quoted above and the Astronet classified software. The contributed software is left to the responsibility of the authors. Any software running on Astronet but non contributed to the Astronet Documentation Service is considered as private and is not accessible since it is not known to the documentation retrieval facility.

The activity of the Astronet Group for Astronomical Application Software will start on October 1984 after the definition of the priorities within the next three years plan 1985-1987 for italian astronomical research.

ASTRONOMICAL DATA BASE

Astronet is presently using the data base facility available in UK Starlink, which contains a Starlink 'ad interim' version of part of Strasbourg CDS catalogs.

The Astronet Group for the Astronomical Data Base defined and realized a preliminary version of an Astronet standard original software for the implementation and maintenance of a network oriented astronomical data base.

The above activity includes the network connection to national and international data banks of interest to Astronet, with particular regard to Strasbourg CDS, ESA ST-ECF at ESO, and ESA ESRIN.

CONCLUDING REMARKS

Astronet is a complex project and it is not possible to detail it here. However, this presentation should be sufficient to give an overview to the general structure, current status, and planned developement of Astronet. More information is available on request through the Astronet Documentation Service.

ACKNOWLEDGEMENTS

Astronet integrates the coordinates efforts of many scientists operating in the national and international astronomical community, acknowledged here as a whole.

V.Castellani, L.Guerriero, F.Pacini, G.Setti, and G.Tofani are due a special acknowledgement for their fundamental action in the various councils of MPI and CNR concerned to the Astronet project.

REFERENCES

Astronet 1982, Ed.G.Sedmak, Mem.SAIt 53/1(1982).

Astronet 1983, Ed.G.Sedmak, Pubbl.SAIt (1984).

Astronet Documentation Service Contact :

Dr. Fabio Pasian
Osservatorio Astronomico di Trieste
Via G.B.Tiepolo 11, 34131 Trieste
Italia

THE CDCA PACKAGE : PAST, PRESENT AND FUTURE

Albert Bijaoui

Observatoire de Nice
B.P. 139
06003 Nice Cedex (France)

Antoine Llebaria

Laboratoire d'Astronomie Spatiale
Traverse du Siphon - Les Trois Lucs
12012 Marseille (France)

Diego Cesarsky

Institut d'Astrophysique de Paris
98 bis, Boulevard Arago
75014 Paris (France)

From 1975, in the "Centre de Dépouillement des Clichés Astronomiques", at Nice Observatory, an Image Processing System, for astronomical purpose, has been built around a PDS microdensitometer and a PDP 11/40 computer.

This system has been realised by interaction between the users, coming from all the french astronomical laboratories to process their photographic plates, and the staff, mainly A. Bijaoui, J. Marchal and Ch. Ounnas.

The main features of this processing system are :

- Program written in FORTRAN IV language, with the RT11 System.
- Full independent programs to realise one operation.
- Interactive dialogues or batch sequences (no command language).
- Image visualisation using graphic displays (vectorial mode).
- File with a defined structure for images and catalogues of

objects, using direct access. These files allow the communication between the programs.

This system has been installed on other computers, PDP 11 or not. To transport easily this software, some basic routines has been defined to create and open image file, read and write part of an image, access to the graphic operations.

With small modifications, the programs have been installed on a T1600 (Telemecanique), an IBM 360/65, an HP1000, an IBM 3381, an HARRIS 700, a solar 16/65....

At Meudon Observatory, D. Cesarsky has mainly realised an adaptation of the software on a VAX 11/780.

More than ten laboratories used or have used a part of the system, leading to exchanges of programs and documentation.

More than 600 programs corresponding to more than 200 000 Fortran instructions allow to process many kinds of astronomical processing :

. General image processing : visualisation, operation, transformations, geometrical corrections, photometrical corrections, noise reduction (linear and non linear), deconvolution, numerical compression....
. Software for the catalogue of objects files.
. Application to stellar or nebular spectrography.
. Analysis of echellograms.
. Analysis of objective prism spectrograms (radial velocity, spectrophotometry, classification).
. Determination of stellar magnitudes and positions.
. Photometry of astronomical extended sources.
. Fully automated analysis of astronomical fields....

For each application we have developed sequences of programs, allowing to extract all the informations.

Now, some VAXs have been bought in French astronomy, so we are transforming all the software.

The philosophy of independent programs is always kept. With the WSM command language DCL it is easy to transform the dialogue to obtain the data in the programs in a command instruction. So, many kinds of command languages can be installed independently of the programs.

The advantages of independent programs are evident for programmation, transport, link with the operating system, errors management....

A today analysis, with the computer science laboratory of the Nice University leads us to oversee that more rational structure of the system will use the new ADA language.

ADA has the advantages of the PASCAL language (definition of the types, the procedures and the structuration). ADA has many interesting features (parallel process, genericity, abstract types,...). The ADA is also an universal language, with a precise norm. There will not have problems for the transport. Unfortunately, it is to soon to realise a fundamental transformation of the software in ADA.

DESIGN OF A PATTERN-DIRECTED SYSTEM FOR THE INTERPRETATION OF DIGITIZED ASTRONOMICAL IMAGES

Piero Mussio

Dipartimento di Fisica

Universita´ di Milano

INTRODUCTION

Data impose constraints in the definition, design, implementation and use of an automatic system for the analysis of digitized astronomical plates. The aim of this paper is to report some exploratory steps performed in the design of such a system (the Automatic Assistent: AA) in support to this often claimed statement. Therefore we discuss the rationale which is behind the design of the whole system and, due to the lack of space, a diary of the experiments which lead to the present definition of the image segmentation tools.

GENERAL DESIGN ISSUES

The AA is designed to help the astrophysicist in his activity of interpretation of astronomical plates, increasing his overall effectiveness as a human problem solver. The AA has therefore to mimic some clerical activities usually performed by the astrophysicist in his data analysis. Following Ledgar, Singer and Whiteside (1), the search for an overall effectiveness even in those cases where it decreases efficiency, was therefore pursued by a friendly and controllable design and implementation (2). Moreover we ask that AA allows a user insight into the performed analysis. To this end a second point has to be cautiosly explored: the astrophysicist is often used to work with analog images while the AA works on digitized data. Digital data are quantized in their nature and affected by digitization noise. A plate once digitized, results into a two dimensional array of integer numbers which vary in a finite range. Each number (called 'gray tone') describes a well defined property of an elementary area (picture element: pixel) of the plate. The astrophysicist's action to be imitated by the AA must therefore be carefully translated from a real/integer numerical environment to a digitized one, affected by a not-yet-sufficiently-explored kind of noise (3).

Moreover, an astrophysicist explores and interprets his data not only on the basis of formal rules, but mainly on the basis of his experience that is in analogy with past decisions, taken in similar situations and memorized as his own 'mental' data base, the whole depending on his own skill. An example of this behaviour is the measurement of the diameters of star images, which, among others, King and Raff (4) describe as follows: " We estimated the diameter of each star image at a point where the image was a fairly dark gray but not quite black. Clearly this judgment will differ from individual to individual, but there did not seem to be a significant systematic difference between the two of us.". Note that how this is an example of "imprecision" as defined in Prade (5). We take into account this imprecision by the use of multivalued logic as a tool of decision (6) (7). In our approach, this tool is used as follows: the AA has to accumulate hints and indications about any structures in the image. Hints and indications are combined into a classification by mean of a multivalued logical tree. This classification is expressed by scores, which memorize the present state of knowledge about the structure. The classification of a structure and the corresponding score may be modified at any stage of the interpretation process and in general cannot be considered established until the last step of the computation.

Scores, hints, indications and multivalued trees can be established and defined empirically, by the observation of the astrophysicist's activity. We do not require that AA automatically classifies every structures present in the plate. On the contrary, AA classifies the identified structures as: 'real' objects, from which galaxies and stars are thereafter extracted, 'uncertain structure' which may be a signal or noise, and 'noise'. 'Uncertain' structure may only be judged interactively by the astrophysicist or, in the future evolution of the system, with the help of the information from different sources (e.g. from a survey, a catalogue of different plates). 'Uncertain', 'real', 'noise' are the translation into the scores assigned to the structures of the person-machine communication language. Until this last classification, a structure is called candidate-objects (CO).

Last but not least, actions performed by the AA must be understood and criticized by different astrophysicists and consequently improved. To this end, it would be necessary to describe every action, performable by the AA in an algebraic form, understandable even by non-programmers. The APL notation is used for the required formal and synthetic algorithmic description of the actions to be performed by the AA on the data (8) (9). The APL programs can be used even for empirical studies and tests on real data.

DESIGN AND IMPLEMENTATION

The design of AA is based on the elicitation of the astrophysicist knowledge and requires two stages: exploration and confirmation. In the exploration phase, sample data are examined by the use of the interactive system ISIID (10) and a prototype system of APL functions is defined and implemented. These functions are defined without any 'a priori' assumption about the nature or distribution of the data. Particularly, it was avoided any reference to continous model, such as the disputed assumption that a digitized star image has a gaussian profile.

In the confirmation phase, the prototype system is used on a new set of data, whose interpretation is unknown. The obtained results are checked against an interpretation performed by an astrophysicist, who does not know the results of the automatic classifications. If the check is not satisfactory, according to some predifined criteria, the programs are submitted to a new refinement or redifinition exploratory step. If the check is positive, the system is accepted and the design ends. The APL functions memorize the knowledge of the astrophysicist and are used as a long-term memory in a pattern-directed-inference system (PDIS) (11). A PDIS is characterized by its organization based on data or event-driven operations (11). PDIS responds directly to a wide range of (possibly unanticipated) data or events rather than operating on expected data using a prespecified and inflessible control structure, as conventional programs do. Since there is no explicit control structure associated with the declared knowledge, new or modified assertions, obtained in the refinement or redefinitions steps by the astrophysicist, can be added easily to the system.

THE DIARY OF IMAGE SEGMENTATION DESIGN

The present implementation of AA follows a three step strategy. AA first segments (12) the digitized images into parts, detecting the candidate objects. Then it describes the shape of each found CO, and tries to classify it on the base of its structure (13). This first level description does not allow the interpretation of nearly round CO., which may be generated both by a star or by a galaxy. AA matches this last problem by studying the fuzzyness of the CO in hand, a star beeing associated to a sharp track, a galaxy to a more spread one (14).

Here we discuss only the first step. The goal is to develop a device to detect in a digitized image the same objects, that an astrophysicist finds in the plate, from which the image is drawn.

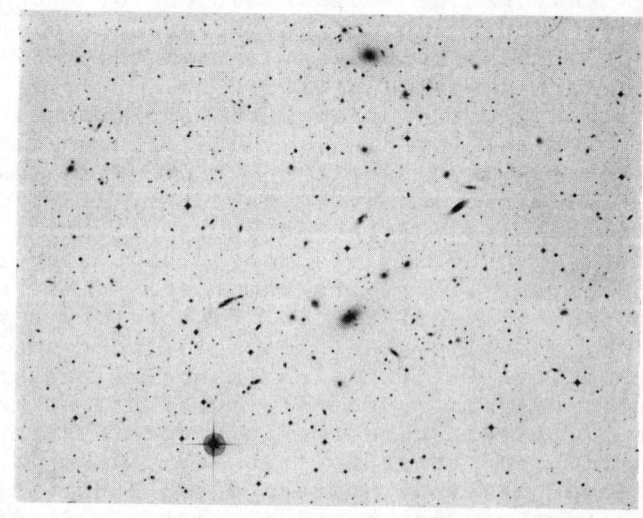

Fig.1 Cluster of galaxies of the southern sky near α 4h 30m, δ o -54 Plate 157J of the ESO Sky Survey; ==67"/mm, mg.lim. =20-21; Digitized field = 20' 20" x 20' 20" (1000 x 1000 pixels); Digitization: Eso Garching microdensitometer; The first subfield used in exploration is marked.

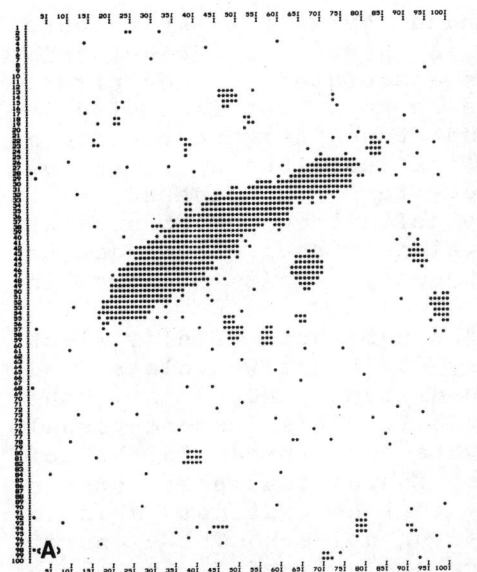 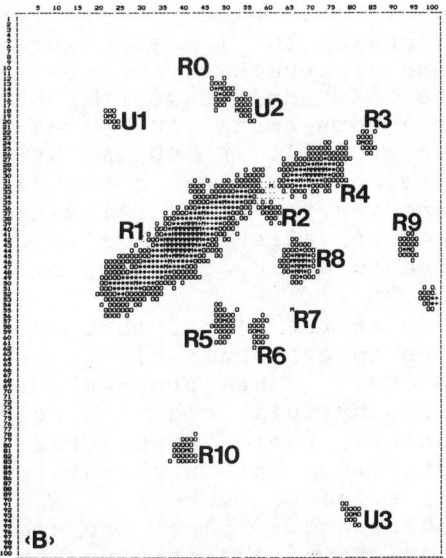

Fig.2: Digital map of the initial subfield obtained: (A) by the application of a threshold which leaves the 90% of the pixels below it. (B) by application of the APL functions for the detection of ordered sets (last definition). The characters identify the visual classification, where U means uncertain, R real. R7 is a bright object lost by the first definition, R2 is classified uncertain by AA, U3 real.

Data for each needed exploration-confirmation cycle were obtained by the field shown in fig.1. The field was divided in subfields. A first subfield (fig.2) was used for the initial definition of the structures of interest. At each successive confirmation step, a new subfield was-used.

A widely used method for the segmentation of digitized images is thresholding (12). But some practical tests showed that the different thresholding methods, known from literature or from our experience, have problems with the detection of faint and overlapping objects (e.g. R1, R2, R3, and R4 in fig.2)

As to faint objects it was first argued that the astrophysicist takes into account structured excesses from the background, which appear 'homogeneus' within a certain field of the plate even when it not intense. From the study of some examples, it was noticed, that this kind of blobs were translated into 'ordered' set of pixels of similar gray level. An 'ordered' set of pixels is defined to be a connected set in which at least one internal pixel exist, that is in which at least one pixel is surrounded by eight neighbours belonging to the structures in hand. To detect this kind of structure, a code is associated to each pixel. The code univocally maps the sign of differences in gray tone among the pixel and its 8 neighbours into a number (15). A CO is therefore described as a set of pixels of given codes which form an ordered set. Moreover the code allows the definition of structures such as valleys and slopes which in their turn can be used to divide overlapping objects, as it is showed in fig.2b .

At the first test on the same data field, (last step in exploration), no noisy structure was classified as CO. The program found some CO, that the astrophysicist did not point out, at a second visual control those structures were confirmed to be of interest. The astrophysicist found that some of the signalled CO were so faint, that he was not able to understand if they are signal or noise. The class of 'uncertain' objects was therefore defined. But three objects, which in the astrophysicist's classification were considered as meaningful CO were lost (tab 1.a). The device was considered unreliable by the astrophysicist, and a new refinement of the

	(A) EXPLORATION				(B) CONFIRMATION				
	R	U	N	T	R	U	N	T	
Real (R)	15	1	0	16	135	1	15	151	
Uncertain (U)	1	2	0	3	5	6	3	14	Automatic interpretation
Noise (N)	2	0	2	4	3	8	7	18	
Total (T)	18	3	2	23	143	15	25	183	

Astronomer's interpretation

Table 1: Confusion matrices for the exploration (A) and confirmation phase. (B). On the main diagonal, one reads the number of cases in which the two classifications match. In the last row, the integral classification of the astronomer by class, in the last column the AA integral classification. In (B) 135 structures out of 143 which are classified as real objects by the man, were classified as real by the AA, 5 uncertain and 3 noise.

proposed definition of CO was required before any confirmation. The new definition was tested again on the same data, but was again considered unreliable because, while catching all the astrophysicist's defined CO, it added some noisy blobs to the class of real objects. It was now observed that CO which were selected by the astrophysicist and lost by the first definition, appear in the plate as 'bright' objects of limited area. Digitalization effects reduce them to a one bright pixel structure. A new definition of CO was proposed: a CO beeing either an ordered set of pixels or a 'bright ' pixel structure. An adaptive thresholding algorithm was now proposed as the definition of 'bright' pixel structure (see 8 chapter 2). The threshold is a statistic from the set of values of the maxima of the ordered sets, which have been previously identified in the field. The new definition of CO was accepted by the astrophysicist and a confirmation phase was executed. The results (tab.1b) are interesting but problems arise with digitized objects surrounded by a strong alho. Therefore a new refinement step is required, and at present it is under study.

CONCLUSIONS

We have discussed the general items of the AA design and an example of their application. Our approach to design suffers from beeing empirical: it is highly dependent on the experience and skill of the astrophysicists, who collaborate to the design, and specialized to a specific class of data.

The design itself is based on an interactive, interdisciplinary knowledge elicitation activity, performed together by the astrophysicist and information analyst. In contrast with other approaches (e.g. 6), we study every single case in which the programmed tools and the astrophysicist differ in their classification. Disagreements are examined by the discussion of the two classifications, none of which is considered 'a priori' better than the other.

It is worthwhile to note that the AA is a formal specification of the astrophysicist's interpretation activity: therefore it may be criticized, changed or updated by everybody interested. Furthermore AA may be used, as it was done with other systems (17), to check the self-consistency of the classification criteria used by the astrophysicist.

If the class of data is changed (that is the kind of plates or the digitization process) AA must be in its turn adapted, checked against the new data. The PDIS structure seems particularly attractive for AA implementation, because any updating or tuning action results in the only modification of the knowledge base, which is constitued by indipendent APL functions (12), without any action on the structure of the system.

REFERENCES

(1) H.Ledgar, A.Singer, J.Whiteside, Directions in Human Factor for Interactive Systems, Lect. note in Comp. Sc.103, Springer Verlag, Berlin, (1981)
(2) B.Garilli, P.Mussio, A.Rampini presented at Int. Work on data anal. in astronomy, IFCAI, Palermo, (1984)
(3) T.Pavlidis, in Pattern Rec. in Practice, pg.123, E.S.Gelsema, L.N.Kanal ed., North Holland, (1980)
(4) I.R.King, M.I.Raff, Pubblications of the Astronomical Society of the Pacific, pg.120, 89, (1977)
(5) H.Prade, Quantitative methods in approximate and plausible reasoning, Rap. L.S.J. 185, (1983)
(6) R.S.Michalsky, R.Chilanski, in Fuzzy Reasoning and its Application, E.M.Mamdani, B.R.Gaines ed., pg.247, Academic Press, (1981)
(7) S.Garribba, P.Mussio, F.Naldi in Synthesis and Analysis Methods for safety and Reliability Studies, G.Apostolakis, S.Garribba, G.Volta ed., pg.154, ASI, Plenum Press, (1978)
(8) K.Iverson, Elementary Analisys, APL Press, Swartmore, (1976)
(9) U.Cugini, P.Micheli, P.Mussio, M.Protti: in Digital Image Processing, S.Levialdi ed., Pitman (1984)
(10) U.Cugini, M.Dell´Oca, P.Mussio, in Digital image Processing, S.Levialdi ed., Pitman (1984)
(11) D.A.Waterman, F.Hayes-Roth, , in Pattern-Directed Inference Systems, D.A.Waterman and F.Hayes-Roth eds., pg.3, Academic Press, (1978)
(12) S.Bianchi, A.Della Ventura, M.Dell´Oca, P.Mussio, APL Quote Quad, Vol 12, No.1, pg.54, (1981)
(13) A.Rosefeld, A.C.Kak, Digital Picture Processing, Academic Press, (1976)
(14) M.Balestrieri, P.Mussio, Proceeding of the 5th colloquium on astrophysics, G.Sedmak, M.Capaccioli, A.J.Allen ed., Oss. Astr., Trieste, (1979)
(15) S.Bianchi, G.Gavazzi, P.Mussio, Mem. S.A.It., Vol 53, No.1, (1982)
(16) G.Bordogna, Tesi Universita´ di Milano, (1984)
(17) P.Battistini, F.Bonoli, A.Braccesi, F.Fusi Pecci, M.Malagnini, B.Marano, Astron. Astrophys. Suppl. Ser, Vol.42, pg.357-374, (1980)

AN ASTRONOMICAL DATA ANALYZING MONITOR

Dieter Teuber

Astronomical Institute of Muenster University
Domagkstrasse 75, 4400 Muenster, West-Germany

INTRODUCTION

The need for exchange of programmes and data between astronomical facilities is generally recognized (e.g. Sedmak et al., 1979, and papers therein), but practicable concepts concerning its realization are rare. Standardization of data formats through FITS (Wells et al., 1981) is widely accepted; for (interactive) programmes, however, identical hardware configurations seem to be the favoured solution. As an alternative, a software approach to the problem is presented.

GAME - GENERIC APPLICATIONS AND MONITORS ENVIRONMENT

GAME was designed to allow the construction of virtual operating systems (VOS). While a real operating system (OS) drives a hardware configuration, the VOS drives the real OS. The VOS presents all data and control information to the various participants in the most efficient way. A simple example: An OS gives access to data by means of filenames which are restricted in the number of letters describing the data, whereas the rest is needed to indicate information relevant to the OS only. The VOS introduces object-oriented identifiers for the users and logical-unit-like identifiers for the application programmes in order to access the data. The first method allows a notation suitable for human use, while the latter method is faster on the machine. Both methods are independent of the filename structure of the particular OS and of the structure of data storage in the system. To accomplish accomodation of the various methods, all participants are connected to the VOS through tables which describe their respective requirements and thus enable the translation.

Several properties of the proposed system will be discussed for the monitor of the VOS, i.e., the control centre of interaction.

Underlying Principle

The GAME concept involves two branches: subjects and administrators. They take the general form of table files and interface routines, respectively. The definition of subjects includes all devices (color display systems, measuring machines, etc.), application programmes, data frames and users. The devices, programmes and a set of data files (standard tables, etc.) shall be called the Astronomical Data Analyzing System (ADAS). All members of ADAS are described in tables.

The basic unit of all tables is named 'element' and is defined to be a field of 16 consecutive bytes. Two kinds of elements are used: vectors and scalars. All vectors have a name field and a pointer field. The latter holds the basic properties of the table, that is protected by the keyword in the name field. In conjunction with the previously gathered information the respective administrator finds its way to the protected table and the associated scalar elements. The scalars contain information, that is not used for table system purposes, but for the applications (parameter values). The tables are edited by the administrators, which are subroutines interfacing the application programmes to the table system. On the lowest level this means simple input/output operations. On the medium level a typical operation is a keyword seek sequence. On the highest level resides the monitor structure.

The Monitor

The monitor structure is based on the assumption that communication between the user and the monitor needs either a menu type interaction or a keyword oriented interaction. In the latter case the monitor structure does not process specific commands, but interprets a whole phrase which is restricted to three kinds of semantic elements: keywords, delimiters and parameter values. Therefore, the monitor structure possesses no inherent command language to interact with the user. A syntax and a dictionary must be defined to turn the structure into a true monitor. The version used at Muenster is the Astronomical Data Analyzing Monitor (ADAM) (Teuber, 1984). But the monitor structure may emulate other systems with respect to their response to the user.

As can be seen in Figure 1, the monitor structure has two dominating function blocks: director and executer. The director interprets and obeys instructions from the director routines which are stored in the users' registers. These routines are written in a proper low-level control language. The instruction set allows modification of specific parts of the users' registers which are used by the executer as described below. The director routines thus determine the syntax of the command language.

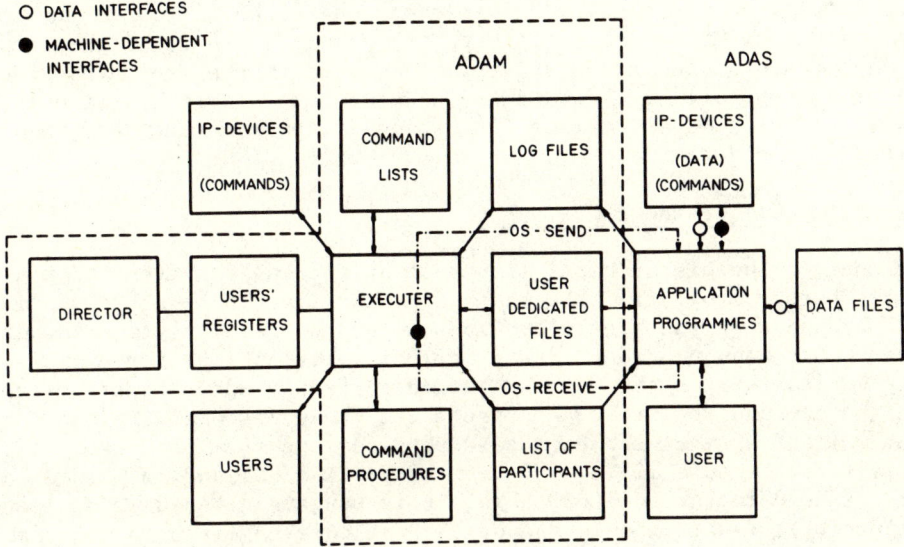

Fig. 1. The monitor (ADAM) surrounded by the image processing participants (users and ADAS).

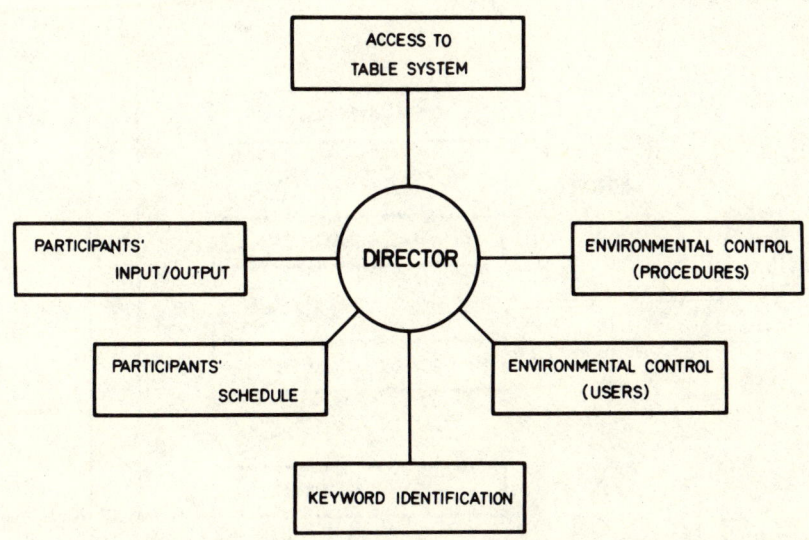

Fig. 2. Functional groups of the executer that may be addressed by the director.

On completion of an instruction sequence the director transfers control to the executer which then performs the selected function using the arguments stored in the respective locations of the users' registers. The major functional groups are shown in Figure 2. All basic table operations may be performed under the direct control of the director. This feature is especially helpful in developing new administrators. The administrators contained in the remaining functional groups permit standardized access to the table system only, i. e., interfaces are called with standard argument list and in predefined order.

Exemplary Use of The Table System

Because of the hierarchical structure of the table system the monitor would become slow, if it tried to find the right path testing always all branches of the hierarchy. To shorten the response time keyword search has been provided with dynamical memorization. For example, the monitor stores the command phrase currently under construction in the command registers. An associated group of commands is held in the command scratch table. Access to a command from that group is very fast, because it is in core memory. If a command is issued that is not held in the scratch table, it is sought in the user dedicated file. If it is not there, one or several command lists may be examined until a match occurred. In case of a match the command and its associated information is copied to all the levels above. Thus the next time the command is issued, access to the associated information is much quicker. To avoid an overflow of the upper levels, commands which have not been used for a level-dependent period may be overwritten.

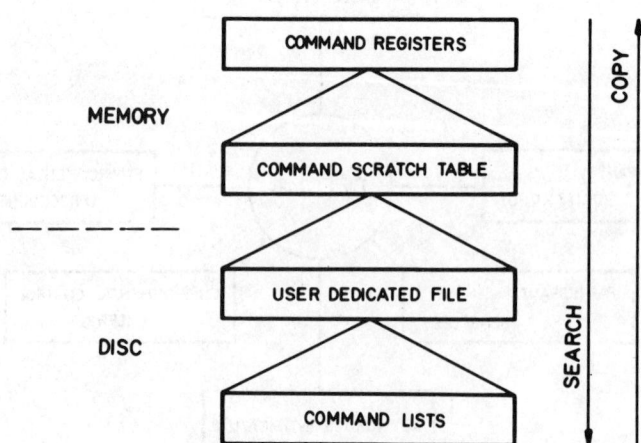

Fig. 3. Command storage hierarchy of ADAM supporting dynamic memorization.

Conclusive Remarks

The basic principle of the proposed VOS is the separation of action and administration which eases the integration of different hardware components, command languages and programme packages. The layered structure of the system allows the implementation of functions on different levels with a smooth transition from one level to the next. An overview of some other properties of the monitor is given in Table 1.

Table 1. Properties of ADAM

Specification	Remarks
multi-tasking	concurrent interactive programmes, batch jobs
multi-user	-
multi-lingual	simultaneous response to different command languages
error response	error checking, tracing and control at all monitor, application programme and interface levels
maintenance	built-in maintenance facilities
table system	for control purposes and parameter storage; hierarchical keyword-oriented tables with redundant routes to provide accelerated access
structure	four major layers: 1. FORTRAN-77, 2. administrators, 3. director routines, 4. command language

REFERENCES

Sedmak, G., Capaccioli, M., Allen, R. J., eds., International Workshop on Image Processing in Astronomy, 1979, Osservatorio Astronomico di Trieste

Teuber, D., ADAM Operator Reference Manual (working edition), 1984, Astronomical Institute of Muenster University

Wells, D. C., Greisen, E. W., Harten, R. H., FITS: A Flexible Image Transport System, 1981, Astron. Astrophys. Suppl. Ser. 44, 363

RIAIP - A SYSTEM FOR INTERACTING WITH ASTRONOMICAL IMAGES

R. De Amicis and G.A. De Biase

Istituto Astronomico, Universita' di Roma la Sapienza

Via Lancisi, 29
00161 Roma, Italia

ABSTRACT

RIAIP (Rome Interactive Astronomical Image Processing) is a software system developed at the Institute of Astronomy of the Rome University, pole of the ASTRONET network, to provide an efficient tool of interactive analysis for astronomical images obtained from ground-based and from orbiting observatories.

The software is written in FORTRAN 77 and has been developed for the DEC VAX computers, under VMS operating system, with DEC VS11 (512x512 pixels) image subsystems.

The main RIAIP features are high modularity and strong interactivity, driven by menus; an usable log-file is also available allowing modes of system use other then the fully interactive one. The main algorithms needed for astronomical images processing are yet implemented and, due to the high system modular structure, new and user defined applications can be easily added.

1. INTRODUCTION

A software system for astronomical image processing must have the following features [1,2,3,4]:

- avoiding the need of constructing a reduction procedure for each particular problem, the consequence is that the system must be as flexible as possible, applicable to a wide range of

problems and must include the fundamental modules for pre-elaborations on images and the algorithms of most common use;
- managing a set of related astronomical images with all the connected data;
- allowing interactive experimenting with algorithms in the construction of reduction procedures, which can be used finally in automatic mode.

At the Institute of Astronomy of the Rome University the software system RIAIP (Rome Interactive Astronomical Image Processing) has been developed[5] on the basis of the previous concepts and of the following ideas, which were given the greatest importance during the system planning and development:

- many users do not necessarily need to have specific programming experience and they are mainly interested in extracting, as easily and quickly as possible, usefull informations from their images,
- easiness of system use,
- unlimited freedom for the user in the construction of the data reduction procedure,
- quickness of system response time,
- easiness of adding applications and algorithms.

The answer to the above constraints was the research of a system with an efficient man-machine colloquium, this fact leading to a strongly interactive software. The use of menues was found to be the most efficient method to drive the interactions and to give warranty for a sufficently self-documenting system. In what follows the RIAIP structure is described in some details.

2. RIAIP ARCHITECTURE

RIAIP is a very modular software system based on three levels and operating with three modalities. The man-machine colloquium is ensured by a structure with menues, which are related to the various operative levels and operations.

2.1. Levels

In order to have a system architecture as functional as possible for the above described purposes, the main operations needed for image processing were identified and grouped into classes. The consequence is a very simple structure based on three operative levels:

- first level, selection of operation classes,
- second level, selection of operations,

- third level, selection of operation details.

As it can be seen from fig.1, the levels are organized in a tree like structure, which has the features of being very simple and of allowing a good modularity.

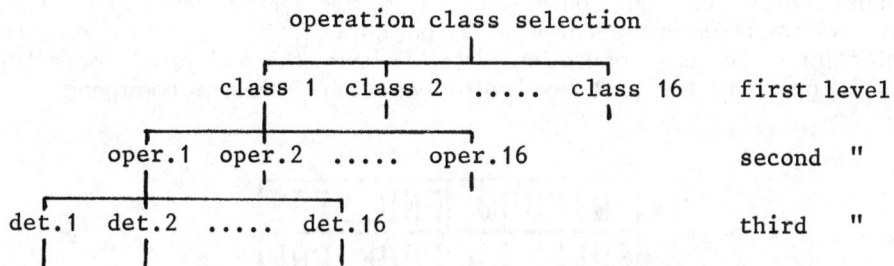

Fig.1. RIAIP structure (modules organization).

2.2. Modalities

The system structure is based also on three modalities:

- full screen,
- four partitions,
- window.

In the full screen modality the display on images is executed on 512x512 pixels. In the four partitions modality the screen is divided in four portions each of 256x256 pixels. In the window modality the display is executed into an user defined square portion of the image. All the elaborations are always executed at the maximum resolution, 512x512 in the first two modalities and in the window dimensions for the last one. The window, with typical dimensions of 128x128 pixels, can be set on any point of the image. The window modality has been introduced to allow elaboration tests in very short computing times. The modalities can be mixed allowing mixed displays.

2.3. Interactivity

A very fast and simple system use, which also minimizes the possibility of errors, can be obtained by inserting commands on menues. The use of menues for the interactions has been selected for its simplicity, fastness and immediacy for the user due to the fact that all the options of the actual operation are shown at the same time. Moreover, with a suitable choice of the menues labels (corresponding to an operation to be performed), a sufficently self-documenting system, which can be successfully used also by inexpert users without the need of an users guide, is obtained. To each menu 16 choices are available so that, taking into account that we have three levels, 16 choices of possible classes, 256 of operations and 4096 of operation details are allowed. Selection by joystick on the menues and the ⟨tab⟩ ⟨return⟩ command are used respectively to go down and up in the system levels (see fig.1). The system structure permits it to go down from the operation class selection to any operation detail within three joystick operations and to go up to the first level with two ⟨tab⟩ ⟨return⟩ commands.

WINDOW	END_SESS
FULL_SCR	FOUR_PAR
	AIDED
	USER_DEF
N_STD_OP	MAKE_UP
ARITM_OP	COMP_OP
IMAGE_OP	
FILE_OP_	GRAPH_OP

Fig.2. First level menu (operation classes).

Fig.2 shows the first level menu, which allows a fast selection of the operation classes and of the display modalities. As it can be seen the operation classes actually included are:

        FILE_OP   operations on files (I/O),
        IMAGE_OP operations on images,
        ARITM_OP arithmetical operations,
        N_STD_OP non standard operations,
        GRAPH_OP graphical operations,
        COMP_OP computational operations,
        MAKE_UP make-up operations,
        USER_DEF user defined operations,
        AIDED   activates the aided mode.

The menu permits it also the modality selection:

FULL-SCR activates the full screen modality,
WINDOW activates the window modality,
FOUR-PAR activates the four partitions modality.

All the other interactions with the system are done by joystick (JS) for graphical interactions and by prompts for numeric values of operation details.

Figures 3a and 3b show two examples of second level menus. Fig.3a shows the menu related to the image operations, which include: selection of the image (from 0 to 3), selection of the display zone, hardcopy, selection of the look-up table, pseudo-color, display af the content of all the four partitions, display of image, levels, peculiar operations of the window modality, maximum and minimum insertion for the pseudo-color, zoom, selection of a ground color for the display zone. Fig.3b shows the graphical operations menu, which includes: selection of the image, selection of the display zone, hardcopy, mean intensity profile on a strip, contours, display, store on disk and load from disk of the graphical files associated to the images, intensity profiles, intensity profiles perpendicular to the image boundaries, selection of a ground color for the display zone.

SL2 F	SL3 F
SL0 F	SL1 F
DISP_ZON	
HARDCOPY	LEVELS
LOOK_UP	WINDOW_T
PSEU_COL	PARA_INS
DISP_ALL	ZOOM
DISPLAY	GRAPH_FL

SL2 F	SL3 F
SL0 F	SL1 F
DISP_ZON	DISP_GRA
HARDCOPY	STOR_GRA
	LOAD_GRA
STRIP	PROFILES
CONTOURS	ORT_PROF
	GRAPH_FL

Fig.3. Two examples of second level menus: a) image operations, b) graphical operations.

It is foreseen, the related work being in progress, to use subprocesses and special devices (like array processors) for some operations on images requiring long computing times when a quick interactivity is not possible. This fact offers also the advantage for the user to do more interactive operations while waiting for the subprocess termination. A facility allowing the system use even without menues (aided mode) will be presented afterwards.

3. STORAGE ORGANIZATION AND ALGORITHMS

RIAIP can work on four images of 512x512 pixels (INTEGER*2), all residing in the central memory at the same time, this allowing it to make direct comparisons between different images and to use all the included algorithms without the need of intermediate operations of loading and storing between the mass storage and the central memory. To each image a label and related graphical files are associated. A work storage area, always residing in central memory, of 2048x2048 pixels is also available for operations on images with dimensions greater than the standard ones (512x512). This storage area also solves some problems related to the loading of 'foreign' images and to the storage of intermediate results of some algorithms.

At present RIAIP includes the main algorithms needed for the Astronomical work, such as pseudo-color, graphical operations (contours and profiles), levels, arithmetical operations, statistics, roto-translation and so on. Owing to the higly modular system architecture, new algorithms can be easily implemented and added. A facility is also available (USER-DEF, in first level), which permits each user to add its own modules and menues.

Some examples of work are shown in fig.4(a,b,c,d,e,f).

4. THE AUTOMATIC MODE

The non-interactive or semi-interactive use of the system is of particular interest. During any session RIAIP generates a log-file on which all the executed operations and the related parameters are recorded. The log-file is written in a suitable very mnemonic code, which is interpreted by the system when the aided mode is activated. The log-file can also be modified or written by the user itself. Various degrees of interactivity are possible up to the construction of a fully automatic procedure:

- fully automatic (no display), RIAIP reads all the commands and values (JS, prompts) from the log,
- fully automatic (with display), same as above but the user can follow on the display all the execution phases,
- semi-automatic (no display), The operations and JS values are read from the log, whereas the answers to prompts are given by the user,
- semi-automatic (with display), the operations and prompts values are read from the log, JS is given by the user and the

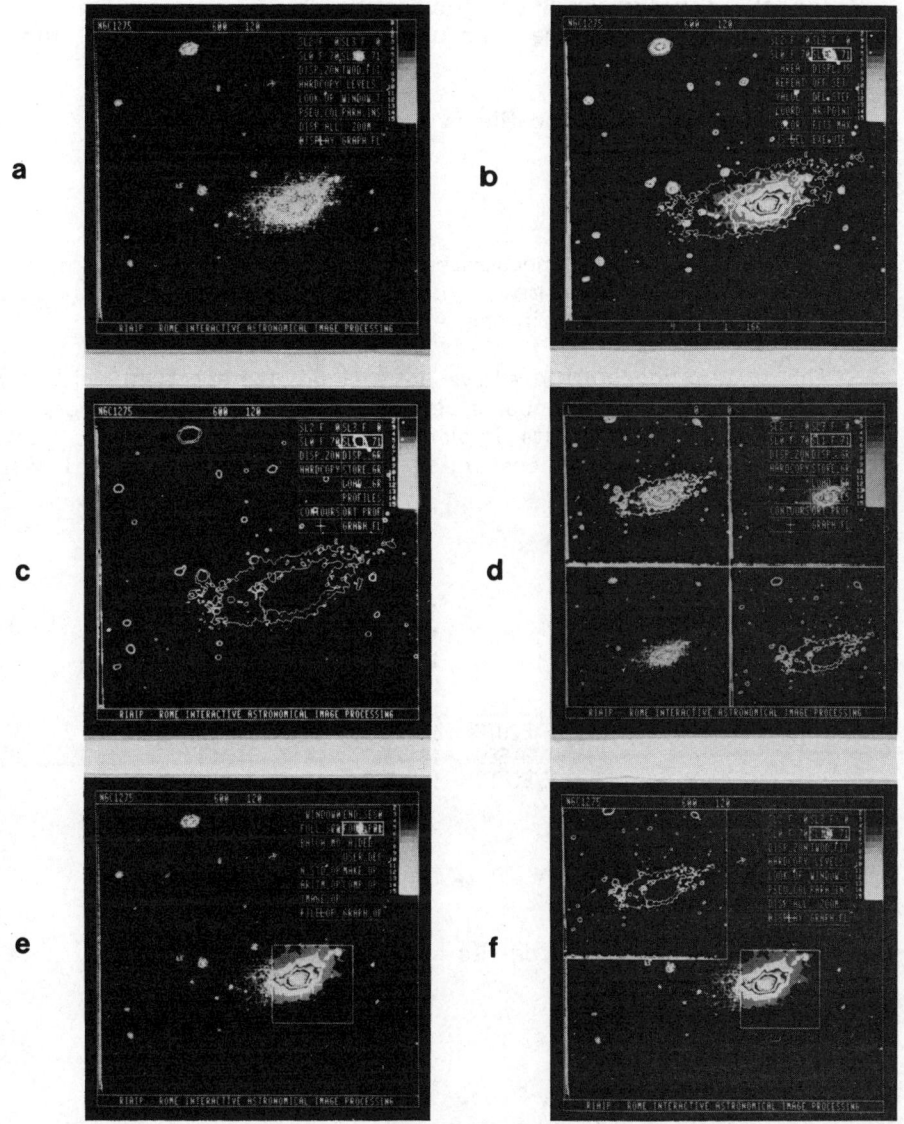

Fig.4. Image of NGC 1275 (obtained from a FOC prototype). a) Full-screen modality: original displayed with pseudo-color technique; b) isophotes on the same smoothed image; c) display of the isophotes only; d) four-partitions modality: the screen content is the same of a),b),c) plus the smoothed image alone; e) window modality: full-screen display of the original image, in the window the portion of the smoothed image can be shown; f) mixed display with the same content as e) plus the isophotes of c).

image subsystem is used for display and JS,
- automatic-interactive, only the operations are read from the log, the answers to JS and prompts being given by the user,
- fully interactive, the menu is enabled and the log is excluded.

A short example of log-file is shown in fig.5.

5. CONCLUSIONS

So far RIAIP has demonstrated to be very simple and easy to use and that its features correspond quite well to the basic requests of the researchers in the Astronomy field.

The system modularity allows an easy implementation with new basic and applicative modules, even by the users themselves. The software structure allows its implementation even on other image subsystems with few changes which do not affect the overall system structure.

```
FILE_OP_, LOAD    ,         , 1,  70,   0
        ,PARA INS,          , 1, 600, 120
IMAGE_OP,PSEU_COL,           , 1
        ,DISPLAY ,           , 1
GRAPH_OP,PROFILES, JS SEL , 1,  62, 142
        ,         ,EXECUTE , 1,  62, 142,  1,  15,  2
        ,         , JS SEL , 1, 313, 164
        ,         ,EXECUTE , 1, 313, 164,  1,  15,  3
FILE_OP_,SL1 F   ,          , 2
        , LOAD   ,          , 2,  71,   0
IMAGE_OP,PSEU_COL,           , 2
        ,DISPLAY ,           , 2
END_SESS
```

Fig.5. Log-file structure example.

REFERENCES

1. P. Crane, K. Banse, Proc. ASTRONET Nat. Meeting, Mem. S.A.I., 53, 1, 19 (1982).
2. F. Macchetto, T. Westrup, Proc. ASTRONET Nat. Meeting, Mem. S.A.I., 53, 1, 31 (1982).
3. G. Sedmak, Proc. ASTRONET Nat. Meeting, Mem. S.A.I., 53, 1, 9 (1982).
4. K.P. Tritton, Proc. ASTRONET Nat. Meeting, Mem. S.A.I., 53, 1, 55 (1982).
5. G.A. De Biase, V. Lieto, R. Lombardi, Proc. ASTRONET Nat. Meeting, Mem. S.A.I., 53, 1, 155 (1982).

EXPERIENCE REPORT OF THE TRANSPORTATION OF THE SOFTWARE
SYSTEMS IN CONNECTION WITH THE CREATION OF MOTHER, THE
VAX/VMS VERSION OF THE GRONINGEN HERMES

> P. Steffen, C. Crezelius, P.M.W. Kalberla and R. Steube
> Radioastronomical Institute University of Bonn, FRG
>
> P.T. Rayner
> CSIRO Division of Radiophysics, Epping, Australia
>
> J.P. Terlouw
> Kapteyn Astronomical Institute, University of
> Groningen, The Netherlands

1.0 INTRODUCTION

The "Bonner Astronomisches Bildrechner System", BABSY (Steffen et al., 1983), was planned as a system based on software exchange in order to minimize costs, man power, and installation time. This idea determined the hardware configuration of the system; considering its main components it is similar to a STARLINK node (VAX 11-780 plus SIGMA ARGS image display).

This "STARLINK-like" hardware configuration was chosen since in 1980/81 it seemed that STARLINK (Disney and Wallace, 1982) would become a major "trend setting" system for astronomical software standards.

2.0 THE INSTALLATION OF ASTRONOMICAL "STANDARD" SOFTWARE

Indeed, installing STARLINK software was nearly as trivial as expected. In the first phase of creating the BABSY software environment, besides other external software (AIPS, from the VLA; NOD2, radiocontinuum software from Bonn; TVS, the Tololo-Vienna System in its STARLINK version) some STARLINK packages were installed: ASPIC and subpackages, PICPRO, and SPICA. Also right from the beginning, we used standards for documentation. These included producing VMS supported "Online Help Libraries" and keeping track of all software included in the system via a software index.

3.0 MOTHER, THE VAX/VMS-VERSION OF HERMES

The second phase consisted of creating a VAX/VMS version of HERMES, the master task of the Groningen GIPSY system and adapting the GIPSY application software itself (Allen and Terlouw, 1980). We consider HERMES, for our applications, to be the best software driver for "real" interactive data analysis, since it provides user control of several tasks at a time, task communication, keyword substitution etc.. Building MOTHER (short for Most Of this is HERmes) showed the complexity of the software exchange problem. Due to the fact that the PDP-11 HERMES version uses many operating system functions, we could perform this work in a reasonable time scale only with personal help from Groningen. Central parts of the internal data file management had to be adapted to the operating system of the VAX. The installation of single tasks of the GIPSY system was easy, though a few subroutines coded in Assembler had to be rewritten and emulations of parts of the PDP 11-70 and of the image processing operating systems were necessary.

MOTHER is like an operating system that initiates and controls application programs, also known as servant tasks. We list the most interesting features of MOTHER in the following table:

1. user controlled multitasking.

2. task-to-task communication; one task can provide another task with input (also possible under user control).

3. book keeping of all user activities and saving output from servant tasks in a logfile as well. Paging and string-searching through this logfile allows retracing of all user actions.

4. keyword management: provides keywords with defaults; prompts are in a standard syntax; position independent input to servant tasks.

5. simple mechanism for rerunning of servant tasks. this allows for changing of one or more keywords without retyping the full command

6. chaining of tasks via a REPEAT facility

7. only minimal effort is required to include user software in the system.

8. improvements of the mastertask are available to all servant tasks.

9. more efficient use of the hardware because of
   a) using only one terminal for a number of parallel running interactive applications.
   b) tasks can run in background to minimize idle time.

4.0 MOTHER, A DRIVER FOR OTHER SOFTWARE PACKAGES

To ensure that users need to deal with only one "astronomical" command language, in the third installation phase we started running applications from other software packages under the control of MOTHER. This is, of course, not convenient for all packages, but, for example, is possible for ASPIC programs without even relinking object code. Limitations grow where applications use special features of their corresponding driver.

Figure 1 explains which packages can be run under MOTHER in the BABSY system. MOTHER acts as a link between these packages. The user can use the same syntax for different packages and will have all the advantages of MOTHER controlled tasks. Besides the

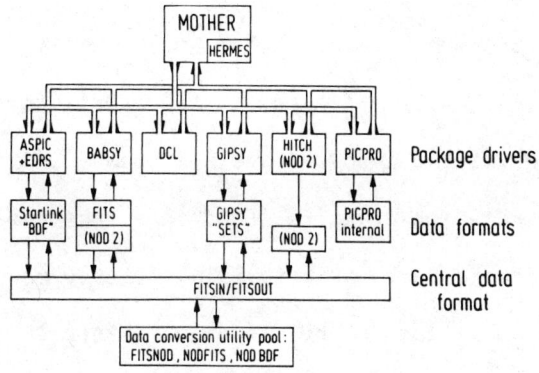

Fig. 1 Mother and related software packages which can be run under MOTHER's control.

astronomical software packages the user also has full access to commands of DCL, the Digital Control Language. This, for instance, allows program development (editing, linking etc.) within MOTHER while astronomical reduction is running simultaneously.

5.0 CONCLUSIONS

Our experience clearly shows that building an astronomical software system based on software exchange in a reasonable time is possible and worth doing. It is not only for the benefit of the "importer" because MOTHER now has been exported to other astronomical institutions.

Two main problems remain when using different external software packages: first there is the well known problem of display software, if the imported package uses a different image processor. A solution can be found in a reasonable time scale only when the package to be adapted is a strictly modular one.
The second problem is due to the different internal data formats of the packages. Our solution requires user interaction, (s)he has to run data format convertors when switching from one package to the other.

6.0 ACKNOWLEDGEMENTS

We thank the Sonderforschungsbereich Radioastronomie for support, R.J. Allen, Groningen, and P.T. Wallace, W.F. Lupton, M. Irwin from STARLINK for their generosity in providing us with software and for fruitful discussions.

7.0 REFERENCES

Allen, R.J. and Terlouw, J.P. ,1980. A multi tasking operating system for interactive data reduction. In IUE Data Reduction, Ed. W.W. Weiss, A. Schnell, R. Albrecht, H. Jenkner, H.M. Maitzen, K. Rakos, E. Mondre (Wien), p. 193

Disney, M.J., Wallace, P.T., 1982, Q. Jl R. astr. Soc. 23, 485

Steffen P., Crezelius C., Rayner P.T., Steube, R. ,1983. Mitt. Astron. Gesell. 60.

SAIA: A SYSTEM FOR ASTRONOMICAL IMAGE ANALYSIS

B. Garilli (1), P. Mussio (2), and A. Rampini(1)

1) IFCTR, CNR, Milano

2) Dip. di Fisica, Universita' di Milano

INTRODUCTION

SAIA is an image processing system designed to analyse astronomical data allowing a complete and detailed control of all the operations the system performs. It is implemented on a HP/Ramtek system. The most urgent purpose of SAIA was to analyse the data collected by Exosat, and this part is now currently operational. Afterwards it has been expanded to allow the analysis of data from different instruments. The performances of the system have been presented elsewhere (Chiappetti et al., 1983). Here we concentrate on its formal aspects and the overall structure.

THE 'RATIONALE' BEHIND THE SYSTEM DESIGN

The two main features of our system are the friendliness of the interactive language and the possibility of evaluating the outputs step by step. Among the other characteristics (Bormann and Karr, 1980), a friendly interactive language should be easy to learn, remember and use, with natural names and sentences.
In order to help users in exploiting the capabilities of the system, we have grouped commands corresponding to actions performed on the same type of data and/or doing similar computations. Commands of the same group have a similar structure, so that the user is automatically driven to the correct syntax by the

analogy in the operations he is performing on the data.

Our aim is to handle both digital and digitized data. We note that the digitation process adds a further noise to the data. Both digital and digitized data, however, are quantized when they undergo the analysis. This quantization means that some algorithms, like rotation or gaussian smoothing, yield different results depending on the adopted technique. Therefore the documentation specifying the kind of technique used in such cases, must be built in the system, under form of warnings or written messages. Thus, the user is able to choose the best strategy, and within given a strategy, the most suitable input parameters.

To summarize, we have extended the concept of 'friendly system' in order to include a) an easy control on numerical results b) warnings relative to operations which may be improper on digital/digitized data c) the possibility of monitoring intermediate results d) a logbook, so that the strategy followed to reach a result is automatically recorded.

OVERALL DESIGN OF SAIA

The design of SAIA is geared around three characteristics: modularity, interactivity and documentation.

Modularity is a logical consequence of the presence of different kinds of data and of the different purposes the system can be used for. Moreover, it allows a greater adaptability to new hardware configurations (Klinglesmith, 1975). Programs dealing with similar data are grouped together in one module. Homogeneity between the different modules is granted by a superinterpreter which manages all the data-independent features of the system. This structure has allowed us to include a whole system (in this case IHAP, Middelburg and Crane 1979) as one of the modules (fig. 1). Documentation is thought to be on three levels (Bianchi and Mussio, 1982): user-oriented, system manager-oriented and programmer-oriented. The interaction between programmer, manager and user is very important for a further development of the system. A good documentation is one of the means by which they can communicate with each other.

1) THE SUPERINTERPRETER

It is composed of some co-routines with the common feature of being used by any of the modules. The module scheduler is the only component really essential to the system to be operative; the others are requested to make

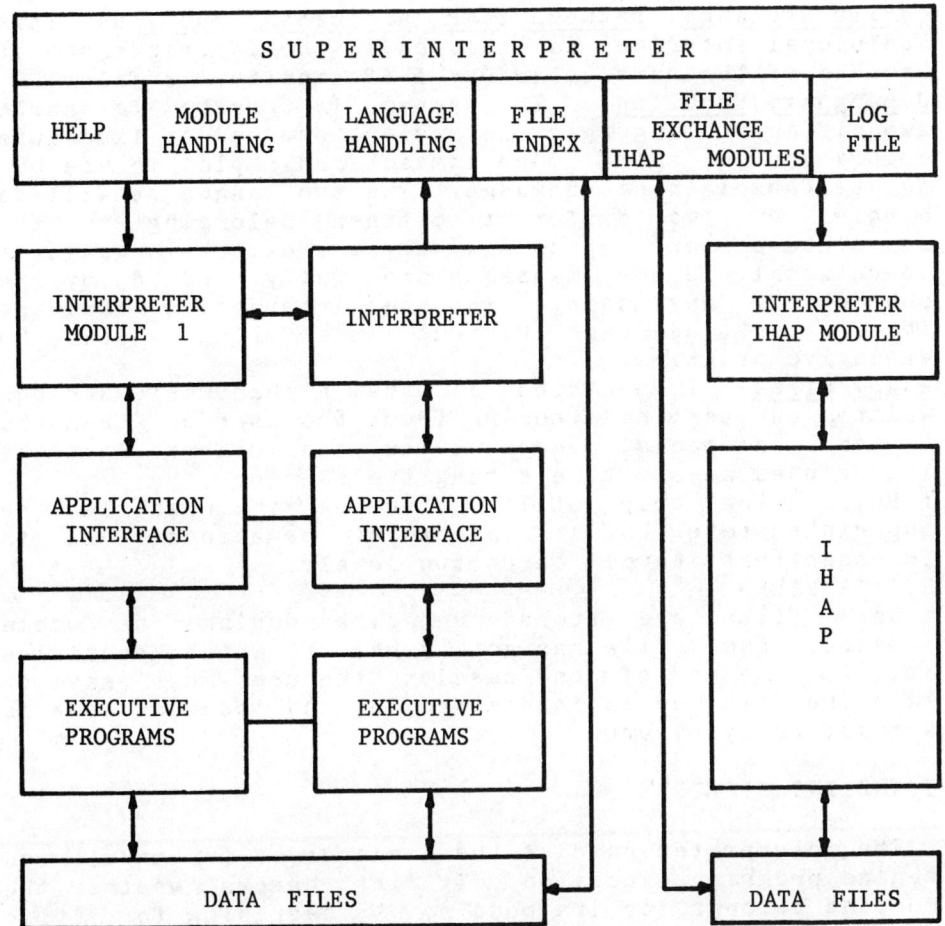

Fig. 1: Schematized overall structure of SAIA

the system a unique friendly structure.
a) Module Scheduler. It passes the user's command to the appropriate module by searching for the command in the different command tables, one for each module. Once the command is found, the appropriate module retrieves the command string and goes on with the execution. If the command is not found in any of the tables, an error message is given.
b) File Index. A command operating on a data file first checks the file existence and then the feasibility of the requested operation. Therefore the file index contains, for each file created during the session, its main characteristics.

c) **File Exchange between IHAP and other Modules**. This additional interface is included in the superinterpreter because of the particular way IHAP handles its files.
d) **Language Handling**. The system is thought to handle two different languages: an easier one and of immediate comprehension, and a more compact and rapid to use but not self-explaining language. The two languages will be handled by two different routines belonging to the superinterpreter. Up to now only the expert language is operational. Since messages are fully handled by one subroutine, any change in the interactive language itself can be made without interfering with the executive programs.
e) **Log File**. It contains all the alphanumeric strings written on terminal (coming from the user or from the system). A special command allows the user to have the log printed at any time during the session.
f) **Help**. The help utility can be included in the superinterpreter but it is not yet operational. It is foreseen that it will be on two levels.

2) DATA FILES

Data files are stored on disk during the whole session. Each file has got a header which identifies it. At the end of the session the user must save on tape the files he is interested in, and the disk area is automatically cleared.

3) THE INTERPRETERS

The interpreter handles the commands of its module and begins program execution. It first checks whether the command belongs to its module, by searching for it in the command table, which is made in an antecedent-consequent form (Waterman and Hayes-Roth, 1978). Once the command is found, its consequent is executed, otherwise the control is passed back to the superinterpreter.

4) APPLICATION INTERFACES

The application interface of a given module is a collection of subroutines, one for each command name, with the function of making all the error checks. These checks are made either by comparison with the file index, or the subroutine itself contains the appropriate restrictions and defaults.

5) EXECUTIVE PROGRAMS

All programs are grouped in segments. After the execution, control is passed back to the interpreter together with the output message, if any.

PROGRAM ADDITION TO THE SYSTEM

A program to be added to the system must be written following a few specifications. No read and write operations from terminal must be included in the program itself. The subroutine belonging to the application interface must be included in the module library. All messages associated with the command must be added to the proper subroutine, located in the superinterpreter library. Eventually, the command name must be included in the command table. A special program accomplishes this task.

CONCLUSIONS

SAIA has been designed and realized to be a friendly and interactive system. Continuous inputs from prospective users contributed to the design fitness and flexibility of the system. Since manpower problems were present, we have built a structure that has allowed the system to be operative in a reasonable time, even if uncomplete in some less essential parts. Updating and addition of extensions or modules is a foreseen and easy operation which does not change the overall structure.

REFERENCES

Bianchi, S. and Mussio, P. (1982), Mem.S.A.I.T., 53, 195

Bormann, L. and Karr, R. (1980), Evaluating the friendliness of time sharing systems, SIGSOC Bull. 12

Chiappetti, L., Garilli, B. and Rampini, A. (1983) 'An interactive system for Exosat Image Processing', 7th European Regional Astronomy Meeting, Florence

Klinglesmith, D.A. (1975) in Image Processing Techniques in Astronomy, ed. de Jaeger and Nieuwenhuijzen, Reidel Publishing Company, p. 271

Middelburg, F. and Crane, P. (1979) in 'Image processing in Astronomy', eds. Sedmak, G., Capaccioli, M. and Allen, R.J., Osservatorio Astronomico di Trieste, p. 25

Waterman, D.A., and Hayes Roth, F. (1978) An overview of Pattern-Directed Inference Systems, in 'Pattern Directed Inference Systems', ed. D.A. Waterman and F. Hayes Roth, Academic Press

A SYSTEM FOR DEEP PLATE ANALYSIS: AN OVERVIEW

P. Di Chio[1], G. Pittella[2] and A. Vignato[2]

[1] Centro Scientifico di Roma, IBM Italia
Via Giorgione 129, 00147 Roma
[2] Osservatorio Astronomico di Roma
Viale del Parco Mellini 82, 00100 Roma

ABSTRACT

A software system for digital processing of astronomical plates, specially devoted to the study of clusters of galaxies, is briefly described. In this processing, typical problems are the large amount of pixels per image and the large number of high magnitude objects to measure for each plate. These problems along with the architecture of a system able to cope with them are described as well as the system functions to process, display and archive image data.

INTRODUCTION

The interest in very distant galaxy clusters is related to relevant astronomical problems such as the test of cosmological models, the structure of the clusters themselves and their dynamical history, and the spectroscopic evolution of galaxies[1]. The observational material usually consists of deep plates obtained by large telescopes and digitized by a microdensitometer. A complete study of a cluster implies the reduction of several plates taken in different color bands, and eventually in different epochs.

Since the size of a typical plate in digital form is up to $10^4 \times 10^4$ pixels and the number of objects present in the field is of the order of 10^4, the traditional visual inspection of such plates is practically inhibited, and then an automated system must be utilized for the analysis; such an approach gives also the possibility to obtain a better reliability of the results by controlling changes of processing parameters.

The image analysis system here described is capable of storing and managing bulk of data of different types, to perform fast processing on large images and provides a set of image display and graphic tools to derive the astronomical parameters of interest.

SYSTEM OVERVIEW

One of the major design aims of the system is to obtain a high level of data independence: this is achieved by an accurate isolation of all the machine dependent routines and the implementation of a user interface completely independent of the host hardware. In this way the portability of the system is made easier, also through the use of a high level language for user-oriented programs. Moreover the maintenance of the whole system is much more comfortable, and routines do not need to be rewritten as a consequence of data structure changes.

A further goal is to realize an open design system in such a way that the user has the possibility to add his own routines, besides the built-in functions, commands and algorithms.

A particular care has been devoted to the end-user interface: it has been realized using a hierachical menu structure, in such a way that also a non-experienced user can access the system immediately: a set of functional keys, defaults and help functions have been designed for this purpose.

The availability of some peculiar hardware, like high resolution color display and array processor, lends toward the use of specialized software, thus decreasing portability but greatly increasing the overall system performance, in terms of reduction of processing time and terminal response time.

The system is designed and implemented to run on IBM systems 43XX and 30XX, under the control of VMSP/CMS[2] operating system. The language used for processing and high-level routines is VSFORTRAN[3] (IBM version of ANSI FORTRAN77) in view of portability, while machine dependent and low-level I/O routines were written in ASSEMBLER H[4] to achieve higher throughput. Furthermore the APL[5] language is used because of its ease of use, language power and interactive capabilities for the analysis of the results through graphic functions and for algorithm tests.

The system is made up by four subsystems: Display, Processing, Data Handling, and Data Analysis and Graphic subsystems.
Let's now briefly describe the functions of each component.

Display Subsystem

This component allows it to display on a high resolution color device (IBM 7350[6]) several kinds of images like plates, sky background, galaxy surface density and so on, and by taking advantage of the processing capabilities of the 7350, to perform image manipulations.

Among them, the most useful functions are: zooming and scrolling of images, handling of color look-up table to enhance faint structures or objects, and local filtering operations to visually validate filter design.
Moreover, the possibility to visualize the superposition of several plates of the same field gives the ability to obtain synthetic information as: the reality of very faint objects, if we compose plates in the same color band; or qualitative knowledge of the nature and distance of objects, if different bands are taken into account.

Processing Subsystem

This component contains computation subprograms to perform the reduction process through the following steps:

Sky background evaluation: to determine the reference value map used for filtering, segmentation and photometry.

Filtering: to smooth photographic noise and achieve signal to noise ratio equalization.

Segmentation: to detect objects with an iterative multi-threshold algorithm to avoid the merge of nearby objects.

Nearby objects detection: to perform a more accurate photometry of each object, by changing the values of those pixels that belong both to the object and its neighbours.

Plate calibration: determination of transparency-intensity relation using both calibration spots or standard star profiles.

Photometry: calculation of object magnitudes and others physical quantities.

Star-galaxy separation: based on morphological parameters.

The processing subsystem can be used both in interactive and batch mode; some of the most time consuming routines can run on the IBM 3838 Array Processor[7].

Data Handling Subsystem

Every data access is made through this subsystem which includes a set of device dependent I/O routines and a set of file management routines. These routines are used to handle internal data structures, by performing operations like creation of control tables and system status inquiry; various types of utility operations are also possible like loading images from tapes, data set dump-restore and so on.

Data Analysis and Graphic Subsystem

This part is written in APL and gives the possibility to access processed data, by producing plots and histograms using the IBM 3277 Graphic Attachment[8]. This facility is used to perform checks on results of processing and develop new algorithms, by taking advantage of the interactive capability of APL.

DATA STRUCTURES

Three types of data are defined in the system:
- Images,
- Catalogs (both single and multiple),
- Miscellaneous data.

The internal structure of data is transparent to the user and the data archive management is performed by means of directories. For image and catalog types of data a descriptor is available in which both user retrievable and system reserved information is reported.

All these descriptors are automatically updated by the Data Handling Subsystem as a consequence of a user action (e.g. creation of an image file from tape, generation of a background map, generation of a catalog by the segmentation program).

Let's summarize the characteristics of each data type.

Images

They are essentially two dimensional numeric arrays $I(x,y)$, in which coordinates are represented by indexes. Under this form, microdensitometer scannings, background maps, object surface density maps and so on are stored.

Taking into account the application for which this system is designed, a typical size for an input image ranges from 4000×4000 to 8000×8000 pixels with a disk space occupation between 32Mbytes to 128Mbytes.

Catalogs

These data sets contain the information relative to each object detected on a given image and are created by the segmentation procedure and updated during subsequent phases, mainly by photometry and classification. The information stored in a catalog is relative to the geometrical and photometrical characteristics of each object.

The geometrical quantities can be divided into two classes: positional information, like x and y boundary coordinates and opacitance centroid coordinates, to easily retrieve each object; shape parameters, like ellipticity and major axis length and direction, used for instance to recognize different types of objects and find their overall spatial distribution.

The photometrical information is the main part of the catalogs because every quantity and relationship of astrophysical interest can be derived from it. As an example it is possible to derive the luminosity function of the galaxy cluster, the cosmological log N vs m relation and the galaxy surface distribution as a function of their types and magnitude classes. Essentialy the photometrical information is made up of a sequence of magnitudes computed inside circular diaphragms of increasing radius centered on the intensity centroid of each object. Using this set of magnitudes it is possible for instance to derive the probability density function of stars and galaxies in a suitable range of magnitudes, by assigning to each object a probability of belonging to one out of these two classes.

Catalogs can be either single-plate or multiple-plate. The first one refers to information relative to the objects present in just one plate, while the second one is created, for the sake of access speed, combining the information coming from several single-plate catalogs.

The types of operation that catalogs undergo are essentially: updating, during plate processing phase; and consulting or displaying, during the data analysis phase. The disk storage occupation of a catalog related to a rather crowded plate can get up to 35Mbytes. In general, more than one catalog is produced for one plate during the processing phase, depending on different parameter values (e.g. segmentation threshold levels).

Miscellaneous data:

This class comprises all auxiliary user data not belonging to the above mentioned classes. Two kinds of data exist in this class: alphanumeric descriptions and numerical quantities, like parameters and coefficients.

In general, these quantities are stored in the descriptors of data they refer to: as an example, in the descriptor of a scanned image, the telescope name, the photographic emulsion type, the observing conditions and so on are reported, while in the catalog descriptor the origin and size of the subimage it refers to, the segmentation threshold levels, the calibration curve coefficients, the type and parameters of the point spread function can be found.

CONCLUSIONS

The system has been completed and tested as far as the routines of the processing subsystem and most of the image display part are concerned; a subset of the data handling subsystem is also operational. Several plates have been already reduced and, in particular, a set of plates from the 4m KPNO telescope and two plates from the CFH 3.6m telescope[9] have been completely analyzed with satisfactory results.

System performance achieved, for the complete reduction of an 8000×8000 image containing 15000 objects, was of about 36 hours (CPU time on an IBM 4341).

The development of this system is done in the frame of the digital image processing activities of the IBM Rome Scientific Center in cooperation with the Rome Astronomical Observatory.

REFERENCES

1. D. Gerbal and A. Mazure, ed., "Clustering in The Universe", editions Frontieres, Gif Sur Yvette, (1983)
2. IBM Corp., "VM/SP: Introduction", Form GC19-6200

3. IBM Corp., "VS FORTRAN Application Programming: Language Reference", Form GC26-3986
4. IBM Corp., "Assembler Language", Form GC33-4010
5. IBM Corp., "APL Language", Form GC26-3847
6. IBM Corp., "IBM 7350 Image Processing System. System Overview", Form GA19-5433
7. IBM Corp., "IBM 3838 Array Processor. Functional Characteristics", Form GA24-3639
8. IBM Corp., "IBM 3277 APL Graphic Attachment Support: Program Reference and Operation Manual", Form SH20-2138
9. P. Di Chio, E. Hardy, and A. Vignato, "A catalog of galaxies in the direction of A2319 cluster", (in preparation).

THE GRONINGEN IMAGE PROCESSING SYSTEM

R.J. Allen, R.D. Ekers* and J.P. Terlouw

Kapteyn Astronomical Institute
University of Groningen
Postbus 800, 9700 Av Groningen, The Netherlands

"The obsolescence of an implementation must be measured against other existing implementations, not against unrealized concepts".

Brooks, 1975; p.9.

SUMMARY

An interactive, integrated software and hardware computer system for the reduction and analysis of astronomical images is described. After a short historical introduction, some examples of the astronomical data currently handled by the system are shown, followed by a description of the present hardware and software structure. The system is then further illustrated by describing its appearance to the user, to the applications programmer, and to the system manager. Finally, some quantitative information on the size and cost of the system is given, and its good and bad features are discussed.

---

*Present address: National Radio Astronomy Observatory, P.O. Box 0, SOCORRO, N.M. 87801, U.S.A.

I. INTRODUCTION

In this paper we describe the current state of an image processing system which has been under continuous development since 1971. This system was originally installed as a single interactive program in a PDP-9 computer with a single-user operating system and modest hardware resources, as described previously by Ekers, Allen, and Luyten (1973). The next version ran as a single program in a time-sharing CDC 6600 computer during the period 1976-78. The present version, commonly known by the acronym GIPSY, is an organized set of independent, interactive programs which runs in a PDP 11/70 computer with advanced image display facilities and a multi-user, multi-tasking operating system.

The design and development of GIPSY has been strongly driven by the needs of its users, and it is the users themselves who continue to add most of the new software to the system. Details of the investment in manpower and of the current size of the system are given later in this paper; a summary can also be found in the paper by Allen and Ekers elsewhere in this volume.

Many aspects of GIPSY software architecture have influenced the designers of newer systems. The original STARLINK project managers decided to include a number of GIPSY features into the definition of their first-generation "software environment" in 1980 (SGP /7). The major new element in their design was to take specific advantage of the facilities offered by the VMS operating system in the VAX computers used on STARLINK. Although this plan was largely abandoned in 1981 in favour of a new scheme (SGP/18.1), several other groups had already begun the design of systems based on the first-generation STARLINK software environment. In this indirect way, GIPSY has had an effect on MIDAS and FIPS (both described in this volume). Also, the development of PANDORA (Simkin, Bosma, Pickles, Quinn, and Warne 1983) at Mount Stromlo in Australia grew out of an early version of GIPSY installed there in 1979.

At the start in 1971, application programs written by the users for GIPSY concentrated on the analysis of two-dimensional radio astronomy map data obtained from the Westerbork Synthesis Radio Telescope (WSRT). With the advent of spectroscopy at the WSRT, the architecture and application programs were expanded beginning around 1976 in order to allow processing three-dimensional data. Presently, the users of GIPSY analyse observations made with a variety of radio, infrared, and optical detection systems. Some examples of the kinds of data currently being processed are given in section II.

Our description of GIPSY will follow the general scheme suggested by Allen elsewhere in this volume. Section III describes the current hardware and software structure of the system. Section IV discusses the appearance of GIPSY to the user, to the programmer, and to the system manager. Finally, in section V we present some statistics about the system and comment on its good and bad features.

II. SOME EXAMPLES OF ASTRONOMICAL DATA

In our paper describing the first generation system (Ekers et al. 1973), we showed a number of examples of the radio continuum aperture sythesis data which was being analysed at that time. Six years later, multiple-image radio astronomy data sets were being accomodated in GIPSY, and methods of displaying such enormous quantities of data were being explored (Allen, 1979b). Presently, multiple-image optical data cubes and infrared observations are also routinely analysed.

We have chosen a few recent examples from current research programs; for the most part these results have not yet been shown elsewhere. Figure 1 illustrates a multiple-image optical data cube, and Figure 2 shows the results of some analysis on that data. Figures 3 and 4 have been obtained by a similar analysis of a radio data cube. Finally, Figures 5 and 6 show exaples of radio and infrared continuum maps.

III. IMPLEMENTATIONS

1. Hardware Configuration

The configuration of computer hardware in which the present version of GIPSY is implemented in Groningen is shown in Figure 7. The host computer and its standard peripherals are provided by the Groningen University Computer Center as a general facility to the university research community. Astronomical image processing is the major activity, but the computer is also used by several other groups. For image processing and display, three additional devices are also connected as peripherals to the host computer. The function of these devices is described below.

Image Computer and Display. This consists of a commercially-available device[a] to which a few minor local modifications have been made. It is used for the manipulation and display of digital images in a variety of sizes and data formats. The internal organization consists of six image memories of 512 × 512 8-bit bytes, but these memories can be re-configured for multiple-byte words and for larger or smaller images. Besides the standard display (e.g.

---

a. Model 70E, International Imaging Systems, 1500 Buckeye Drive, Milpitas, CA 95035.

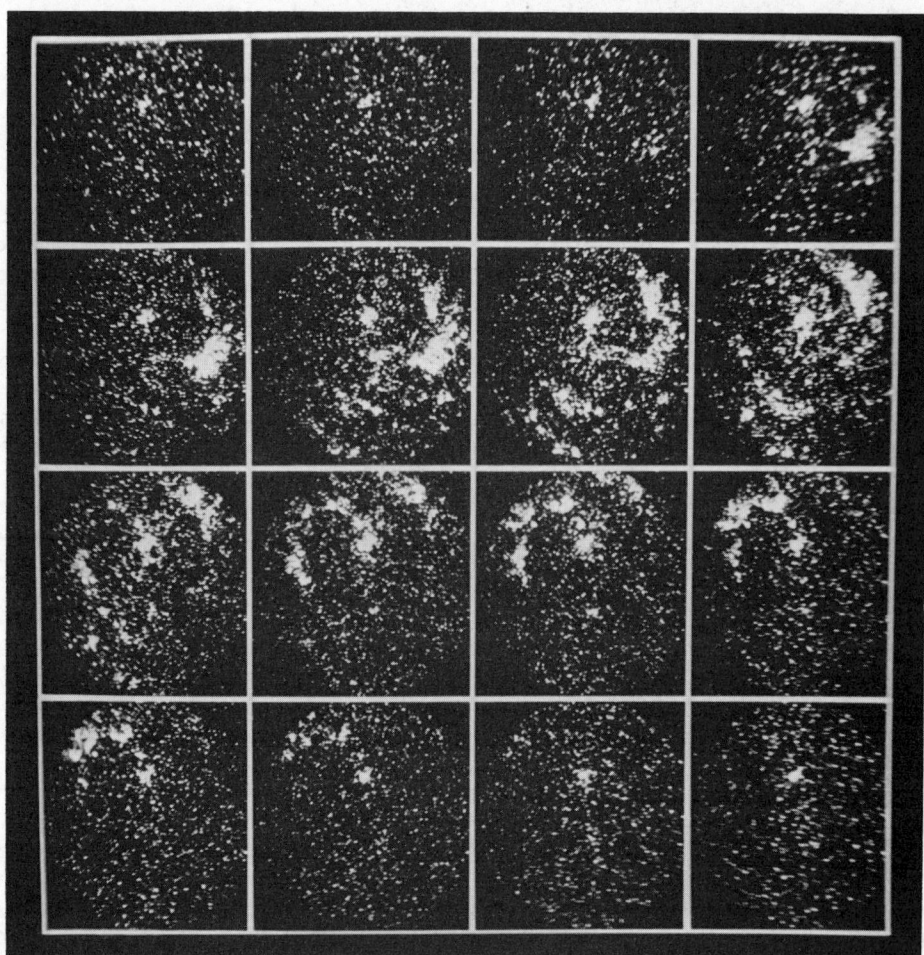

Fig. 1 Portions of a seeing-limited data cube of optical Hβ line emission from a central 5.5 arc minute field on the southern spiral galaxy M83 = NGC 5236, as observed with the TAURUS imaging Fabry-Perot spectrometer and the Image Photon Counting System detector on the Anglo-Australian Telescope. The full data cube is 256 × 256 pixels × 98 channels. The channels selected for this mosaic have a velocity width of about 27 km s^{-1} and are separated by 20.7 km s^{-1}, beginning at 670 km s^{-1} in the upper left corner, and decreasing to 360 km s^{-1} at the lower right. The continuum emission has been estimated from channels outside the range of the Hβ line and subtracted from each remaining channel; however, owing partly to intrumental effects, the bright nucleus of the galaxy remains visible just to the north of the center on each channel map. From unpublished work by R.P.J. Tilanus; see also Allen, Atherton, and Tilanus (1985).

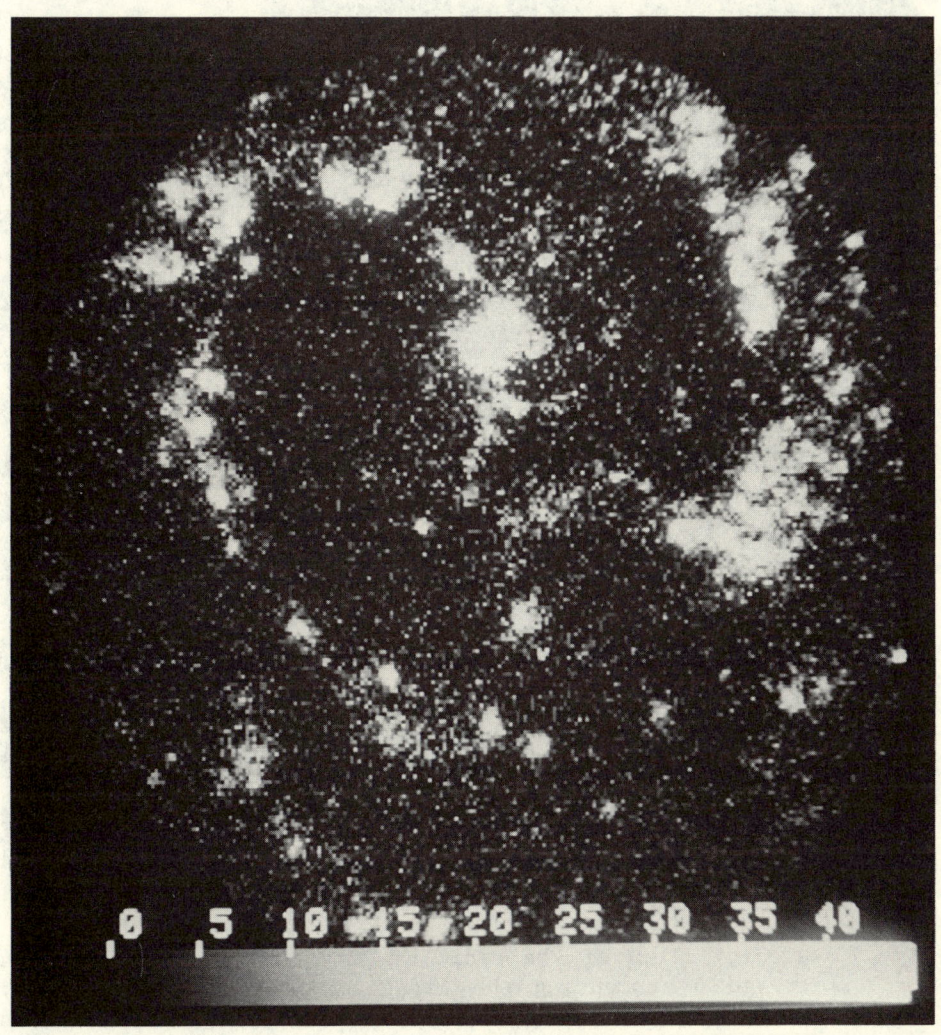

Fig. 2 Image of the total Hβ emission of M83, obtained from the data cube illustrated in Figure 1 by computing the zero moment along the velocity axis at each pixel position. The gray scale along the bottom of the figure indicates the total number of photons detected per 1.3 × 1.3 arcsecond pixel. From unpublished work by R.P.J. Tilanus.

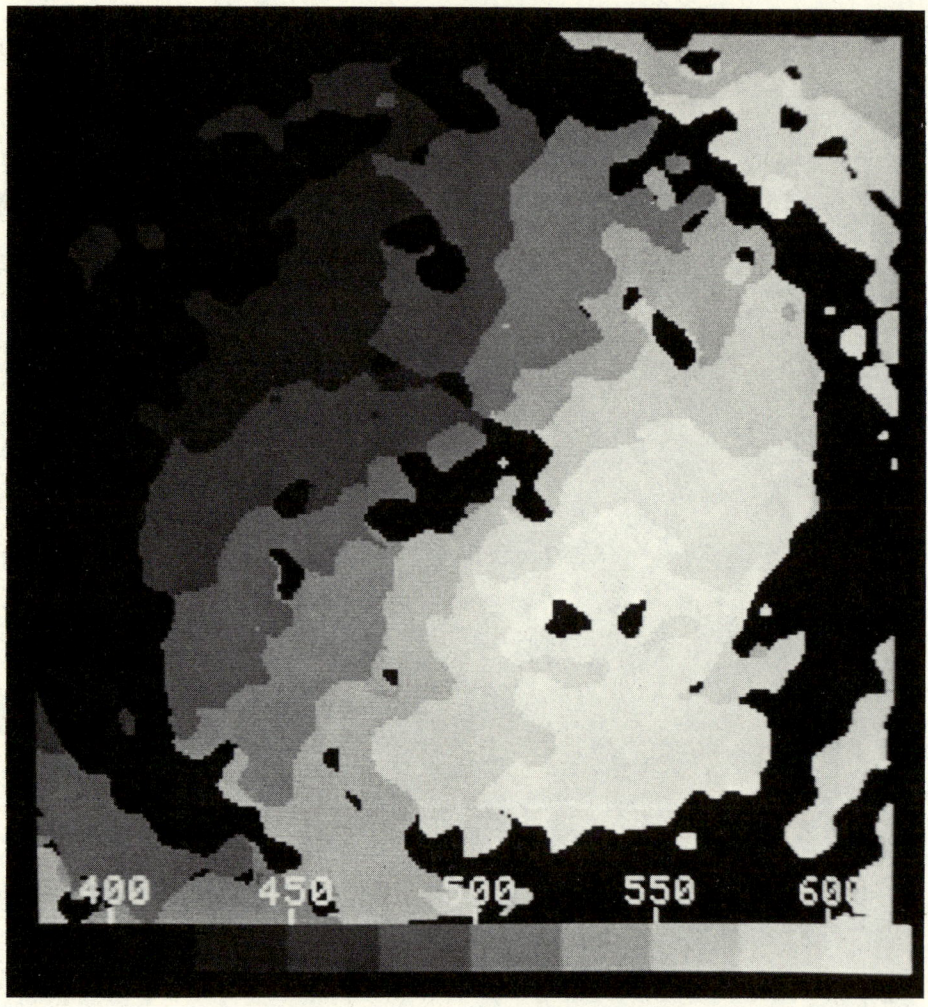

Fig. 3 Velocity field of the radio HI line emission from M83, as derived by computing the first moment along the velocity axis at each pixel position in a data cube obtained with the Very Large Array in the United States. The original data cube size was 512 × 512 pixels × 31 channels; it has been smoothed for this analysis to a resolution of 15 × 15 arcseconds × 20 km s^{-1}. A threshold technique has been used to exclude regions of insufficient signal from the velocity calculation. The field size of 12.8 arcminutes encompasses more of the galaxy than the Hβ image of Figure 2, but it still largely excludes a faint outer HI ring which is partly visible in the corners of the picture. The grey scale along the bottom is calibrated in km s^{-1}. From unpublished work by R.P.J. Tilanus.

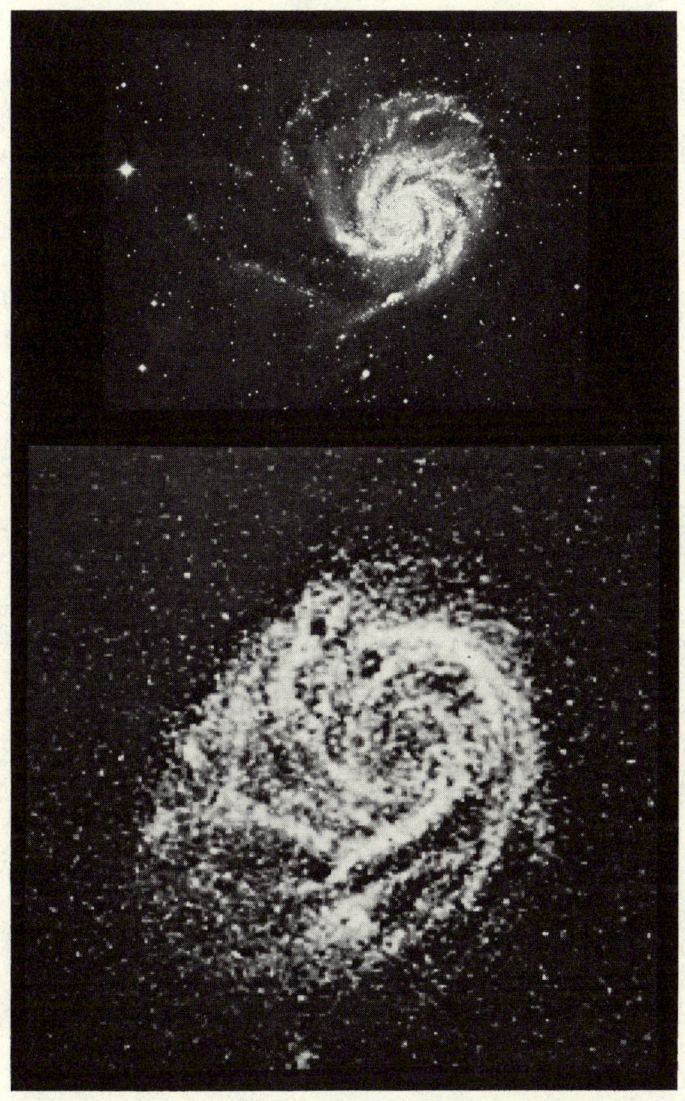

Fig. 4 Distribution of the radio total HI line emission in the northern spiral galaxy M101 = NGC 5457 (lower panel) compared to its optical image (upper panel) on the same scale. North is to the left. The original data cube was 512 × 512 pixels × 16 channels, observed with the Westerbork Synthesis Radio Telescope at a resolution of 24 × 30 arcseconds × 27 km s^{-1}. The HI picture shown here was obtained by computing the zero moment of the data cube along the velocity axis at each pixel position, using a thresholding algorithm in order to reject channels with insufficient signal. From Allen and Goss (1979).

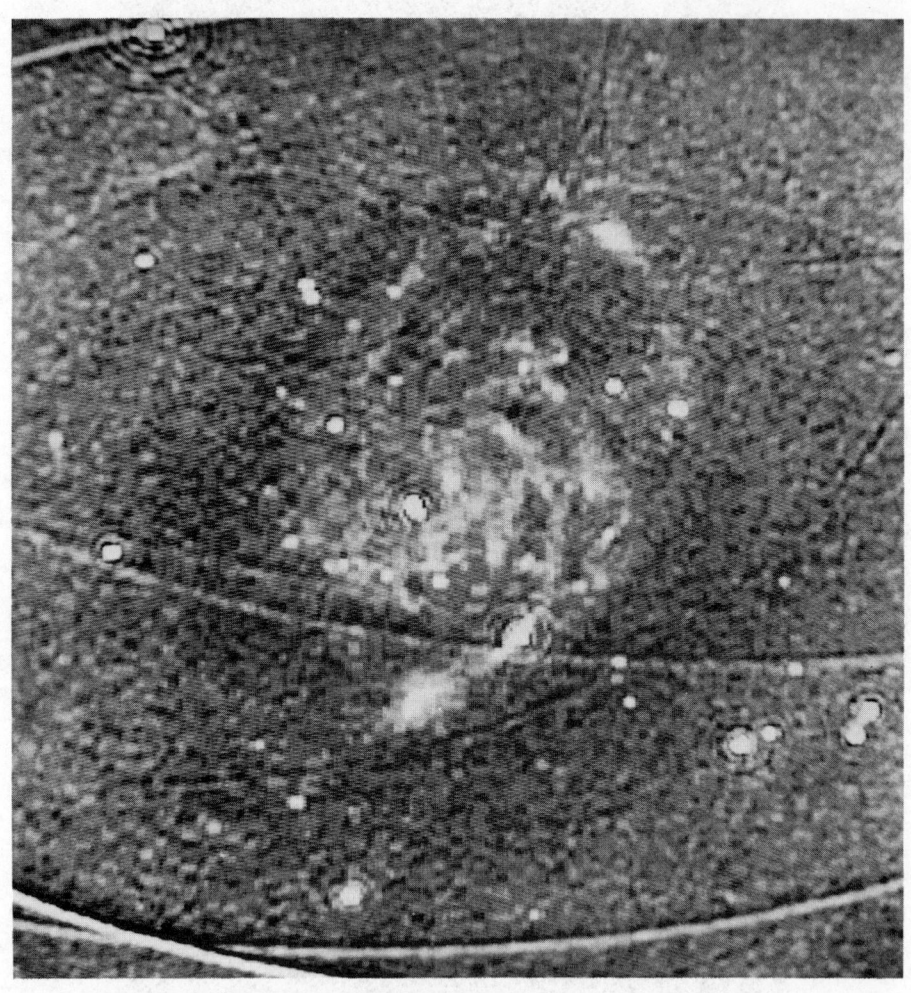

Fig. 5 Distribution of the radio continuum emission from M101, as observed with the Westerbork Synthesis Radio Telescope in the Netherlands at a wavelength near 21 cm. North is to the left. The resolution is 13 × 16 arcseconds. The arc-shaped features crossing the image and the small concentric rings surrounding discrete sources are instrumental effects which can be removed by further processing. This image can be compared with the optical picture in Figure 4, and with the old, much less sensitive result shown in Fig. 5 of Ekers et al. (1973). From unpublished work by Viallefond and Goss.

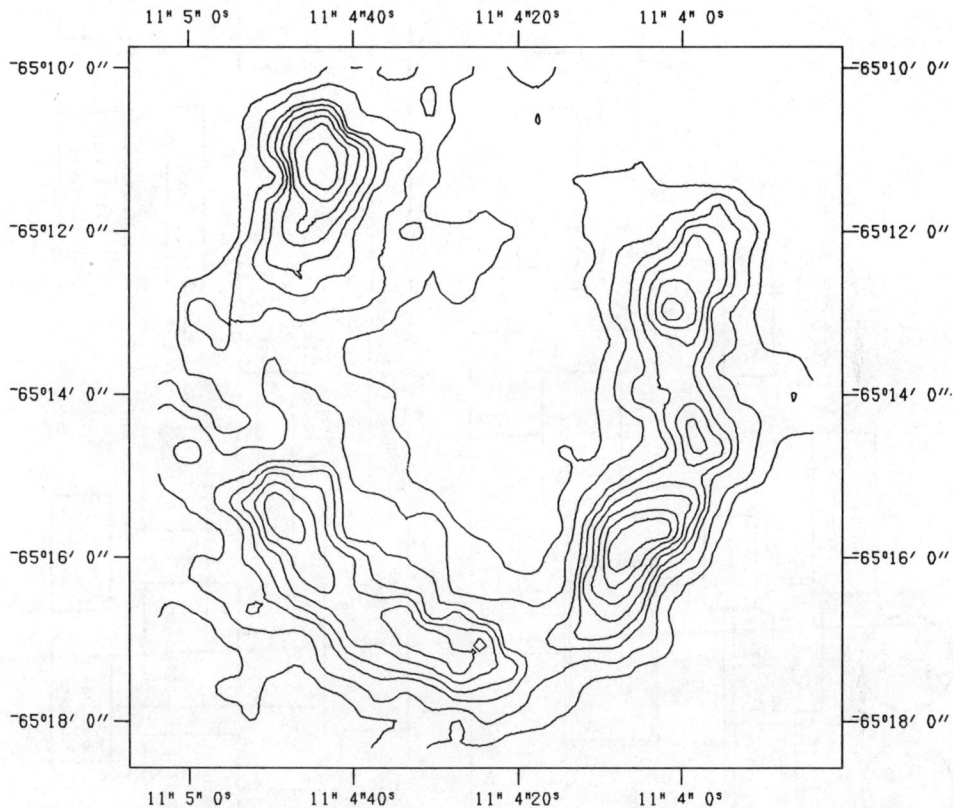

Fig. 6 Distribution of the infrared continuum emission at a wavelength of 50 microns from the galactic ring nebula RCW58, as observed with the Chopped Photometric Channel on board the Infrared Astronomical Satellite. The resolution is 1.5 × 1.5 arcminutes. The contours show emission by dust, which has a colour temperature of about 30 to 40K. There is a good correspondence with Hα features of the emission line images published in Chu (1982). The radius of the nebula is about 2.4 pc. From unpublished work by van der Hucht, Jurriens, Wesselius, and Williams.

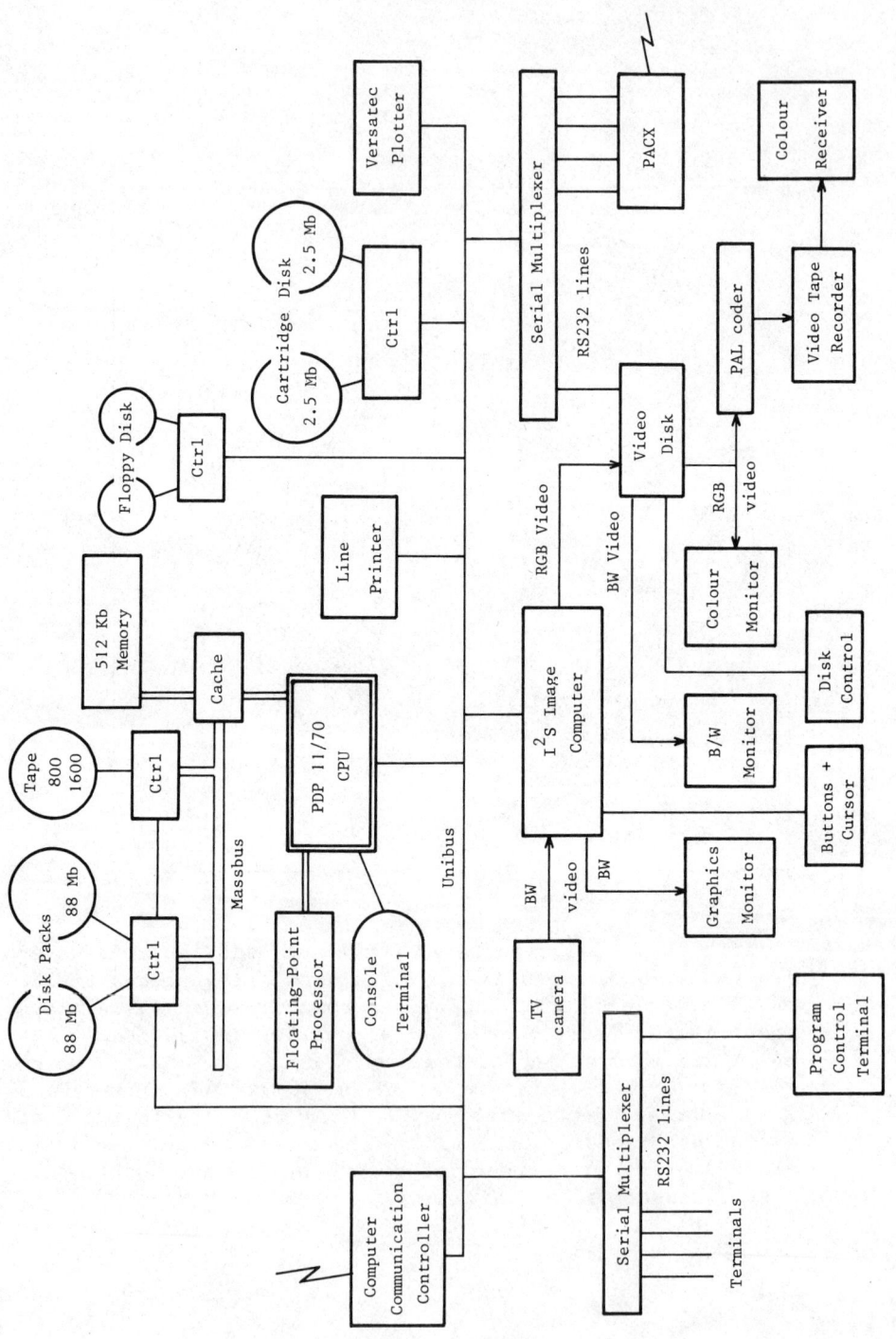

Fig. 7 GIPSY hardware (1983). Arrowheads denote analog signals.

colour, zoom, etc.) and cursor interactive functions, the device is capable of performing arithmetic operations on 512 × 512 8-bit and 16-bit signed and unsigned pixels at video rates (100 nanoseconds per pixel), and of feeding the results back into its own memories. The performance of the machine for such calculations has been evaluated previously (Allen, 1979), and an implementation of the CLEAN algorithm commonly used in radio astronomy has also been described (Allen, 1983). Storage of graphic images is provided in six memories of 512 × 512 1-bit pixels. Graphics can be overlaid on colour or grey pictures, or displayed on a separate screen.

Video Disk Recorder. This is an analog device[b] used for storage and cinematographic playback of processed images. It has a capacity of 100 colour or 300 grey pictures. Controls are provided for displaying sequences of pictures at various speeds up to the full rate of 25 frames per second. The video disk is used for:

- reconnaissance of large 3-dimensional data bases, by coding one of the principle axes as a time sequence;
- temporary storage of intermediate results for visual inspection during a lengthy data-processing session;
- blinking between two stored images in order to discover differences; and
- storage of processed images for later recording on film or on video cassette.

Electrostatic Plotter. This device[c] provides high-quality hard copy for graphics images such as contour plots and profiles, and working copies of half-tone images. Colour hard copy is obtained by photographing the display screen.

2. Software Architecture

The general structure of GIPSY software is analogous to (but simpler than) the combination of operating system, utility programs, and compilers found in any general-purpose computer. The major elements are:
- a master control program to handle user terminal commands and to initiate and monitor the application programs;
- a three-dimensional data base;
- a set of input-output interface subroutines to allow application programs to communicate with the user terminal and the data base;
- a structured high-level programming language;

---

b. Model RP3332B, Image Processing Systems, 70 Glenn Way, Belmont, CA 94002.
c. Versatec V-80.

- a standardized method for adding new program modules to the system library; and
- a structured documentation scheme.

Application programs. These are isolated from the user, data base, and master control program by a set of layered interfaces. This structure is intended to reduce the level of connectivity of the whole sytem, according to the precepts of the programming "principles" discussed by Allen and Ekers elsewhere in this volume. The general structure of a GIPSY application program is shown in Figure 8. The code written by an astronomer for his particular problem is symbolised by the inner dashed box (number 3) in this Figure; a structured Fortran-like language called Sheltran (Croes and Deckers, 1975) is used for most of our programming. Sheltran forces the programmer to write his code as a set of closed control structures, and provides additional facilities for improving the readability of the program listings. The application program communicates with the rest of the system through a small number of standard interface subroutines (dashed box numbered 2 in Figure 8). Finally, the program along with all of the interface and utility subroutines is built into an independent unit of executable code which requires only the master control program and the host computer operating system in order to run. In this way, faults in one application program are prevented from having a major impact on the system as a whole.

Master Control Program. This control program was introduced into GIPSY in 1979. The motivation was to improve and simplify the interface between the user and the applications program which he wishes to run. The design specifications have been discussed in more detail by Allen en Terlouw (1981). The main activity of this program turns out to be transferring messages between application programs and the user, between application programs and the operating system utilities, and among application programs themselves; hence its name "HERMES". Its services to the user are:
- to relieve users who are not computer programmers from the necessity of learning how to handle the computer manufacturer's operating system;
- to provide a simple overview on the terminal screen of the status of all application programs which are running;
- to standardize the appearance of all application program dialogue; and,
- to keep a "log file" of all activities and to allow the user to present any page of this file at random on his terminal.

The user services provided by HERMES will be further described later in the section on the user interface to GIPSY. We mention here several other features which were not originally basic design elements but which have turned out to be significant:

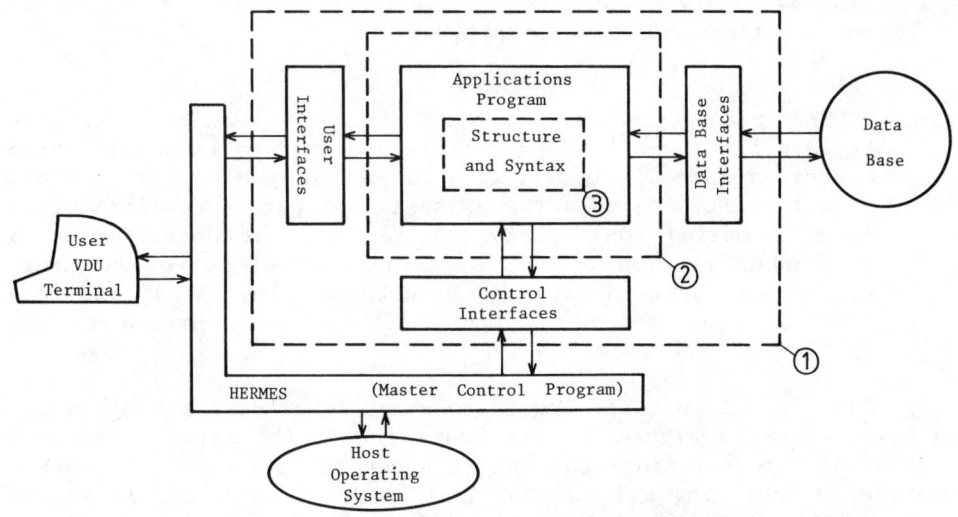

Fig. 8 GIPSY applications program structure.

- HERMES is in a position to keep and update command strings for each application program which it has run. These text strings contain the most recent values for parameters in the form of all KEYWORD = VALUE pairs typed by the user for a particular application program. A macro facility can now be provided by giving this string an abbreviated name. Application programs can then be easily re-run with parameters from the last execution plus any desired current modifications.

- Viewing the message traffic in the other direction, HERMES is also in a position to analyse requests coming from an application program and decide whether to present the query to the user on his terminal screen or to provide a reply from some other source. This opens the way for a rather elaborate scheme of defaults for parameters, which may be conditioned on the data and on parameters obtained previously. These defaults are controlled from within the application program, as will be described in the section on the user interface subroutine to follow.

- Since HERMES interprets user terminal input on a character-by-character basis, it can respond to special single-keystroke commands (which are not echoed on the screen). For example, the "TAB" key causes the display of the previous page in the log file on the terminal screen, while "control F" starts a program to produce a grey representation of the display screen on the electrostatic plotter.

- HERMES provides also a central <u>format conversion</u> facility between the formats of parameters required by the application program and the formats available in the command text string. This facility includes the expansion of parameter lists given in an "implied DO-loop" format, as will be described in a subsequent section on the appearance of GIPSY to the user. Also, user typing errors can be detected immediately and corrections requested before the information is passed to the application program.

<u>Data Base Structure</u>. The three-dimensional data base is shown schematically in Figure 9 as a "data cube". The actual implementation of this structure on disk is shown in Figure 10. The entire data cube including all astronomical header information is stored as one standard disk file using file definition and maintenance utilities provided by the host computer operating system. Data sets are referred to by a "set number", and the map plane slices (z axis) of the data cube by a "subset number". The implementation is a compromise between conceptual generality, speed of access, and simplicity.

<u>Interface Subroutines</u>. GIPSY interface subroutines are divided functionally into three categories: user, data base, and control. We discuss here as an example one major subroutine from each category.

USRINP: This subroutine is used for obtaining parameters from the user. It provides the programmer with the following functions:
- interrogates the user terminal for a (list of) parameter(s) if that parameter (as designated by its keyword) is not otherwise available;
- provides the parameter in a format chosen by the programmer, independent of what the user actually typed;
- handles all possibilities for default values of parameters;
- gives immediate syntax error messages to the user, so that the returned parameters are at least correct in that respect;
- saves "keyword = parameter" pairs for subsequent use in the application program; and,
- adds text strings recording these transactions in the log file.

The subroutine is called by the application program as:
CALL USRINP (ARRAY, NELS, TYPE, LENGTH, DEFAULT, KEYWORD, MESSAGE)

KEYWORD is a text string used as a prompt, as a label for the parameter string given by the user, and as a label to store parameters in a "macro" to be

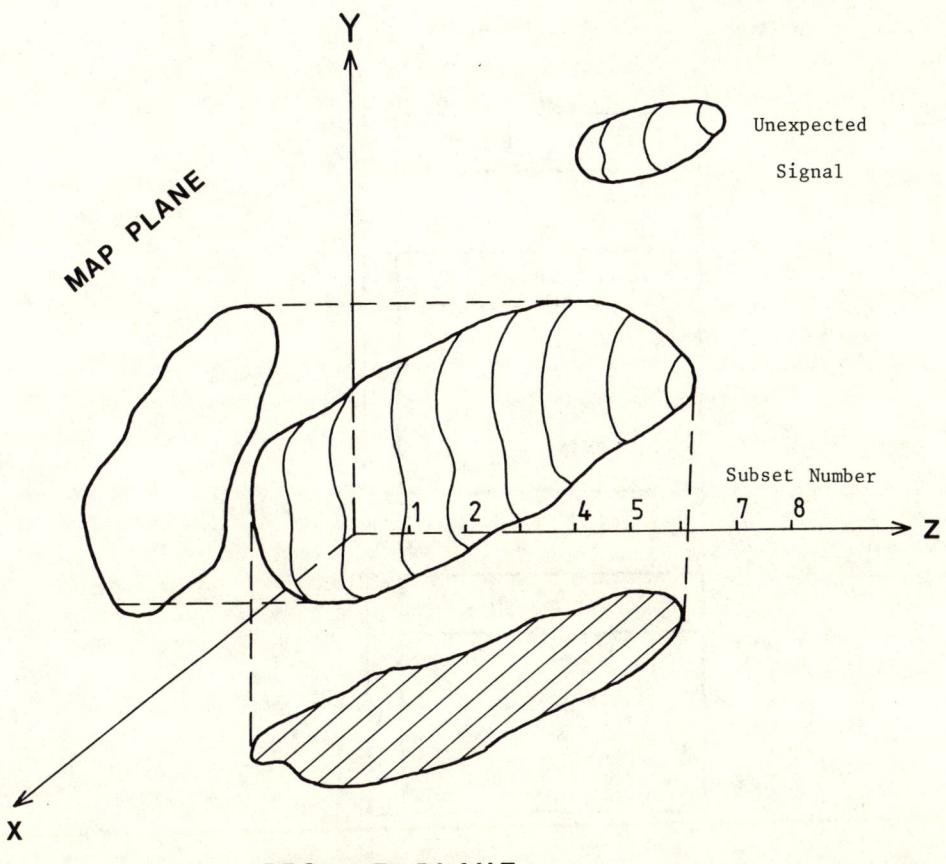

Fig. 9 The three-dimensional data structure or "data cube" of intensity or number of photons.

used as defaults in a future execution of the program. DEFAULT is an integer defining what kind of defaults the applications programmer will accept:

- DEFAULT = 0 means no default is possible. The user is prompted and must reply with a parameter. An example is when requesting the set and subset numbers of an image.

- DEFAULT = 1; the user is prompted, and if he replies with a carriage return the current value of the parameter (preset by the programmer before calling USRINP) will be returned as a default. An advantage of this construct is that defaults can be computed from within a running program, based on previous-

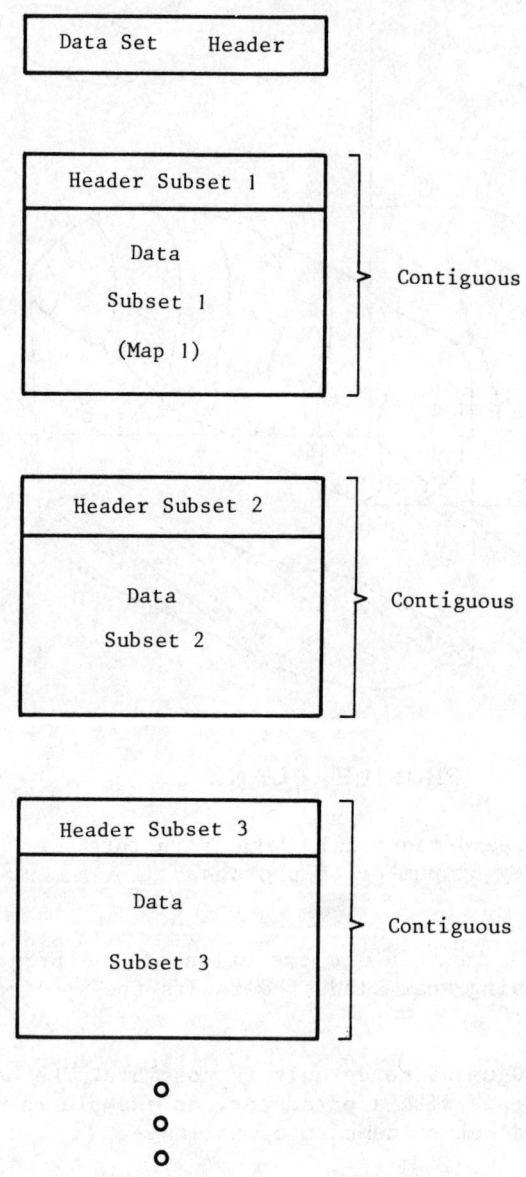

Fig. 10 Implementation of the data base as one file.

ly-entered parameters and on the data itself. An example is to default to the full size of the image (determined by the program from reading the image header) for some numerical operation which may in general be done on a restricted area.

- DEFAULT = 2 is intended for all parameters which the average user does not normally need to know about or to change. If it really is necessary to change them, this can be done by pre-specifying the keyword and parameter list any time before the program arrives at the particular call to USRINP; otherwise the application-program-selected defaults are automatically taken. Keywords with this level of default are often referred to as "hidden keywords", since under normal circumstances no prompts appear on the user's terminal. They are also very useful for testing programs.

Note that the user may give a parameter string at any time. The master control program saves the information as an ASCII string until the applications program requests it.

The default facilities described above, coupled with the capacity of the master control program to save command strings, provide a high degree of flexibility for both novice and experienced users of GIPSY. Novice users usually wish to use only the standard features of an application program; they appreciate being extensively prompted for relevant parameters, and are happy to have the program choose sensible defaults by itself. Also, they do not want to be asked for parameters which are not necessary. More experienced users quickly find such prompting to be a nuisance, and prefer to prespecify parameters or use a private set of defaults defined in a text string when starting a program. In this way they can also use special features of the program accessible via hidden keywords.

Flexibility in the interpretation of defaults is also available to the GIPSY programmer. For instance the "carriage return" default = 1 can be taken to mean different things in different parts of a program even if the keyword is the same. As an example, the program DISK begins by asking the user to provide the keyword SET=; carriage return is interpreted to mean that the user wants an abbreviated list of <u>all</u> data sets available to him. After providing this list, the program then prompts for SET= again, but now a carriage return stops the program. In both cases, if the user replies with a specific set number the program gives him a description of all subsets contained in that data set.

<u>READXY</u>: This subroutine is used to read a piece of the data cube from disk storage in the map plane (see Figure 9). The data set is specified by a SET number, and the location on the z-axis by the SUBSET number. The call is:
CALL READXY (SET, SUBSET, XBEG, YBEG, ARRAY, NX, NY, IER)

The frame to be extracted from the given subset is defined by its beginning coordinates XBEG, YBEG, and size NX, NY. The data is returned in ARRAY and an error code in IER. NY may be 1 if only one line of data is required. The coordinate system used for subsets is the normal, right-handed system which one learns in primary school, with the point (0,0) at the center of the image; XBEG, YBEG therefore refer to the lower left corner of the frame to be extracted. GIPSY application programs normally do not concern themselves with opening and closing files of image data. The interface and control routines handle this automatically; if not already open, a data file is opened at the first access request and closed when the program terminates. Finally, we note that pixel values are stored on disk and manipulated in programs as floating-point numbers. This entirely avoids requiring programmers to pay attention to integer overflow problems (which were a headache in our first-generation system), or to write both fixed- and floating-point versions of their code.

DEPUTY: This subroutine allows one application program to run a second application program, using the parameter list of the first program. The first program may add to its parameter list any keywords which it wishes to pass using the subroutine WKEY before calling DEPUTY. It may be convenient (or even necessary) to rename keywords using the subroutine SUBST before calling DEPUTY. For instance the output data set identified by the keyword SETOUT= can, by this means, be associated with the input SET= of a subsequent program started with DEPUTY. The call is:
    CALL DEPUTY (TASK),
where the one input argument is a text string giving the name of the application program to be started. Execution of the first program is normally suspended until the second program has finished. The second program may transmit parameters back to the first by using WKEY, provided the keywords are then requested (by calls to USRINP) after the first program has resumed.

Notes on keywords and control subroutines. The parameter string for a given application program is private to that program; GIPSY keywords are local. This is done in order to minimize the connectivity of the system. Programs which are started by other programs do share keywords; however, the system treats such "spawned" programs as a single entity with the name taken from the first program. The user is therefore confronted only with that name, and is unaware that the work may be carried out by several different application programs in succession. A program failure at the end of the chain causes HERMES to abort all other programs in that chain.

Updating the software library. Programs which are ready to be made available for general use are inserted into the system library by means of a standard procedure. This procedure, which may be operated by the programmer himself, is automatically started by logging on the host computer operating system with a special password. The procedure is set up in an interactive program which requests the names of the input files, processes these through compilers, builds the task (executable program) if necessary, and writes the results to the program library on disk and to backup media. It also makes entries in a history file where all permutations on the system library are recorded. After this procedure has been run successfully, the programmer deletes all files pertinent to that program which he may have in his own disk space. This saves on space, but also guarantees that when someone else may wish to modify that program he may be confident that the system library contains the most recent, operable version of the source code. The update procedure also provides for deleting programs and documents which are no longer required.

Documentation scheme. This scheme is divided into three levels, designated 1, 2, and 3:
- Level 1 documents provide general information and descriptions of how to operate the various application programs. Users need only be familiar with documentation at level 1;
- Level 2 documents provide descriptions of all interface subroutines to the user terminal, data base, and for control. At this level we also find various utility subroutines and slave tasks, and descriptions of how to write level 1 programs;
- Level 3 documents describe software which is specific to the particular configuration of computer hardware and host operating system currently in use. Programs at this level are usually not explicitly called by an astronomer who is writing an application program; they are, of course, called by the level 2 subroutines which he uses.

Documentation examples are given in Appendices C and E.

IV. HUMAN INTERFACES TO GIPSY

As discussed by Allen elsewhere in this volume, a data processing system presents a different appearance to different people depending on whether the system is viewed by a user, a programmer, or a system manager.

1. The User's View

An astronomer who plans to use GIPSY for his data analysis may begin by reading a set of documents collected in the "New User's Documentation Package". Time on the system for production processing is normally allocated to users in blocks of several hours long. Each major user (or guest) of GIPSY receives a private disk pack which he may use for his own data processing; these packs can hold up to about 80 maps of 512 × 512 pixels. After installing his disk pack on the drive, he sits at the control console facing a standard

video display terminal and keyboard (VDU) and flanked by a number of television screens. Logging on the host computer with a special password initiates an automatic start-up sequence which, besides performing a number of housekeeping functions, presents the user with the latest page of the Daily News. This is a computerized text file containing commentary on software bugs and on the latest modifications to the system. The master control program HERMES then takes over, re-organizing the VDU screen from the conventional line-by-line format to the format sketched in Figure 11. The subdivisions are:
- the "User Command Area", consisting of two lines at the bottom of the screen. All printable characters typed by the user are echoed here;
- the "Task Status Area" consisting of 4 lines showing the status of currently active tasks (i.e. application programs);
- the "Common Output Area" consisting of the top 18 lines of the screen. This section is used to display pages of the log file, showing information sent from application programs.

The Common Output Area (COA). The COA shows one page of 18 lines from the log file. These pages are numbered, and on the right edge of the screen in the Task Status Area are shown the page number currently being displayed on the COA, followed by the current page number on which HERMES is writing messages. There are two modes of displaying the log file information: "page mode", and "non-page mode". In non-page mode (the default on startup) the current- and display-page are always the same, i.e. if a new page is initiated, the old one disappears from the screen. In page mode the information on the screen will stay there until the user instructs the system to change the page. There are two classes of commands to manipulate the COA: control characters; and commands terminated with <CR> (carriage return). The control characters always have an immediate effect, regardless of the state of any line currently being typed, and are not echoed in the User Command Area. These page control characters and commands refer only to the display, and do not inhibit tasks from continuing to write in the log file in the normal way. Commands of both types are listed in Appendix A.

The Task Status Area (TSA). This is the middle section of the terminal screen consisting of four lines which may contain status information of "active servant tasks"; they are the application programs requested by the user, or utility programs initiated by HERMES. To the extreme right of the TSA a time-of-day clock and the current- and display-page numbers are given. Some examples of the status information are (asterisks shown blink on the screen):
- When a task has been activated, but is not running yet;
    taskname WAITING TO BE RUN
- When a task is running;
    taskname RUNNING

```
<USER   >TRANS ,SETOUT=
Output  set and subset(s): 0,1
<USER   >TRANS,SET=6 1,SETOUT=
Input   set and subset(s): 6 1
Output  set and subset(s): 0,1
<USER   >TRANS /WAIT
<USER   >TRANS/GO
<USER   >ROTATE
<USER   >ROTATE,SET=6 1
<USER   >ROTATE,SETOUT=
<USER   >ROTATE,SETOUT=0 1
<USER   >ROTATE,POS=
Set    6, 1 from (-127-127) to ( 128 128)
<USER   >ROTATE,AREA=
 COTRAS:instrument unknown.  WSRT assumed.    **WARNING **
             L=-127 TO 128    M=-127 TO 128
<USER   >ROTATE,ANGLE=10

- ROTATE Set   0,  1;  8 percent of map ready                    12:10
- PRINT    * default is 30 * 50 if only the center is given      549  551
- CPLOT    * Type set and subsets (0,1)
- LOOK     USER ABORT
PRINT ,AREA=_____
_____
```

Fig. 11   User VDU terminal screen format.

        or
        taskname status message (supplied by servant task)
- When a task is pausing;
        taskname *PAUSING (optional message)
- When a task requires input;
        taskname *input request message from that task
- When a task encountered a fatal error;
        taskname error message -FATAL
- When a task has crashed (i.e. stopped without notifying HERMES);
        taskname *CRASHED
- When a task has finished processing normally;
        taskname +++FINISHED+++
- When a task has been aborted by the user;
        taskname USER ABORT

The last four messages will disappear as soon as the user initiates another servant task or when the user types ↑R. The command ↑R can also be used to recover the TSA if it is accidentally destroyed.

The User Command Area (UCA). The User Command Area (UCA) consists of two lines located at the bottom of the display screen. It is logically one long line, with 148 typing positions available for user input. Some examples of the commands that can be typed will be described later. HERMES itself can also "type" in this area. It does this for instance in the following cases:
- when it is likely that a "taskname,keyword=" combination will have to be typed, HERMES types it for the user;
- when the user types ↑Y or ↑T, a cursor processor is activated. When the cursor position has been read out, the coordinate values are typed in the UCA and can therefore appear as keyword parameters for input to another (active or still to be activated) task.

Text present in the UCA which is either erroneous or otherwise unwanted is simply deleted with the DELETE key (or ↑U to erase the whole line). For instance in this way a task waiting on input via a "taskname,keyword=" request in the UCA need not be satisfied before initiating some other task. ↑U erases the text, and a new command can be given by the user. If the UCA is empty, striking the space bar causes HERMES to cycle through presenting the currently unsatisfied keywords of active tasks.

At the end of the UCA, HERMES displays error messages which are related to the line typed by the user (e.g. BAD SYNTAX, ALREADY ACTIVE, NOT PRESENT, etc.). When a line is accepted by HERMES, it will disappear from the UCA; if it is rejected (as can be seen from the error message) it stays on the display, but will disappear as soon as the user starts typing a new line. Rejected lines are not displayed in the COA and do not appear in the log file.

Finding the documentation. Copies of all user documentation are kept in loose-leaf binders at the control console. This documentation includes general information as well as descriptions of all Level 1 application programs. Copies of these documents can be produced on the VDU screen and printed in the log file by entering the command:
    HELP   TASK= program name.

Several frequently-used tasks can be initiated using the (programmable) function keys located across the top of the VDU keyboard.

Giving Parameters to Tasks. If the user does not pre-specify any parameters when initiating a task, the task will prompt him for the necessary information as it proceeds or else take default values as indicated in its documentation. As the user becomes more experienced the screen dialogue and delays involved in answering these prompts can be circumvented by pre-specifying parameters, for example as follows:

    PRINT SET= 12,4   AREA= FC DG 20 20
This command will initiate a task to print the pixel values from an area of 20 × 20 grid points centered on the field center of the map in set 12, subset 4. The KEYWORD-PARAMETER pairs may be given in any order, separated by commas or blanks. HERMES will save the keywords, giving the associated parameters to the application task when it requests them. Several frequently-used keywords such as SET= and AREA= are available as single-key commands by means of the row of function keys across the top of the terminal keyboard. A task which has been previously run can be repeated with the same parameters by typing @TASKNAME. Any selected parameters the user wishes to change can simply be appended; for instance if the user wishes to repeat the execution of PRINT described above but on a different map he need only type
         @PRINT SET= 6,3
which will print the pixel values from the map in set 6, subset 3, over the area given in the immediately previous initiation of the task PRINT. Note that the KEYWORD=PARAMETER pairs for a given task apply only to that task and not to any another task, even though the keywords for different tasks often have the same names.

    Formats for Input Parameters. The format for input values is completely free; HERMES takes care of converting the numbers into the formats actually required by the application task. Any of the legal Fortran formats are accepted; otherwise list elements are assumed to be character strings. If a character string has been requested by a task, no conversion is made. For real and integer values the user may give a repetitive list in the form s:e:i where s = starting value, e = ending value, and i = increment. For instance the keyword for contour levels
    CNT= -3:10:2
means the list of contour levels -3, -1, +1, +3, +5, +7, +9.
Coordinates which have been requested as parameters can be given to any task in a variety of forms, as specified by a prefix followed by a blank. For instance
    POS= * 13 27 6.4   23 34 10
designates a position at right ascension (RA) = $13^h27^m6\overset{s}{.}4$, declination (Dec) = +23° 34' 10". The prefixes currently available are:
   G   grid units in RA and Dec with respect to the map center,
       e.g. G 10 -12. This is also the default if no prefix is
       given;
   *   celestial coordinates RA and Dec in hours, mins, secs and
       degrees, mins, secs;
   D   RA and Dec in degrees;
   O   RA and Dec in degrees offset from the map center;
   RP polar coordinates from map center in degrees;
   FC celestial coordinates of the map center; and
   LB galactic coordinates in degrees.

A rectangular framed area of a map may be specified either as the lower left and upper right corner or as a center and size; HERMES knows which of the two you mean. For instance:
   AREA= -10 -12 +4 +6.
   or
   AREA= -3 -3 DG 15 19
both refer to the frame with lower left corner X,Y = -10, -12 and upper right corner +4, +6. Two further prefixes are available for areas:
   DD   size in degrees of RA and Dec; and
   DG   size in grid units.
For instance the command
   AREA= FC DG 11 11
is equivalent to
   AREA= -5, -5, 5, 5
and refers to an area of 11 × 11 grid points located at the map center.

Using the Cursor. A very convenient way of obtaining position and amplitude information from an image is to use the cursor, moving it by means of the trackball and selecting the desired parameters with the function keys on the cursor box. The cursor task can be started with ↑Y or ↑T. The first cursor is a small cross which comes equipped with the grid coordinate values of the pixel in the image under the cursor; the second is a "full screen" cursor to be used with vector plots such as contour diagrams. Information about the coordinates and amplitude of the pixel under the cursor is continually updated in the task status area. Depressing function keys on the cursor box causes parameters to be "typed" in the user command area, as follows:
- button A gives the current grid coordinates of the cursor in the order X, Y;
- button B gives the coordinates in right ascension and declination;
- button C gives the pixel amplitude; and
- buttons A and B together terminate the cursor task.

Note that the parameters entered into the UCA in this way are the same as if the user had typed them himself; characters may be erased (with the DELETE key) and new characters inserted until the desired values are obtained.

Finishing the session. Since the results of his data analysis are kept on his own private disk, the user may easily interrupt his computer work and return to it in the future. The session is closed by initiating the task BYE, which also will take care of saving, or printing and deleting, the log file accumulated during the analysis session. If kept, the log file remains on the user's private disk; at the following startup, the system will continue at the same place it left off.

Feedback from the users. It is virtually impossible to test each application task for reliable operation under all combinations of input parameters and data. Failure of specific tasks, or suggestions for improvements of their operation or documentation must be brought to the attention of the person currently responsible for that task (see the documentation of the task) by using a Software Failure Report. A stack of blank forms is available at the GIPSY user console. The original, along with all supporting documentation (e.g. relevant pages from the log file), is to be given directly to the person concerned; the carbon copy is to be added to the stack kept in the accompanying folder at the user console in order to indicate to the System Manager and to subsequent users that the problem has been reported. All active users of GIPSY are strongly encouraged to attend the weekly Software Lunch Meetings, where suggestions for improvements and current problems are discussed.

2. The Application Programmer's View

Although there are currently about 180 application functions available in GIPSY (concentrated in 155 programs, cf. Appendix B), the methods of data analysis become more complex with the experience gained and there is continual pressure to modify and improve existing programs. In addition, GIPSY is now being used for the analysis of optical and infrared astronomical data as well as the radio astronomical observations for which the first set of application programs were designed, and new software is required for this. Many users therefore become programmers; in this section we describe GIPSY from that point of view. The first step for a new programmer is to read over the contents of "The New Programmer's Documentation Package".

Presenting the Problem. It's important that a proposal to write a new program be discussed with other astronomers using and programming GIPSY. The reasons for this are:
- the required function may already be available, perhaps as a combination of other functions;
- someone else is already working on the problem; or
- others need such a program too, and would like to offer suggestions about how to write it.

The regular weekly Software Lunch Meeting is the occasion to present a proposal for a new program. Such a proposal is best made in the form of a draft of the user documentation for the program, describing its function and all the keywords. This gives others a chance to propose improvements before such changes would require rewriting the code.

Software Conventions. Programs written for GIPSY are expected to conform to a number of conventions and standard practices which have been adopted in order to improve the maintainability and

durability of the software:
- use a structured programming language. For those who are used to programming in Fortran, the Sheltran language is an excellent alternative which is quickly learned and strongly supported in GIPSY; and
- become familiar with the function and features of the standard level 2 subroutines which are used to interface GIPSY application programs to the data base, to the user, and to HERMES.

In addition to these general practices, there are a number of specific syntactical conventions to be observed. Some examples of these are:
- make frequent use of subroutines, or of Sheltran procedures if subroutines are not convenient. They improve the modularity of the program and make it easier to read;
- avoid the use of COMMON to transfer information to private subroutines;
- subroutines which are private to the program should be kept in the same source code file as the main program itself;
- use PARAMETER statements to define array dimensions, buffer sizes, FOR loop limits, etc. This facilitates modifications in the future;
- avoid EQUIVALENCES;
- declare the TYPE of all variable and arrays explicitly; and
- do not use multiple ENTRY or RETURN points in subroutines.

Documentation Conventions. As we have suggested earlier, the user documentation for a new application program should be the first thing to be written, before the program is coded. The standard format for the user documentation is given in Appendices C and E. Descriptions of how the program works belong in the code of the program itself; see Appendix F for an example. As a good rule of thumb, there should be as many lines of description in the program as there are lines of executable code.

Coding and Testing. An applications programmer receives a private password for the computer and a modest allocation of disk space on which to develop the software. Three source files have to be constructed:
- the application program SHELTRAN code;
- a command file with the instructions for building the task;
- the final user documentation.

When the program has been compiled and the first task successfully built, a test is needed. This takes priority over production data analysis during working hours. Getting HERMES to run a test program is easy; the program name is merely preceded by the programmers own identification under which the test version of the application task resides. If full error message reporting is desired in order to catch difficult bugs, this can be switched on at start-up by choosing the appropriate answer to the start-up question.

<u>The Referee System</u>. Software written for GIPSY is subject to peer review; a referee can be selected by the program author. Referees examine the source code and the accompanying user documentation for comprehensibility; they are not required to check the program operation at the user console.

<u>Updating the Software Library</u>. This topic was described previously in section III.2 on the software architecture. After the procedure has run successfully, the programmer is given the opportunity of adding some prose to the DAILY NEWS in order to inform others of the changes and additions. Since all the source versions are now safe on the backup disks and the task itself is in the GIPSY library, the programmer can delete all files pertaining to that program under his own identification. This will be required anyway, since the number of blocks available to him on the system disk is too small for more than a few software projects.

3. The System Manager's View

The main job of the System Manager is to promote communication amongst the users and programmers. There are a number of mechanisms in GIPSY for accomplishing this:
- the list of current software projects;
- the Daily News;
- software failure reports;
- the weekly software discussions; and
- the referee system for new software.

Except for the first item, these topics have been mentioned earlier. The list of current software projects is an important working document for the system manager. It is a computer text file which he maintains, with a brief entry describing the nature, current status, and name of the persons responsible for new programs under development and for improvements to existing software.

In the course of the development of GIPSY over the years we have found it useful for the system manager to have the close assistance of two colleagues with rather different specializations:
- a professional programmer familiar with the operating system of the host computer but not concerned with analysing astronomical data; and
- a senior user intimately involved with astronomical applications but not concerned with writing software. This person acts as the "friend of the users".

V. AN EVALUATION

As we have stated in the introduction, we view the present version of GIPSY as the product of more than 13 years of continuous development. We are now on the verge of moving to our fourth host computer and operating system, and it is an appropriate moment to sum up the current status.

1. System Size and Cost

An impression of the size of the system software can be obtained from Tables 1 and 2, which provide a breakdown according to level and type of the programs and documentation. We continue to invest manpower at the rate of 3 to 4 man-years per year, a typical value since the start of the project. The average maintenance and development capacity is about 35,000 lines of source code plus documentation per man.

The hardware and software costs are summarized in Table 3, integrated over the expected 8-year lifetime of one generation of computer hardware. In this time span, the hardware and software costs are about equal.

2. Good Features of GIPSY

From the experience of users and programmers we note the following positive points about GIPSY:

- there is a large number and great variety of application programs, and the system has the flexibility to allow them to be combined in novel ways;

TABLE 1

Breakdown of GIPSY software on the host computer system disk as of 10 October 1984 by level, including binary tasks, (level 1 only), source code, object libraries, and documentation:

Level	Description	Number of Programs	Total Size (Megabytes)
1	Application Tasks	155	13.2
2	Interface and Utility Subroutines	116	0.8
3	Hardware and Operating System Subroutines	96	0.3

TABLE 2

Breakdown of GIPSY software on the host computer system disk as of 18 October 1984 by type. The typical packing density for source code is 38 lines per kilobyte and for documentation 25 lines per kilobyte.

File Type	Number of lines
SHELTRAN source code	83,227
FORTRAN source code	13,001
MACRO source code	2,548
TOTAL	98,776
Documentation — Level 1	16,597
Level 2	8,205
Level 3	3,063
TOTAL	27,865

TABLE 3

Rough Costs of GIPSY in millions of Dutch guilders over 8 years (1977 - 1985).

Hardware items	Cost
PDP 11/70 with peripherals	0.90
I^2S image computer and IPS video disk	0.42
Total Investment (1977)	1.32
Contract maintenance/yr	0.1
Local engineering help/yr (1 day per week)	0.02
Total maintenance costs (8 yrs)	0.96
Total Hardware cost over the lifetime of the project	2.28
Image processing software development costs, 3.5 man-years/year over 8-year lifetime of the project (3.5 × 0.08 × 8). Total software costs	2.24

- simultaneous operation of several application programs is quite useful, and the system provides a good overview of what is going on;
- the use of private data disk packs is much appreciated by the users (it avoids a lot of tape operations and gives them added flexibility in carrying out lengthy data processing jobs), and relieves the system manager from the odious task of allocating and recuperating space on the system disk;
- the software "environment" of interface subroutines eases the job of developing new application programs;
- it is easy to test new programs;
- system maintenance is simplified by the standard updating scheme with automatic backup;
- the standardization on floating point numbers simplifies the software; and
- the adoption of common programming standards has led to an integrated system with a high level of modularity and internal consistency.

3. Present Inadequacies

There are a number of areas for improvement in the future:

- the implementation of the data base structure on disk is not sufficiently flexible. For instance, the astronomical headers are of fixed length and cannot be extended, and subsets cannot be separately deleted from a data set;
- many users are of the opinion that it would be helpful to their bookkeeping to have a history file attached to a data set. This file lists the processing operations carried out on that set in chronological order;
- many application program tasks are so large that they must be built up of overlaid segments in order to fit in the 56 kilobytes of address space available (after subtraction of 8 kilobytes shared region required for inter-task communication). Constructing overlays requires an unnecessarily detailed knowledge of the computer hardware and operating system and is prone to errors;
- the image computer is not fully integrated into the data base structure; and
- the application programs currently available are quite heavily oriented to the analysis of radio astronomical images, both maps and spectral line profiles. More application software is needed for handling optical images, both surface photometry in the map plane and spectra in the profile plane. The software currently available is useful for radio data cubes from the Westerbork telescope and the VLA, for optical emission-line data cubes obtained with TAURUS, and for sky maps obtained with IRAS. GIPSY is not yet able to conveniently handle IPCS spectra, galaxy photographic surface photometry, or radio

single dish line profiles or interferometer visibility data.

Finally, we mention several points occasionally presented as inadequacies of GIPSY, but which we consider debatable:

- the master control program does not allow user access to operating system utility programs;
- the programming language (Sheltran) is not widely known;
- communication between tasks is limited to "Keyword=parameter" pairs. Large data buffers cannot be transferred, and synchronous simultaneous operation of tasks is not supported;
- the system does not support "minimum match", so that task names and keywords must be typed in completely; and
- the "HELP" documentation is limited, so that a new user has to know quite a lot about the system before he can get started. There are not enough "Documents about Documents".

ACKNOWLEDGEMENTS

We owe a great deal to the graduate students and staff members who have used, programmed, and criticized GIPSY over the years. Many good aspects are due to them, and they have helped us understand why many of our own ideas were not as good as we thought. Several staff members have made major contributions to the application program library and deserve special mention: Miller Goss, Ulrich Schwarz, and Seth Shostak. Renzo Sancisi has offered many excellent suggestions for improvements (many of which are unfortunately still pending), and has acted very effectively for quite some time as the "user's friend".

The manuscript and illustrations were prepared with the assistance of Ineke Rouwé, Gineke Alberts, George Comello, Wiebe Haaima, and Geert Tamminga.

The PDP 11/70 computer is provided and maintained by the Groningen University Computer Center. We are grateful to the director and his staff for their continued cooperation.

REFERENCES

Allen, R.J. 1979a, in Image Processing in Astronomy, ed. G. Sedmak, M. Capaccioli, and R.J. Allen (Osservatorio Astronomico di Trieste), 233.
Allen, R.J. 1979b, in Image Formation from Coherence Functions in Astronomy, ed. C. van Schooneveld (Reidel, Dordrecht), 143.
Allen, R.J. 1983, in Three-Day In-Depth Review of the Impact of Specialized Processors in Elementary Particle Physics, ed.

Istituto Nazionale de Fisica Nucleare Sezione de Padova (Padova, Italy), 323.
Allen, R.J., Atherton, P.D., Tilanus, R.P.J. 1985, in The Milky Way Galaxy, ed. H. van Woerden et al. (Reidel, Dordrecht), 275.
Allen, R.J., Goss, W.M. 1979, Astron. Astrophys. Supp. 36, 135.
Allen, R.J., Terlouw, J.P. 1981, in Proceedings of the Workshop on IUE Data Reduction, ed. W. Weiss (Observatory of Vienna), 193.
Brooks, F.P. Jr. 1975, The Mythical Man-Month (Addison-Wesley).
Chu, Y.-H. 1982, Astrophys. J. 254, 578.
Croes, G.A., Deckers, F. 1975, Informatie 17 (Netherlands Society for Computer Science, Paulus Potterstraat 40, Amsterdam), 109.
Ekers, R.D., Allen, R.J., Luyten, J.R. 1973, Astron. Astrophys. 27, 77.
Simkin, S.M., Bosma, A., Pickles, A.J., Quinn, G., Warne, D. 1983, Wisconsin Astrophysics Series (Univ. Wisconsin).
SGP/7, Starlink General Paper 7, 1 July 1980 (Rutherford and Appleton Laboratories Computing Division).
SGP/18.1, Starlink General Paper 18.1, 11 September 1981.

APPENDICES

Owing to space limitations, the appendices to this paper are not reproduced here. They are available from the authors.

PANEL DISCUSSION: SYSTEMS FOR DATA ANALYSIS

WHAT THEY ARE; WHAT THEY COULD BE?

        Chaired by P. Crane

        Participants: P. Benvenuti,
                         R. Allen,
                         P. Wallace and
                         D. Wells

CRANE: I'd like to pose a couple of questions: (1) Command Languages - A tool for the astronomer or for the programmer? (2) Portability - Holy Cow or Red Herring? I propose that we start with the first one and see how far we get. If we don't get past that, fine. If we get on to the question of portability, this is also fine. Let me just open up the discussion by asking Rudi Albrecht to make a comment.

ALBRECHT: Of course, command languages started up as a tool for astronomers. It's a convenient way to get the system to do what you want, in terms of data analysis, in terms of calculating magnitudes, equivalent widths, and so on. A recent development is to use the command language of the image processing system as a tool for the development of application programs. This developed out of a user driven need. What happened was that users asked the system developers to provide them with a tool to string up those commands, to put them in a file, to edit the file, to submit them as a batch process, and that led into command languages that were programming languages in their own right.

ALLEN: I'd like to make a drawing (Fig. 1). What's the matter with the original idea of saying Run X? That's the way you would do it, just in a normal operating system. I would call it a very high level of aggregation of an application program or a system. In other words, it doesn't have command language at all, it just

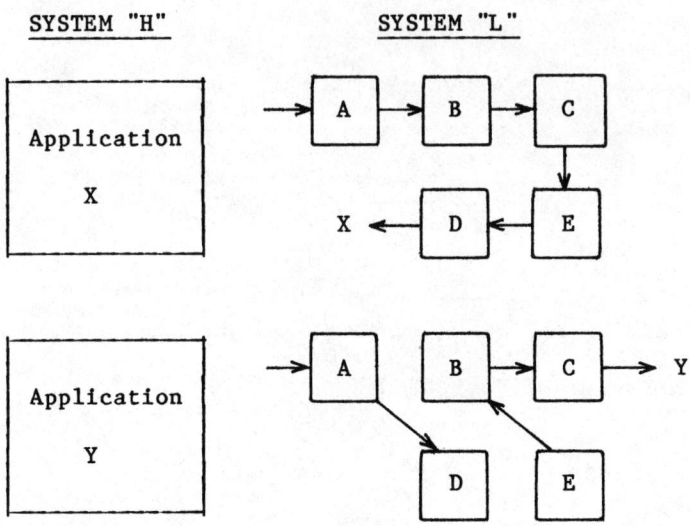

Figure 1: <u>Left side</u>: system with a high level of aggregation
<u>Right side</u>: system with a low level of aggregation

uses the command language of the computer. And the other one is where the modules are very tiny. You hook them together with a very sophisticated command language. These two are very extreme examples. To illustrate they're very extreme you could say, with this model we might write four programs: add, subtract, multiply and divide that would operate on an image and around that a complicated language which would allow me to compute sine and cosine by series expansion. People are smiling already because they know that the implementation would be rather expensive. What we would do if we had that? We would want to build a compiler so that once the astronomer had gotten his procedure the way he wanted it, he could compile and boot it over to the system on the left in Figure 1. So the whole purpose in that case with a high level command language in the extreme model would be destroyed somehow. Now I wonder where we want to be? Somewhere in between these two extremes. There are systems going which look like Run X, and we're talking about systems which, now and in the future, are close to this kind, very small fractionated things. I think what we need to find out is what are the advantages and disadvantages of these two approaches. And if we stick to these two extremes then we're not treading on anybody's toes because there is no system that works right now and nobody would dare to offer an image processing system which requires you to enter run x, run y, run z under the normal control language. That's much too simple minded. So I would like to use the coming discussion to think what the advantages of these two approaches are and to understand whether command languages really are useful at all.

WELLS: Well, I think that's a good subject to discuss and clearly there is a wide range of opinions. What Ron Allen has presented is the way it appears everywhere in astronomy known to me today. You're somewhere in that spectrum. The people who build state-of-the-art computer systems today for interactive use see the problem differently. What we see in Figure 1 is the outgrowth of thirty years of computing in the traditional style. In one form or another all systems existing today in the commercial market look like that, with a few minor exceptions. Some people think that there is a different model of the future. In particular the Japanese think it, companies like Apple think it, and you should understand that comes out of Xerox technology. You could catch some of it in Tanimotto's talk a few days ago if you listened closely to the details. The model there is that the traditional approach isn't the way the human brain works. That's the problem. It's the way ancient computers work and even today's computers mostly work. So some people think we ought to work with objects, and we are to pick them up and move them around. That is, we ought to have data objects and we ought to have process objects and essentially grab them with graphics and move them through processes and put them places. The best example of this style is the Apple Lisa machines, and more recently the Mackintosh. These are graphics oriented machines which pick up objects and move them. It's pictures, and, of course, this is a very natural concept for people. The other model which many people believe is the future in commanding computers is the speaking voice. I don't know the answer but anyone who is really thinking about the design questions - for software, for doing data analysis and research and making computers more useful to people has to understand the distinction between our classical, computer language type and a very flexible human oriented approach. I can't tell you what is going to happen in the future in that area. I'm sure, though, for the rest of the decade, we will all be somewhere in the spectrum of Figure 1. I don't want to urge any designers to rush out and try to build anything for astronomy, it's too soon.

TANIMOTO : I wouldn't go quite so far as to say that I don't think anyone should experiment with these things because it is only by trying these things that we find out that we can use them or we can't use them. I haven't thought very much about what might be good graphical representations for astronomical image processing procedures but there might well be some. Certainly, the data objects could be represented by icons pretty effectively. At least you could categorise objects like galaxies and stars with a little iconic representations but I don't know whether people would feel comfortable working with a system in which all the data and operations are represented on a screen by little pictures.

WELLS: If you want to know what this kind of system feels like you only have to go to a computer store that sells Apples and look at a Lisa or a Mackintosh, particularly the Lisa system. It is just a different way of thinking about computing. I can't see how to do astronomy with it either, but it's so innovative an idea that I can't imagine that we will be able to ignore it for ever.

BENVENUTI: Well, I think we will always have to take into account that this system finally has to be used by astronomers and by a large variety of astronomers. I think we have to take the average astronomer and put him in front of these different options and try to imagine what would be his reaction to these different possibilities because I think in that area we can learn a lot if you try to see the evolution or the interaction of the astronomer with the existing systems. At least in ESO, there is the experience of people working with IHAP and developing the way of using IHAP which led them to the building-up of MIDAS. I think that if one bases the classification of advantages and disadvantages purely on the system quality and the possibility of what the skilful computer science people could do, we might make a mistake, because if later on the astronomer will not use it because it is too complicated or too difficult to learn, then we will fail. I think one has always to take that into account and try to build up these advantages and disadvantages directly from the use of the system.

CRANE: I think that the typical astronomer doesn't want to learn any image processing system. He doesn't want to know any commands. He wants to be able to do it by pure thought if he can and without having to touch a keyboard or a wand or even say anything.

WALLACE: I'd like to also comment on the question on Ron Allen's proposed listing of advantages and disadvantages. I think that one of the advantages of the very fragmented architecture on the right there is that it gives the opportunity of using features which are provided by the operating system which in the left hand of Figure 1 architecture is not really possible. For example, on a VAX, you might want to mix DCL commands with image processing commands. I really don't think it is possible at this stage to limit astronomers just to those features which the environment people have thought of and made portable. There are so many nice things that you can't afford to hide them. That's why I think that the portability of applications programs is far more important than portability of environments at this early stage. On the subject of whether command languages are for astronomers or programmers, I've got a couple of remarks about that. The first is that we do have users who use IDL which offers quite a high level command language and there is no doubt at all that the users enjoy it and use it quite fully. But you've got to have a very efficient mechanism for invoking program fragments for that to work. If it takes five seconds to fire up a program it's not going to work in that style of command language. Now, the other thing about whether it's still a tool for program development. There is a danger with anything to do with languages because it is a very attractive area for programmers. If you mention the phrase "command language", immediately you get reactions from programmers who think "Ah, what an opportunity to write a compiler". So you've got to be very careful and make sure that you establish how important the command language is, because ultimately it is the applications that really do the work and the command language is a convenience.

RIDGWAY : I think that it is probably a mistake to try to pick out where you want to be in the range between those two extremes in Figure 1 because, in fact, "Run x" is what you want to do and the completely fragmental scheme is what you usually wind up doing to start out. In other words, if you have only a run "x", then traditionally what the astronomer has to do is get into "x" and change "x". Now, moving that kind of flexibility into a command language environment may be an adaptation that's necessary for a complex system like this in which, perhaps in spite of one's intentions, it becomes difficult to get into "x" and change "x". So, I think you have to look at providing that range of availability of operations, and at making a smooth transition between the fragmented and integrated extremes. It's difficult, and that's perhaps where your current research and software development problems lie.

CRANE: Can I try and summarize this? What I see people saying is that really what the astronomer wants is on the left (a high level of aggregation) but what he really needs is on the right (a low level of aggregation) and somehow we have to be able to provide at some level what we see on the right but we certainly have to provide what we see on the left.

TANIMOTO : I get the impression that the astronomers do not want to do programming and that's why they want to have a finished product thing on the left rather than have to compose the sequences on the right. From a programming language point of view, what you really want is a system in which you can look at things at any level that you may need to. Once you have completed the description or procedure or anyone else has done that, then you want to run "x" or you want to actually use "x" as a component of another sequence at another level. So it seems to me that there is no answer to this question of where should we be targeting our system between these two levels, but we should be building a kind of system with which we could freely move between different levels. There are obviously more than these two levels. There is really a continuum of different levels assuming that these procedures like "x", whether it's a cosine or something like this, could be used in a higher level. You want to always be able to build on that or stand on the shoulders of the people who come before you.

DISNEY: When one decides to buy a computer for some specific task these days, one usually tests out all the various alternatives against a benchmark and then one takes into account all the other factors which may be political and economical and so on. What we really need is a way of benchmarking or testing all these various command languages and, in fact, of all these high level systems environments. I only say this very hesitantly but this seems to be beautifully summarized in this picture which many of you have seen.

Figure 2: Data Analysis System Bazaar: What it should be!

There are only two ways of doing it that are practical. One is by essentially laying down the law so that there is one group which is much more powerful than any other group, which might for example be the Space Telescope Science Institute, and there are certain advantages in that, at least the dictatorships get the trains to run on time. On the other hand, from all astronomers' point of view, it doesn't always work out that way because those dictatorial decisions are often taken by non-astronomers and I do have worries about a system that's imposed on us essentially by NASA, where there is an enormous bureaucracy. The alternative, which is not terribly easy to set up but nevertheless it should be possible, is when all these possibilities are offered to us, that we actually run our own benchmark on it, either formally or informally. I see that the ST/ECF should have a role to play in this and that is to get together a group of typical astronomers with some typical needs, from some relative novices through to people who are real experts, and let them try to do their astronomical programs on the various different systems, and then mark them and put the pros and cons in and then try to make some sort of a decision. Because short of that we are going to go on proliferating or we're going to have a decision imposed on us. I'd like to ask the panel whether there is some possibility of actually going this way?

WELLS: I think too much diversity in users interfaces is bad for astronomy because astronomers are travellers and they have to cope with this diversity and it's simply counterproductive for research work. I do think that you need some degree of diversity just to protect against error and to have an insurance policy essentially. How much is a good argument. I have a hard time believing that one unified design will ever be the final answer for astronomy. But I think today we have too much diversity. The Space Telescope Institute may drive the whole matter into a standard. I think that that's just a special case of a general rule that I've advocated for several years which is that the large institutions which have large instrumentation systems and are big data producers should take the responsibility to produce a quality product and make it widely available. I have hoped that in this way we might at least reduce the number of user interfaces and software packages to a more manageable level since the number of large institutions is quite finite. That's just a purely practical answer which requires no grand coordination or committees or reviewer, or anything of that sort. I'd recommend to the major sites that they take far more responsibility for their data and make their software widely available to their user community in an effort to help their users by reducing the diversity.

BENVENUTI: Well, I think that if we compare this system that we have seen in this by now famous viewgraph (Figure 2) with the level of aggregation that is predominant by Ron Allen, I don't find they are so different if we take out probably IRAF and maybe on the other side Starlink - but all the others I don't think they differ too much in the level of aggregation internal to the system. So I think we have really to discuss the future trend, not the existing system. Now, what worries me is when I hear that you want both, that is, you want a low level of aggregation, because that will help to build up new software and at the same time you want high level of aggregation for the astronomers because they want to run a specific task without bothering too much about programming. What bothers me is what we have to pay for that. It is not clear to me that going in that direction we will have a very efficient system. And that's a question that I can't answer. I would like to see if some computer science people could give me a hint on that.

KEGEL: I've said this many times before, I don't think, at first, that there is a conflict between run x and the low level of aggregation that Ron Allen is proposing. There is no conflict because both of them can exist in the same system without influencing each other too much. I think the only problem you have is when you have built out of the fine grains, as Ron has drawn on the right. The thing you want is to transform it into a run x type of thing and you have to be able to do so very smoothly. And that's where Piero Benvenuti's efficiency considerations come in. I think that it will require a considerable effort either into making the language efficient enough to leave it in the command language or a considerable effort to produce an efficient compiler for your command language to do an automatic conversion from the right to the left. And I'm not quite willing to give an estimate of how much effort is involved in producing such a compiler because I don't think anything of this kind has been made yet. Even IDL does not provide, as far as I know, a real compiler version of its own language. So, we really have to be a little bit careful about it. On the other hand, given a carefully designed language, and given the state of the art in compiler building, it should not be insuperable. If we really could combine all the effort which is going into system software building these days, where many people are trying to reinvent the wheel, each for their own very good reasons because they have their own particular needs which they think are not really catered for in their own system - if we could really come to some kind of cooperation of efforts to concentrate on this issue, I think you could make a very big step forward within a few years.

HAMAKER: I think there is another way how you could implement flexibility in this level of aggregation. You could have many options available within this task X. For instance, as a simple example, you combine two images and you can add them or you can subtract them or you could divide them or anything else. Now, you could have the task X ask what do you want. But you could also have a default mechanism which allows you to specify in advance that you want to do that. So you have both options available. By setting your defaults you could define the level of aggregation and the number of questions which are going to be asked to you. Now, I think this offers you the possibility of tailoring task X to what you want and having a fine granulation there without having to go into all the problems of command procedures and that sort of thing, which I think an astronomer doesn't want.

TANIMOTO : I'd like to mention that command languages may in some sense be doomed eventually because they're sort of a left-over from the batch processing era. When you submitted a batch job in the sixties on punched cards you had to provide job control, all the operating system stuff was separated from the applications. Then, when the first time-sharing systems came along, again the operating system commands were at one level, you talked to the monitor, what was called the operating system, and then your application was written in a separate language, and so forth. A number of the new work stations did not make a distinction between the command language and the programming language that the machine works in. A lot of the problems of trying to figure out what to put where go away as soon as you move to the integrated computing philosophy. Probably the best examples of this are Lisp machines where the machine language of the system is in some sense a high level language. You don't have this duality of modes where you're either at the operating system level or you're at the application level. You're always in the Lisp environment. If you want to reallocate some peripheral devices or files or something, it's simply a command in the language. The language works both in an interpreted fashion and compiled. In general, when you're talking of a system as an interpreter you mean something which scans the input and causes it to be executed. But there is also the possibility of compiling a function. And we're compiling functions now rather than programs and when you're working in the Lisp environment you're usually working with a mixture of compiled functions and some interpreted functions. I think the main advantage of this kind of integrated approach is that you don't have to learn two separate languages; you don't have a command language plus an application language. You have one language, and it is very easy to move things from one to another.

You might want to think about how you would go about moving some of the facilities that are now found in command languages into an integrated computing environment. You really want to have one language and not have to deal with things on so many levels.

WALLACE: Yes, I think I agree with all that. I'd just like to remark that it's interesting that the command languages that are being produced for astronomy are apt to be much less powerful than even languages like FORTRAN. So, they're rather lower level things, and yet the parameter interfaces into the application programs are far more complicated than those allowed of programming languages where you've got very elaborate defaulting possibilities and so on. So there is a curious overlap developing where application program interfaces are getting much more complicated than subroutine interfaces. And yet the command languages are not as powerful as even old languages like FORTRAN. I suspect the astronomers could do quite well with BASIC and I'm sure that they'd love interactive FORTRAN as their command language.

WELLS: Those remarks about LISP are closely related to the ones I made earlier about graphics driven systems, because there are a lot of relationships there. I might also give a historic note for astronomers, since the language FORTH was invented within astronomy and it shares many of the properties of LISP. Anyone who really knows about FORTH is well aware of its ability to do some of these things, not as well as LISP systems. I agree that it is nice to get rid of the command language and talk about a unified programming/computing environment. We have a major barrier, however, as Pat Wallace pointed out, because we've all been trained to think FORTRAN. It's a major problem. And in a practical sense, I don't see how we are going to escape that in this decade, much as we might wish to.

ROTS: Let me sound a warning about a system with a low level of aggregation. One of the examples of such systems was CANDID which was developed in the early days of the VLA and it had a very low level of aggregation. One of the problems was that it was too powerful. It did not really protect the user, especially the inexperienced user, from doing disastrous things to his data. Let me make another comment: I think another element in the whole problem of control languages is that we're catering to a rather wide variety of astronomers, wide variety in level of experience and sophistication and needs. Let me give an example: I think one of the good control philosophies for a novice user could be a menu, because essentially the user does not have to read any or much documentation before he starts out. The trouble is, of course, that it gets in the way of the more experienced user, and it becomes rather inflexible after a while. My feeling is that if we approach the subject from the user point of view we may well have to turn to systems where we actually not only give the user a wide variety of things he can do, but also a wide variety of ways in which he can command whatever he wants to do, different levels of control interaction. Maybe something like menus for the novice and command structures for the very sophisticated user.

TEUBER: What do you think of this approach of the command language independent systems? I think they may support key driven systems, menu-driven systems and the graphics systems. What else is there? You have a menu of graphic symbols that you can put into tables, and graphic symbols normally are represented by display lists, so that is where you generate them. So you can support this also on a system like that. But you are not bound to one kind of system.

WALLACE: I just got to respond to that. The Starlink environment certainly is command language independent and I think a lot of the others we have heard about are. The layers of subroutines are constructed so that the command language can be replaced without the application programs being affected. Now, some of the people who have been developing Starlink are so convinced that that's the right thing to do, they won't even discuss command languages because they are details of implementation and they regard the application program interfaces as the important thing. So we will have different command languages in Starlink, and it won't matter.

CRANE: Maybe we should move on to the second question I've posed here. Portability: holy cow or red herring? When I say "holy cow or red herring", I think most of us would agree that it is perhaps a holy cow. On the other hand, what we really don't know and what we'd really like to have some input on and some discussion of is "what do we have to pay for transportability, both in terms of development costs and in terms of execution performance?" I think, we have at ESO an example of an extremely unportable system, the IHAP system, and yet we have a highly efficient system. It has been optimized for its particular piece of hardware and competes very well on a Hewlett-Packard 1000 system with a VAX-based system where the VAX is an intrinsically faster machine by a factor of at least four. We certainly see that there is very little difference in performance. Of course, the history is such that we also have quite a large user community there and people are not drawn by the factor of four from the Hewlett-Packard system to the VAX-based system. How much do we have to pay for transportability?

KEGEL: I'm going to rephrase the question. "What do we have to pay not to have portability?" Because that's going to cost you a lot of money and I think your own example proves where the problem is. You are now writing a completely new system and you have to redo each and every of your applications to get it working on the MIDAS again. And that's the cost you have to take. Of course there is a gain in efficiency, I'm not going to deny that, but I really think that given the state of the prices of hardware and the prices of software, there can be no question that the cost in software should be the thing you aim at reducing, and not the cost of the hardware.

CRANE: If we are going to move to far more sophisticated architectures such as were discussed by several people here on Tuesday, do we really know how to build something that's transportable to those architectures? Are the programming languages that are going to be required for those architectures going to take advantage of those architectures in the way that we need? It is clear that the array processor software in AIPS is not transportable to other array processors. Is the state of the art such that we can really take advantage of transportability when we want to move to far more sophisticated architectures?

WELLS: One of the key things that we are protecting is our investment in our software if we engage in portability and layered construction in the software. I might add that another major advantage which NRAO has been driven by in the AIPS project is that we wish to be able to do a competitive hardware procurement at any time that we want to do it. We do not want to be constrained by some choice of operating system made some years ago. We want to have the freedom to change if needed and to gain whatever advantages we might gain by changing our hardware. However, when you're doing that, if you are going to want to make a competitive hardware procurement changing operating systems, vendor, software and so on, the underlying system software that the vendor gives you, that's terrible for astronomy, if the operating system user interface provided by the vendor is allowed to show too much in your application code and therefore I think that if you really believe in portability and how to do it, you are obliged to build a virtual operating system to go on top of the vendor's code to give a user interface and that's the high level command languages we've mostly been talking about and that you're obliged to make that command language be independent of the vendor's code. Essentially, you're obliged to provide an interface that an ideal vendor would have provided. Now, in looking ahead to the future, if you wish you can argue that UNIX is that system and indeed in the years to come we may find that that will be a true statement. I can just speak for my institution at NRAO when it had to make its decisions that could not be assumed. In 1979-1980 one could not assume UNIX was the wave of the future. Even today it would be a little risky and so we were obliged to create a virtual operating system. But I believe that that advantage of isolating the users from changes in the hardware and between machine differences is a very valuable thing. In terms of performance, when Phil Crane said, "Well, the AP code is not portable", micro code is not, because that's like writing a machine language for some machine. However, the problem for vector pipeline type processors is to learn how to write high level code appropriately so that it will port and that's clearly an active area of research with all the vendors who make vector pipeline processors. If you look at the notes about what FORTRAN is going to look like later in the decade, you are likely to be struck with the idea that it's not going to be FORTRAN as we know it in the past and, furthermore, that the FORTRAN people must come to grips with vector processing. I think, in fact, portability is possible for vector machines. For the parallel processors, I don't quite see how to do it yet.

WALLACE: Can I just make a few more comments? Wim pointed out a very important point which is that whereas there may be costs of arranging for transportability, there are certainly also costs of not making programs portable. However, of course, you're talking about the long-term versus the short-term. In my experience the funding authorities pay only lip-service to the long-term when it comes to software, they don't pay people or money. So, I agree completely with him, there is a cost either way. If we can accept that FORTRAN will survive, and from what Don Wells said this is true, and if you go look at FORTRAN 8X, you might be a bit surprised. I think it's relatively easy to make application programs portable and I don't think it costs any more. I think an application programmer is going to spend the same amount of time writing a program in one of these environments than without. I think there are very, very substantial costs in constructing the host subroutine libraries, operating kernel and all that stuff. I think a figure of 30 percent was mentioned earlier by someone and I'd put it even higher than that in developing the environment itself. There is also a cost in performance, I'm sure. If you go writing your own virtual operating system it will not be as fast as the one supplied by the vendor and I think it might well be enormously slower. For example, in the DCL/VMS, a program load takes half a second. In the Starlink environment as it is at the moment, it is an order of magnitude more. That's not acceptable in my opinion. I also have very grave reservations about the philosophy of accepting the lowest common denominator in both operating systems and hardware. I think it's unfortunate that you have to make a decision, for example, with image displays that you're only going to use three or four features so that you can go and buy the displays from any vendor. I don't think that's acceptable and a way has got to be found around that. A possible solution is to provide something elaborate like GKS so that you can use all these displays and find out that they've got the features a given program uses. It's very hard to persuade programmers to make all the enquiries. So it's an unsolved problem. I personally think that the way you avoid taking the lowest common denominator in operating systems is by accepting that a certain amount of the machine user interface will show through to the user of an image processing system. For example, it may be possible to have environments where data objects have file names which belong to the host machine and on different machines you would type in slightly different character strings for those parts of the system. If we don't accept that we are going to make use of the vendor operating system at a user level, then the writers of environments are going to have to construct back-up utilities and everything and they are not going to do it as well as the machine vendor.

BENVENUTI: Yes, I would like to comment on the investment that you put in the software and how this will develop with the future. I think that the real investment and the real durability of this investment is in the application program. I think we all agree on that. So, I can see that in the future when you will have to rebuild a host system because your hardware changes and you have a more efficient way of hosting what you have produced in the past, this investment will decrease compared with what you have accumulated from the previous time. So, I would put my priority on the applications programs. Those are the things that should survive for a very long time while a host system is going to be changed because the hardware offers of the future will be so attractive. I notice if I look back in the past that we have already reached a reasonable portability of the application program. It's much easier now to transport an applications program from, say, MIDAS to FIPS, for example, or to the other similar environments and so I think that this is an achievement that should not be lost in the future.

MUSSIO: The user must be guaranteed, not only in the efficiency of the system, but in its effectiveness. I mean, it is not only a question of crunching numbers and getting outputs but a question of crunching numbers and getting results. Results which may be controlled and be controllable by the user. You always speak of changing the hardware and maintaining the programs. To change the hardware and maintain the programs means to change the computation underlying the program. How do you guarantee that the results will be the same and reliable?

WELLS: It's an important technical point. In my talk about AIPS the other day, in one of the little items in our future projects category was to write a test, certification, benchmarking package. One of the principal reasons for that is we were jumping from machine to machine so fast that I began to be worried that some of those machines might not compute arc-tangents correctly or that when they invert matrices they might get slightly different results because the machines round off the numbers differently. And, when you get serious in large systems, very large systems, the problem you raise is very real. How do you assure that two different computers get answers that are consistent within an acceptable statistical error? I might add, practical experience in the AIPS group shows that there is yet another problem you wouldn't dream of. If you deal with the array processors as we have, you might find one day that it doesn't give the same answer it did last week and, indeed, that has happened. Therefore, one aim of a test and certification package is to run the same problem at regular intervals and check that it gives the same answer. Mercifully, with most conventional computers, you never think such thoughts, because if the computer hardware begins to misbehave the operating system soon crashes and you call in the fixer to fix it up.

DISNEY: I was just going to ask a question to the people representing their own systems, the big systems. It's very interesting to know what the weaknesses are as well as the strengths. Some people have told us what they have but I'd certainly like to hear the advocates of the big systems like AIPS, MIDAS and STARLINK, telling us for two minutes what they think the main weaknesses of their own system are and letting people in the audience tell them if that's the only thing that's wrong.

CRANE: Well, I think it's a question of what you're trying to do because, as Ron Allen pointed out to me, he's had plenty of people come to him and say, "Oh, well, MIDAS is a pile of junk"; but that's because they were trying to do something that MIDAS really wasn't capable of doing at that point. The weaknesses of many of these systems are not what they do, but what they don't do. It's the application programs that make these systems strong. If you're asking what the system weaknesses are in MIDAS, I'm not capable of making a professional evaluation of the system level things. On the other hand, I can tell you what we're missing in terms of application programs.

WELLS: I think an objective, a truly objective analysis of relative merits of these different design approaches is practically impossible and I'm aware of two major studies that have been done in recent years by groups attempting to choose from the things marketed by other groups, and each professed to be objective, and I remain unconvinced as to the degree of their objectivity because I looked at the outcome and realised I could predict it from what I knew of the people who made the study.

AIPS fits in the run X category. It's very classical in that sense. It's an operating system for running big programs. It in fact is a portable virtual operating system that runs major application programs. So if you look at AIPS, that box in AIPS is very big for those boxes that were in Figure 1. Those are huge programs. It is not the little bitty programs that you hook together to do things that were not thought of by the designers. It's not very good at that. And so, if you were an experimenter wanting to do experimental data analysis in areas that were not conceived by the system designers, you find that it's an inconvenient, uncomfortable environment, and I don't want to stress that too strongly when I say that I am reflecting what experimenters say. And I might remark from watching this scene for a long time that experimenters are the people who usually criticize these large scale system designs. They have very different interests. They want to play with software, and they tend to demand a level of granularity in the code, and a level of elegance, they want the latest elegant system concept, and they will always criticize any project that attempts to produce ten man-years worth of code and deliver it to a user community. That's a problem area.

ALLEN: These are not unique to AIPS!

WELLS: No, no, these are things that have lasted a long time. The IRAF project, for example, which is the latest thing sneaking in through the door in Figure 2, is based on a long tradition at Kitt Peak, looking at these projects over many years. They hope to do better. They hope to have the best of both worlds. It's a grand dream. It's been a dream for a long, long time. You'd love to have a beautiful integrated environment which yet makes the experimenters happy. I don't think the millenium has arrived.

ALLEN: I would like to return to the question of comparing the systems. I'm not so sure that I would be so pessimistic in making that comparison. It does take a lot of work, it's going to require defining some sort of standard kind of image processing things that people want to do, and perhaps going and actually trying to do them in institutes where the systems are installed and are optimally running, but if that job were to be taken on for a number of different standard kind of projects, and carried out, for instance by the ECF as an example, I think that the community would benefit from having some objective data on the comparison of these systems. It comes back a little bit to what Mike was asking earlier, to have a set of benchmark programs at a level somewhat higher than checking the arithmetic units of computers, a set of benchmark programs which would do some of the standard simple things, not even necessarily terribly complicated. Actually go to the places where the systems that have been installed and are optimally running, and run these benchmark programs and then describe what happened. How did you have to do it, in the sense of the walk-through as I was mentioning a couple of days ago. Give us a user walk-through of how you actually had to do the job, how long it took, and what the results were. That could provide very useful data, and if ECF would consider doing that, they would at least be an independent organism whose results were surely to be respected.

WELLS: You should understand that the Space Telescope study lasted well over a year and went through many revision levels and involved an analysis of projects represented by some of the designers in the group here, and it is quite an elaborate study. It didn't do some of the things you suggested, although they imported systems, had them running on their computers, and attempted an objective evaluation. So, if you want to do better, you're going to work hard.

BENVENUTI: Yes, I really share the opinion of Don Wells on this because indeed, the study performed by the Science Institute took a long time to be performed, and finally, I think it is a fair study of the different systems they analyzed. However, the conclusion they draw is questionable in my opinion, and probably other elements enter into the final decision, and I'm sure if at ECF we do the same, if we decide to invest the same amount of time to make these analyses properly, which certainly will take some time, I'm not sure that the decision we will have to take at the end will be driven by the fair conclusion of the study, because you have other elements to take into account and you can't avoid doing that. So I prefer, in fact, to say that all the systems we have seen and all the systems that are available on the market are more or less equivalent. There is not one of them which is by far the best. However, I agree with you that what one should do is to shift this interesting testing to the applications program. There might be applications programs that can do the same thing approaching the problem in different ways, and there I agree that we should try to test these systems, try to test these application programs or these approaches, and provide a fair answer to these different applications. There, I think we can make something useful to the user, more than testing an environment which is linked to so many things, starting from the hardware to the personal involvement in the development of the system, and many other things, like money which is available for the development of these systems.

ALLEN: To clarify that what I was asking for was not to do another independent study with a view towards choosing product X, but really just independent of the choice of what product you're going to have to choose at the ECF, for instance, I think that you would be doing the user community a favour just to help us understand how one makes the comparison. Even if the choice might be driven by completely different things, it at least provides some data. We don't even have any data at the moment. We don't even have the data which shows us, in a summary, how you do a particular standard simple image processing thing in the number of different systems which are currently available even in Europe today. That's essentially the walk-throughs which I asked for earlier. I would like to view that question independently of making a decision as to which product you're actually going to install, because that's driven by decisions which are not always simply one of efficiency. I would like to divorce it completely from that. But in the meantime it could be a very useful piece of information.

CRANE: I think we've clearly generated a lot of interest here. I'm not sure whether we should stop now since we've already been here for an hour and twenty minutes, and I'm a bit reluctant to cut off people who are jumping up in the audience to ask more questions. So I hope, if I try to cut off the discussion now, that in fact we'll continue this discussion anyway. I'm sure we will, and I just would like to thank the members of the panel, and the audience for their active participation, and Antoine Llebaria for giving us the inspiration to talk together by drawing Figure 2.

SESSION

IMAGE PROCESSING

CHAIRMEN:

S. TANIMOTO
A. GAGALOWICZ
F. ROCCA

ASTRONOMICAL INPUT TO IMAGE PROCESSING

ASTRONOMICAL OUTPUT FROM IMAGE PROCESSING

Edward J. Groth

Physics Department, Jadwin Hall
Princeton University
Princeton, NJ 08544 USA

INTRODUCTION

This paper discusses some aspects of image processing - first with a high level look at image processing, its inputs, outputs and operations. Then, a specific research project is examined to illustrate how various image processing functions are used to derive results from image data. The project is the construction of a catalog of galaxies from photographic plates. The functions illustrated include correcting for the nonlinear plate response, modelling and removing the background with a polynomial, removing bright objects, histograms, smoothing, object detection, object parameter estimation, and manipulation of tabular data.

WHAT IS IMAGE PROCESSING?

The simplest definition of image processing is that it is data analysis performed on arrays of intensity versus two spatial coordinates. However, in recent years, image processing in astronomy has come to encompass many data types and operations. For example, the analysis of long slit spectra involves two dimensional arrays of intensity versus one spatial and one wavelength coordinate. Echelle data, such as those described by P. Crane in this volume, give flux versus order number and wavelength and processing involves extracting the one dimensional flux versus wavelength. Also in this volume, D. Wells describes how radio interferometers can produce three dimensional arrays of flux versus one frequency and two spatial coordinates and R. Fosbury describes how stepped slit spectrographs can be used to produce similar data in the optical band. The raw data returned by the *Einstein* X-ray satellite were a series of events, each tagged with a time, energy and two spatial coordinates and it was from these events that images were constructed. Grism data, such as those described by P. Hewett in this volume, are two dimensional arrays of flux versus a spatial coordinate and a second coordinate containing both spatial and wavelength information. Often, image processing includes operations on sets of images, taken with different polarizations, with different colors, or at different times.

Astronomical image processing can involve operations on one, two, three, or even four dimensional arrays with some combination of spatial, temporal, frequency (or wavelength or velocity), polarization, or color coordinates.

DEVELOPMENT OF ASTRONOMICAL IMAGE PROCESSING

Since the advent of astronomical photography, astronomers have been in the image processing business. Many measuring engines were developed to extract information from photographic plates. These included traveling microscopes, 'Grant' machines, iris photometers, blink comparators, and densitometers, to name a few.

Digital image processing began in the 1960's primarily motivated by the unmanned missions to the Moon and planets which returned the first close up pictures of these objects. Here, the emphasis was on 'fixing up' the images - removing the detector and telemetry artifacts which detracted from the visual appearance of the images.

Since then, many factors have stimulated the development of digital astronomical image processing. These include the development of electronic imaging detectors such as CCD's in the 1970's, the implementation of the large radio arrays such as Westerbork and the VLA, NASA and ESA space astronomy missions, and the use of scanning micro-densitometers to generate digital data from photographic material.

Astronomical image processing is still a young discipline. Much of the available information is not published in the normal astronomical journals, but in conference reports (such as this one), internal documents, and 'comments in code'. Many older astronomers suffer from 'computerphobia' and do not show the same respect for an outstanding image analyst as they might for an instrumentalist or observer. Nevertheless, image processing is an active field undergoing rapid development as it encompasses an increasing fraction of astronomical data analysis.

WHY IMAGE PROCESSING?

Besides the fact that more and more astronomical data are obtained in the form of images, there are many reasons for the popularity of image processing. Perhaps the most important is that humans interact readily with an image. Results are more easily grasped when presented as an image rather than tables of numbers. Viewing data in the form of images makes efficient use of the eye-brain processor for pattern recognition.

In many cases extensive image processing is required to obtain presentable images from the raw data. The previously mentioned X-ray images from the *Einstein* satellite are an example. The gamma-ray images from the COS-B satellite (see G. Simpson's contribution in this volume) where obtained only after extensive reconstruction since the spatial position of each gamma-ray event is only determined to within an annulus. Radio images from the large arrays start out as complex fringe visibilities in the u-v (Fourier transform) plane.

With modern digital image processing, one can do everything that used to be done with photographs - usually more rapidly, more accurately, more flexibly, and in greater quantity. For example, consider the problem of identify variable stars from two images of the same field taken several days or weeks apart. With photographs, one would use a blink comparator. With modern digital image processing, one might still blink the two images (now on a video display), but there are several other approaches. The images can be subtracted and large differences found automatically or by visual inspection of the difference image. A catalog of stars, listing positions, magnitudes, and uncertainties, can be constructed from each image and the catalogs can be searched for statistically significant differences. What's more, image processing allows for adjustment of the registration, zero point, magnitude scale, and so on, before the images are compared. Thus it is possible to blink two images obtained from telescopes with different scales.

But the real power of image processing is not just in doing the same old thing better, but in the new applications that are possible. These include, for example, construction of images from non-image data, examination of the full dynamic range of modern detectors, spatial resolution enhancement, stacking of images for greater signal to noise and the detection of fainter sources, isolation of weak features for detailed study, and generation of images from derived parameters. Many other techniques could be added to this list and many more will surely be developed in the future.

IMAGE PROCESSING FUNCTIONS

Image processing functions are often divided into two (overlapping) groups, referred to as *calibration functions* and *analysis functions*.

In the case of calibration functions, the inputs are raw detector data and calibration data, the operations correct for detector imperfections, and the ideal outputs are linearly scaled intensities versus a regularly spaced two dimensional array of coordinates (for the case of a classical picture). The algorithms tend to be specific to the detector.

The inputs to analysis functions are the outputs from the calibration functions, the operations derive astrophysical information, and the outputs include images for visual interpretation, trends or relations between measured parameters, catalogs of sources and associated measurements, and so on. The algorithms tend to be specific to the scientific problem at hand.

The distinction among calibration and analysis functions is useful since it helps modularize software development and data processing, allows for the incorporation of instrument expertise in the calibration algorithms and scientific expertise in the analysis algorithms, and provides the potential of automated calibration.

On the other hand, these distinctions should not be overdone. The boundary between calibration and analysis is often ill-defined, the required calibrations may depend on the science to be done and in many cases, several iterations of calibration and analysis may be required before the final result is achieved. Also, construction of calibration data suitable for input to the calibration functions may require the application of functions similar to those used during analysis.

CALIBRATION FUNCTIONS

Calibration algorithms depend on the nature of the data to be calibrated, the instrument used to obtain the data, and to some extent on the science to be done. Elsewhere in this volume, D. Wells describes how the *CLEAN* algorithm is used to generate images from radio-interferometric data. To calibrate typical optical data, photometric and geometric corrections may be applied.

Photometric corrections (often called radiometric corrections in other disciplines) may include correction for detector non-linearities (which requires measuring the instrument transfer function), correction for detector offset, and correction for spatial variations in detector sensitivity (also called flat field corrections). The exact manner in which the corrections are applied varies from instrument to instrument.

Geometric corrections are applied to remove distortions in the detector or telescope optics. When the distortions are not severe, measurement of the distortions may be sufficient, as this allows the positions of identified features in the image to be corrected to 'true' positions. If the distortion is more severe or if images are to be superimposed or mosaiced, resampling of the image onto a uniform grid may be necessary. Geometric corrections interact with photometric corrections since distortions vary the effective area of the source seen by each pixel of the detector. Also, when resampling is necessary, the noise properties of the data are altered. In the case of bilinear interpolation, four input pixels contribute to each output pixel, so the output pixels are not statistically independent even when the input pixels are. Some loss of information necessarily occurs whenever an irreversible transformation is applied to the data.

Calibration may involve the detection and identification or replacement of 'bad' pixels. Such pixels may occur as a result of defects in the detector or from external causes such as the impingement of a cosmic ray on the detector. Bad pixels may be identified so they will be ignored in later processing or they may be replaced by an estimate of 'what should have been there. Replacement is risky since it amounts to 'creating data from nothing'.

Another step which often occurs during calibration is reformatting the data from the instrument specific format to a local standard format and perhaps to a transportable format such as the Flexible Image Transport System (FITS) devised by Wells, Greisen, and Harten (1981) and its extension to groups (Greisen and Harten 1981).

ANALYSIS FUNCTIONS

The kinds of functions that come under the heading *analysis* are limited only by the imagination of the astronomer. Thus they are open-ended and more difficult to describe than calibration functions. This section gives some useful characteristics of analysis functions and lists general tools that most well equipped image processing systems will contain.

Astronomical images may contain anywhere from a few hundred pixels to over a hundred million pixels. With this many pixels, it is rare that all of them represent 'good' data. And with 100 million pixels, a six sigma event is not unlikely. Analysis algorithms must be robust in the sense that they handle bad data or statistical outliers gracefully. Algorithms should allow for 'undefined data' - for example, bad pixels identified during calibration. Also, it should be possible to select arbitrary subsets of data on which to work.

Often, many functions will be applied to an image during the course of analysis and it is important that all operations be logged with the image, so that the analysis can be reconstructed at a later date. Of course, an image data set must include more than just the pixels of the image. At the very least, the data set must include a description of the image - its size and data type. In addition, the image description may include other auxiliary data such as the coordinates of the image, the date it was obtained, the telescope and instrument used to obtain the image, and so on. The auxiliary data may also include the operations log.

Important outputs of astronomical image processing are *lists* - for example, a list of all stars in an image, including positions, magnitudes, and colors. A good image processing system will have the capability to handle lists (also referred to as *tables* or *catalogs*) as another data type.

Increasingly, astronomical studies involve comparison or correlation of data obtained at widely disparate wavelengths, e.g., radio and X-ray images. To facilitate these studies, there must be a capability, such as that provided by the FITS tape format mentioned earlier, for image exchange with other sites.

A versatile image display system is mandatory for image processing, especially to support interactive analysis. Desirable features of an image display device include capabilities for contrast enhancement, zoom, pan, cursor input, graphics overlays, and simultaneous or blink display of two images. Color is useful, not just for creating 'pretty pictures', but also for highlighting important features and encoding a third dimension of information in a two dimensional display. As display devices become more sophisticated, additional capabilities will come to be 'required'.

General functions that an analysis system should support include the following:

- Editing an image or its description and generation of numerical listings of a portion of the image or listings of the description.

- Display of images and control of the display device.

- Generation of test data.

- Image arithmetic.

- Detrending, including polynomial fits and subtraction.

- Generation of histograms and other statistics.

- Extraction of subimages.

- Mosaicing of images.

- Smoothing, Fourier transforms, and spatial filtering.

- Resampling onto a new grid (for rotation, scale changes, etc.).

- Removal of bright objects.

- Photometry of objects.

The preceding list is necessarily incomplete as the menu of standard functions in an image processing system depends on the requirements and tastes of the system's designers and users.

The list is incomplete for another reason: if one is to do forefront research, one is necessarily performing research that has not been done before. This can mean observing with a new instrument, observing new objects, or in many cases, developing new software so that more powerful algorithms can be applied to the analysis of the data.

ILLUSTRATIONS VIA A SPECIFIC PROGRAM

The preceding sections have been quite general. Now, a specific research program is examined in order to illustrate some of the previously mentioned functions in greater detail and show how they are actually used in image processing. Although the illustrations will start from raw data, it is not possible to go all the way to a scientific result as the software development for this project is still in progress.

The objective of the program is to construct a catalog of galaxies going fairly deep - to 22.5 or 23 magnitude, covering a limited area of the sky but large enough to examine the spatial distribution of galaxies to 1 degree separations. The plate material consists of two strips of overlapping 4m IIIaJ plates spaced on 40 arc minute centers. One strip, containing 14 plates is located near the North galactic pole, and the other strip contains 15 plates near the South galactic pole. Both strips cover an area approximately 1 by 10 degrees.

Many investigators have developed automated image analysis and catalog construction techniques - see for example Butchins (1981), Carter and Godwin (1979), Godwin and Peach (1981), Grueff et al. (1984), Herzog and Illingworth (1977), Jarvis and Tyson (1981), Koo (1981), Koo and Kron (1982), Kron (1980), MacGillivray et al. (1976), Peterson et al. (1979), Reid and Gilmore (1982), Sebok (1979), Shanks et al. (1984), Strom et al. (1981), and Valdes, Tyson and Jarvis (1983). The unique features of this project are the combination of limiting magnitude and area coverage and perhaps some of the data processing algorithms.

The plates are digitized with a PDS microdensitometer. Due to limitations of the densitometer software, and the capacity of 1600 bpi tape drives, each plate is digitized in four overlapping quarters with about 4000 by 4000 pixels per quadrant (about 8000 by 8000 per plate).

As is well known, plates are quite nonlinear, so the first step is the conversion of photographic density to relative intensity. Each plate has on it a set of sensitometer spots, with known relative intensity, exposed at the same time as the plate was exposed. The densities in these spots are measured and fit to a curve of log intensity versus density. This curve is then used to perform the density to relative intensity conversion.

Figure 1 shows a mosaic of the four quadrants from one plate at low resolution. Each pixel in this image represents an average of 50 by 50 pixels in the full resolution image. Actually, the average and standard deviation for each pixel were computed twice. The second time, pixels further than 3 standard deviations from the first average were omitted from the computations. This illustrates the point made earlier about the need for algorithms to be insensitive to outliers. The goal is to obtain an image in which most of the pixels represent the sky background. The pixel rejection scheme does this except near fairly bright objects. No resampling was required to align the images before joining them since they were all based on the same coordinate grid (established by the microdensitometer). A mosaic of the four 'standard deviation images' is shown in

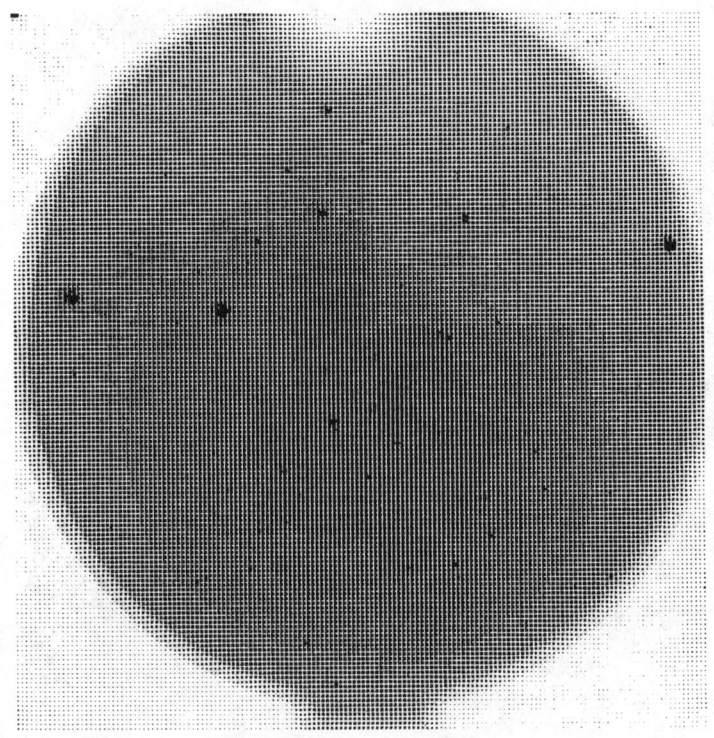

Fig. 1. A mosaic of the four quadrants from one plate. This image is a low resolution image of an entire plate. Due to the use of the averaging algorithm described in the text, most of the pixels represent background.

Figure 2 and indicates that the noise is reasonably uniform over the image.

The next step is to correct for the non-uniform sensitivity over the field of view seen in Figure 1. To do this, polynomials are fit to the low resolution background images. However, each quadrant contains a considerable number of pixels beyond the field of view which should not be included in the fit, so windows are appended to the images to sharply define the edge of the field of view and limit the fit to the good data. Two dimensional, eighth order polynomials are fit to each quadrant separately. The fit is iterated, and pixels too far from the fit are discarded at each iteration. Once the polynomials have been determined, they are subtracted from the data and the result is divided by the polynomial. Assuming the non-uniformity in Figure 1 is the result of non-uniform sensitivity rather than scattered light or other additive (rather than multiplicative)

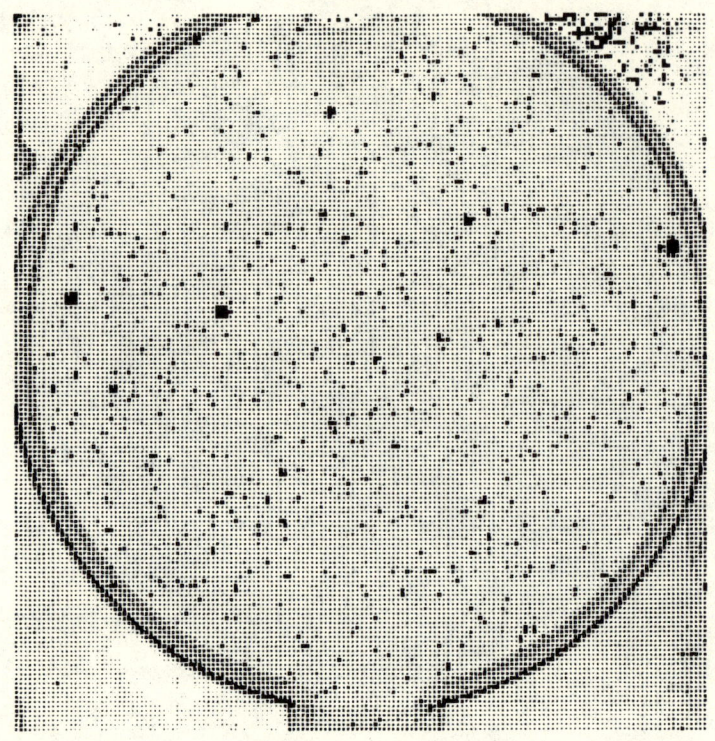

Fig. 2. A 'standard deviation' image in which each pixel represents the standard deviation in a 50 by 50 pixel array from the full resolution image.

effects, this procedure generates data in sky units with the sky removed. The result of this flat fielding applied to the low resolution background image is shown in Figure 3 where the typical variation is a few tenths of a percent of sky. The same flat fielding process is applied to the full resolution data using the polynomials derived from the fits to the low resolution data.

Eventually the images will be searched for objects by looking for pixels above a threshold. Bright stars have wings which can extend several hundred pixels. In order to find faint objects near bright stars, either the detection algorithm must take account of the local background or the wings of the stars must be removed. An example of the former approach is described by M. Malagnini in this volume. The latter approach is adopted here. Figure 4 shows the region around a typical bright star at the full resolution of about one-half arc second per pixel. The method for removing a bright star involves locating the center

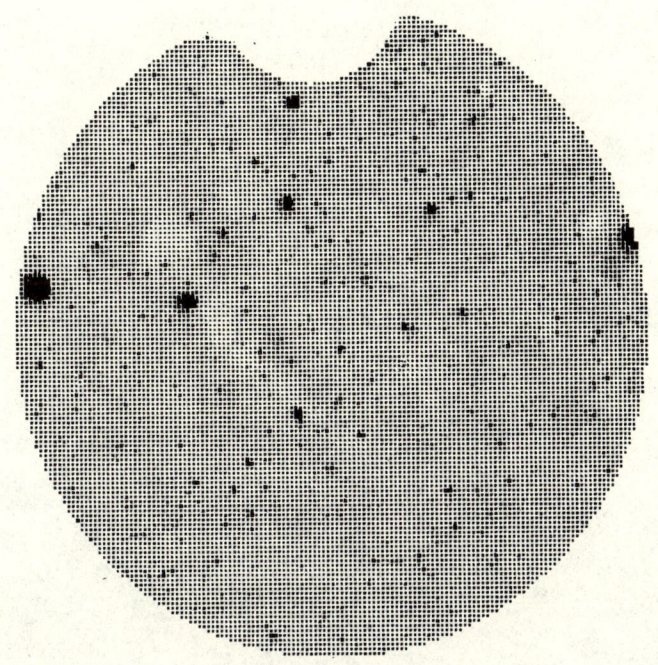

Fig. 3. The data of figure 1 after flat fielding. The typical variation is a few tenths of a percent of the sky brightness.

of the star, constructing the radial profile by averaging the data in annuli concentric with the center, and then subtracting the profile. The profile computation is iterated, again discarding pixels too far from the mean, so that nearby objects do not corrupt the profile. The result of subtracting the profile and eliminating the core of the star and the diffraction spikes with windows is shown in Figure 5.

The goal of object detection is the detection of faint galaxies. The faintest galaxies will be undistinguishable from stars. Those that can be distinguished from stars (without other information) will have images somewhat larger than star images. So, the goal becomes the detection of faint objects somewhat larger than the point spread function. To optimize the detection of these objects the data are smoothed with a circular 'boxcar' filter of radius 3.2 pixels. (That is, each output pixel is the average of all input pixels whose centers are within 3.2 pixels of the center of the output pixel.) For Gaussian shaped objects, this filter is optimum for objects with a radius of about 2 pixels - about 1 arc second and roughly

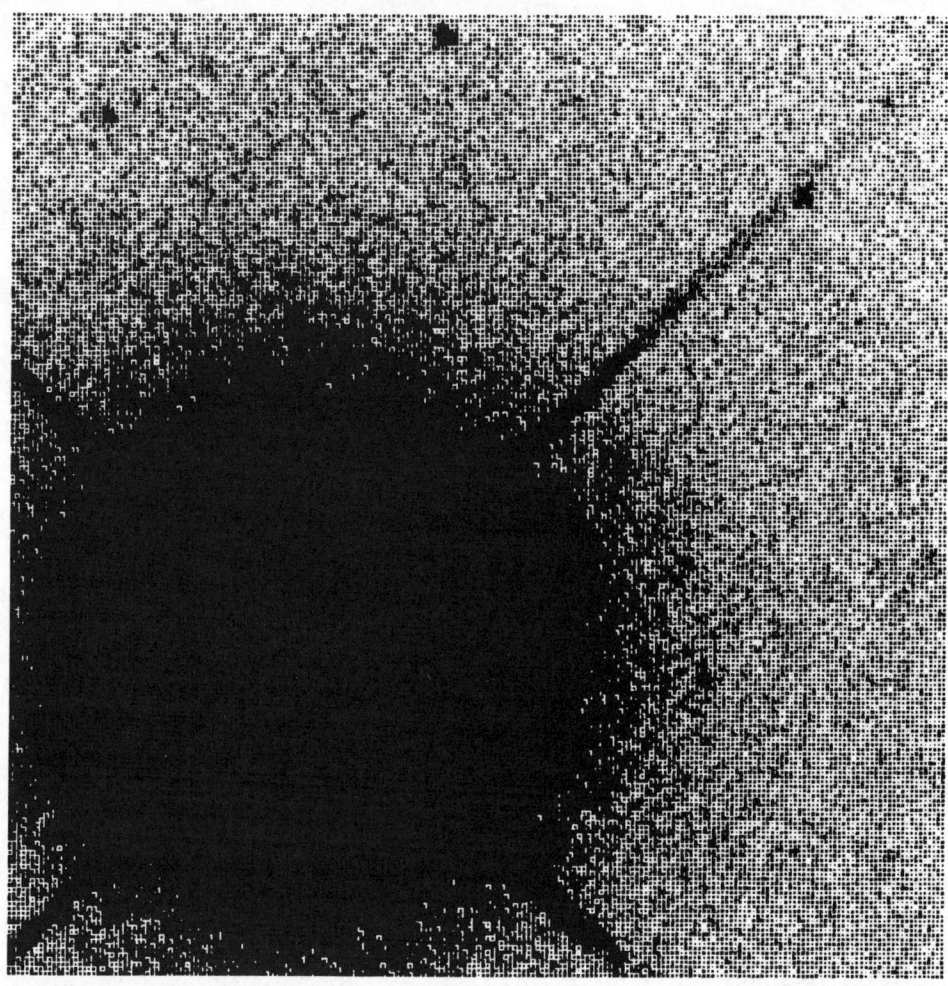

Fig. 4. The region around a bright star. Each pixel is about 0.5 arc seconds square. The wings of the star occupy the entire field.

twice the seeing radius. Of course, the optimum filter would have the same shape as the object, but this filter produces a signal to noise about 90% as that of the optimum filter for Gaussian shaped objects.

The detection algorithm is fairly straightforward. All pixels above threshold in the smoothed image are identified and those which are sufficiently close to each other are identified as belonging to the same object. Objects containing fewer than

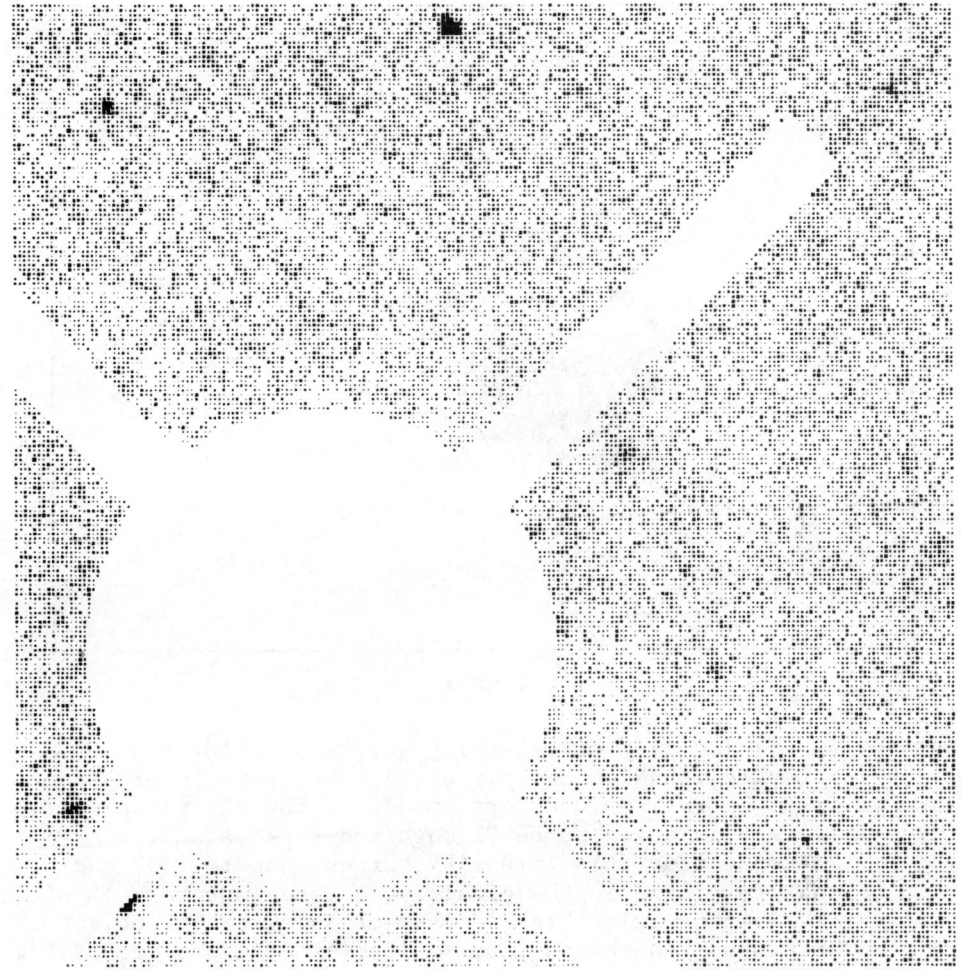

Fig. 5. The star profile has been subtracted and the core of the star and the diffraction spikes have been excluded from further processing by the application of windows.

five pixels above threshold are rejected. Histograms are used to determine the threshold.

Once the objects have been identified, a two dimensional Gaussian plus background is fit to the unsmoothed data for each object. The fit includes the above threshold pixels as well as a neighborhood around these pixels which extends far enough to ensure that the background is well determined. The Gaussian contains six parameters: X coordinate, Y coordinate, total intensity, and three parameters related to the shape, size, and

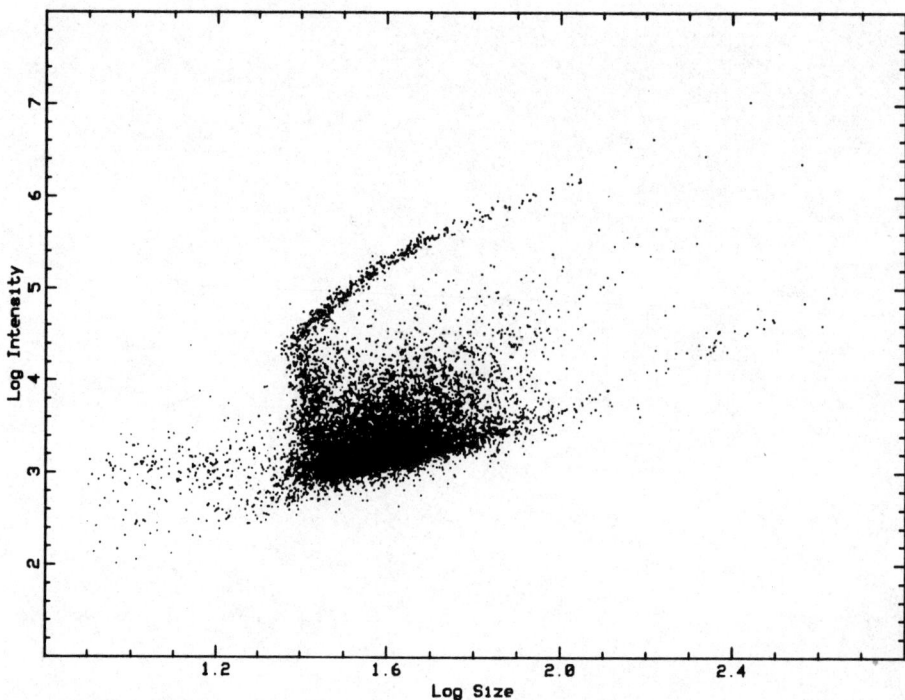

Fig. 6. A log-log plot of intensity versus size for the 12020 objects found on one plate. Intensity units are such that intensity=1000 corresponds to the sky per pixel. If sky is taken to be 22 magnitudes per square arc second, then this intensity corresponds to 23.7 magnitude. Size is the effective radius (square root of semi-minor axis times semi-major axis) of the object in microns. An object with a diameter of 1 arc second has log(size)=1.43.

orientation of the Gaussian. From the geometric parameters, the major and minor axes and position angle of the best fitting ellipse are derived. In the event that several objects are close together and identified as a single object by the detection algorithm, two or three Gaussians are fit simultaneously. Typically, 6% more objects emerge from the fitting procedure than were input.

The output of the fitting procedure is stored in a tabular data structure - each row contains data for one object and each column contains a particular parameter. Some of the columns contain diagnostic data such as the number of iterations while others contain the data of interest such as the intensity or the estimated error in the intensity.

Figure 6 shows a scatter plot of intensity versus size for the 12020 objects found on the plate in Figure 1. There are a number of features in this plot. The vertical band of objects near log(size)=1.4 which bends to the right at brighter intensities contains stars. (Star images all have the same size, so stars should lie along a vertical line. However, as more of the center of the image becomes saturated, the fitting procedure overestimates the size causing the star locus to bend to the right.) The cloud of points to the left of the star locus and and along the very bottom of the main cloud of points comes from objects (or noise) with only a few significant pixels - not enough for a simultaneous fit of seven parameters. The rest of the points are galaxies.

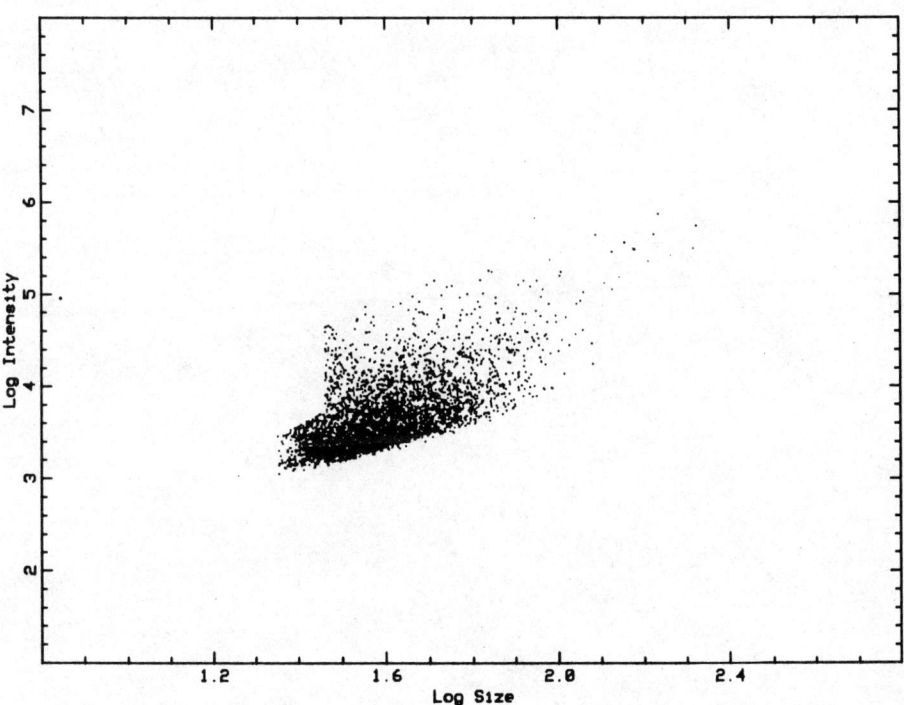

Fig. 7. The data of figure 6 after applying cuts to select high signal to noise objects and 'by hand' cuts to isolate galaxies. There are 4457 objects in this figure.

In order to clean up this plot, one can restrict the data to those objects in which the fit converged, the signal to noise (the ratio of the intensity to its estimated error) is greater than 10, the axis ratio (ratio of minor to major axis of best fitting ellipse) is greater than 0.1 and log(size) is greater than 1.35. When all these constraints are applied to the data, 5406 objects remain. The signal to noise constraint is responsible for most of the 6614 objects that were eliminated.

The remaining objects must be classified as either stars or galaxies. Unfortunately, an automatic classifier algorithm for this project has not yet been written. Eventually, techniques like those described by J. Friedman in this volume may be useful. However, for the purposes of illustration, a crude classification is made 'by hand' by designating all those objects in the previously mentioned star locus as stars and all others as galaxies. At the bottom of the vertical band of stars, the galaxy density overwhelms the stellar density, so all objects in this region are taken to be galaxies. The resulting plot for galaxies, containing 4457 objects, is shown in Figure 7.

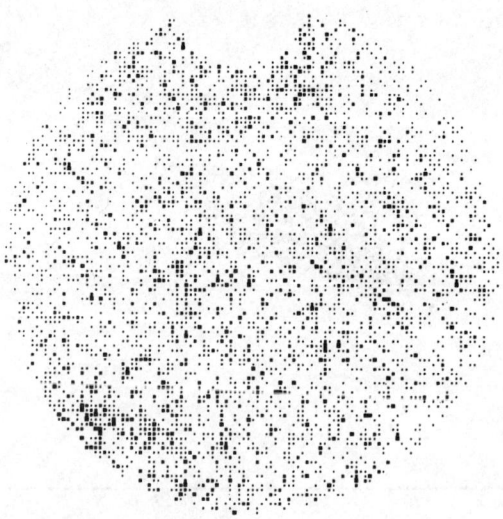

Fig. 8. The spatial density of the objects in figure 7. Each pixel represents the density of objects in a half arc minute square cell.

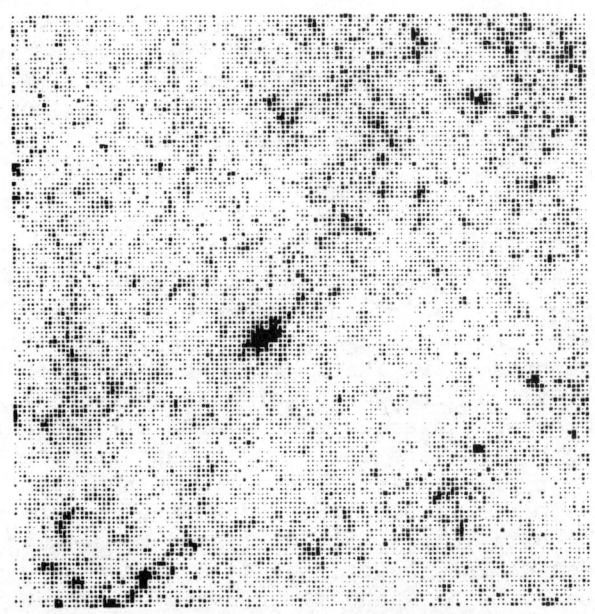

Fig. 9. A portion of the Shane Wirtanen galaxy map for comparison with figure 8. Each pixel represents the galaxy density (to 18.6 mag.) in a ten arc minute square cell.

The spatial density of these galaxies is shown in Figure 8 and for comparison, the density of galaxies in a portion of the Shane Wirtanen map (Shane and Wirtanen, 1967; Seldner et al., 1977) is shown in Figure 9. Note that Figure 8 shows a portion of the sky about 50 arc minutes in diameter, whereas the portion of the Shane Wirtanen map is 640 arc minutes on a side. Although difficult to tell from the visual impression made by these two figures, the distribution in Figure 8 is much smoother than that in the Shane Wirtanen map reflecting the fact that the 4m plate is much deeper than the Shane Wirtanen map so the apparent clustering is smoothed out by averaging over more clusters along the line of sight. Clustering is clearly seen in Figure 8. The Shane Wirtanen map is often described as containing filamentary structure and perhaps traces of similar structure can be seen in Figure 8 as well, despite the fact that it is so much smoother than the Shane Wirtanen map.

Software yet to be developed for this project will convert plate coordinates to astronomical coordinates (this requires generating a star catalog somewhat deeper than the SAO catalog so that there will be several stars on each plate); use the plate overlap regions to generate consistent magnitude scales and determine the reliability of the algorithms; merge the plate catalogs into a single catalog; and provide a more reliable galaxy-star-noise classification than the 'by hand' method illustrated in Figures 6 and 7. Once the merged catalog has been constructed, the problem of the clustering of galaxies can finally be investigated with these data!

CONCLUSION

In this paper, I have tried to show some of the image processing techniques used to generate scientific results from astronomical data. As our image processing capabilities grow, so do our requirements. The Shane-Wirtanen catalog of galaxies was constructed by visual inspection of plates with a microscope. Over 1300 plates were examined and over a million galaxies were counted. The results were recorded in hand written form on log sheets - one per plate - and gave only the number of galaxies in each 10 by 10 arc minute cell. When the results were first published, the data volume was compressed by summing the counts into one degree cells. Later, the data sheets were entered into a computer data base and distributed on magnetic tape. Today, digital image processing gives us the capability to construct large catalogs with much more information than simply the count per cell. The catalog can include positions, magnitudes, shapes, sizes, colors (if several color plates are available), and even crude redshifts (if many colors are available). In the future, even more capabilities will be available and these too will come to be regarded as image processing requirements.

I would like to thank the organizers of this workshop for creating a productive and enjoyable conference. I also thank J. Peebles for inspiration and E. Spillar, R. Fishman, and U. Lindqwister for assistance. This research has been supported in part by NASA through contract NAS5-25084.

REFERENCES

Butchins, S. A., 1982, Automatic Image Classification, *Astron. Ap.*, 109:360.
Carter, D., and Godwin, J. G., 1979, Photometry of the Cluster of Galaxies A1146, *Mon. Not. Roy. Ast. Soc.*, 187:711.
Godwin, J. G., and Peach, J. V., 1981, Photometry of the Cluster of Galaxies A1367, *Mon. Not. Roy. Ast. Soc.*, 200:733.
Greisen, E. W., and Harten, R. H., 1981, An Extension of FITS for Groups of Small Arrays of Data, *Astron. Ap. Suppl.*, 44:371.

Grueff, G., Vigotti, M., Wall, J. V., Benn, C. R., A Deep Radio and Optical Survey Near the North Galactic Pole - II. Magnitudes and Colours of Objects in the Optical Fields of 5C 12 Radio Sources, *Mon. Not. Roy. Ast. Soc.*, 206:475.
Herzog, A. D., and Illingworth, G., 1977, The Structure of Globular Clusters. I. Direct Plate Automated Reduction Techniques, *Ap. J. Suppl.*, 33:55.
Jarvis, J. F., and Tyson, J. A., 1981, FOCAS: Faint Object Classification and Analysis System, *A. J.*, 86:476.
Koo, D. C., 1981, Multicolor Photometry of the Red Cluster 0016+16 at z=0.54, *Ap. J. Letters*, 251:L75.
Koo, D. C., and Kron, R. G., 1982, QSO Counts: A Complete Survey of Stellar Objects to B=23, *Astron. Ap.*, 105:107.
Kron, R. G., 1980, Photometry of a Complete Sample of Faint Galaxies, *Ap. J. Suppl.*, 43:305.
MacGillivray, H. T., Martin, R., Pratt, N. M., Reddish, V. C., Seddon, H., Alexander, L. W. G., Walker, G. S., and Williams, P. R., 1976, A Method for the Automatic Separation of the Images of Galaxies and Stars from Measurements Made with the COSMOS Machine, *Mon. Not. Roy. Ast. Soc.*, 176:265.
Peterson, B. A., Ellis, R. S., Kibblewhite, E. J., Bridgeland, M. T., Hooley, T., and Horne, D., 1979, Number Magnitude Counts of Faint Galaxies, *Ap. J. Letters*, 233:L109.
Reid, N., and Gilmore, G., 1982, New Light on Faint Stars - II. A Photometric Study of the Low Luminosity Main Sequence, *Mon. Not. Roy. Ast. Soc.*, 201:73.
Sebok, W. L., 1979, Optimal Classification of Images into Stars or Galaxies - A Bayesian Approach, *A. J.*, 84:1526.
Seldner, M., Siebers, B., Groth, E. J., and Peebles, P. J. E., 1977, New Reduction of the Lick Catalog of Galaxies, *A. J.*, 82:249.
Shane, C. D., and Wirtanen, C. A., 1967, The Distribution of Galaxies, *Pub. Lick Obs.*, 22:1.
Shanks, T., Stevenson, P. R. F., Fong, R., and MacGillivray, H. T., 1984, Galaxy Number Counts and Cosmology, *Mon. Not. Roy. Ast. Soc.*, 206:767.
Strom, S. E., Forte, J. C., Harris, W. E., Strom, K. M., Wells, D. C., and Smith, M. G., 1981, The Halo Globular Clusters of the Giant Elliptical Galaxy Messier 87, *Ap. J.*, 245:416.
Valdes, F., Tyson, J. A., and Jarvis, J. F., 1983, Alignment of Faint Galaxy Images: Cosmological Distortion and Rotation, *Ap. J.*, 271:431.
Wells, D. C., Greisen, E. W., and Harten, R. H., 1981, FITS: A Flexible Image Transport System, *Astron. Ap. Suppl.*, 44:363.

REVIEW OF METHODOLOGIES IN DIGITAL IMAGE PROCESSING

Andre Gagalowicz

INRIA
Domaine de Voluceau-Rocquencourt
BP 105, 78153 Le Chesnay Cedex
France

ABSTRACT

We present a brief review of some new methodologies in digital image processing. The aim of image processing is to "understand" automatically images and 3-D scenes by computer. The various steps leading to this point are commonly divided in two major parts: image processing, properly speaking, and image analysis. Image processing techniques have been already largely developed and are well known from the astronomers' community so that we will emphasize image analysis techniques and simply mention some new results in image processing which may be of interest to the reader.

INTRODUCTION

Industrial systems and scientific phenomena are more and more controlled, tested and sensed with the use of images obtained from cameras, telescopes, microscopes, etc. The aim of digital image processing was primarily to help a human observer to interpret specific events occurring through these images, but we are more and more inclined to suppress the observer. We desire to ask the computer to interpret images automatically in the same way as this human observer would interpret them. This interpretation may eventually induce actions for the system (actions of a robot for example).

All the loop starting from the image sensing and ending up with its interpretation is of image processing concern for the non specialists community. In fact, "image processing people" separate this field into two parts: image processing and image analysis (see Fig. 1). For them, image processing corresponds roughly to all the techniques designed as an aid to image interpreters. Image analysis

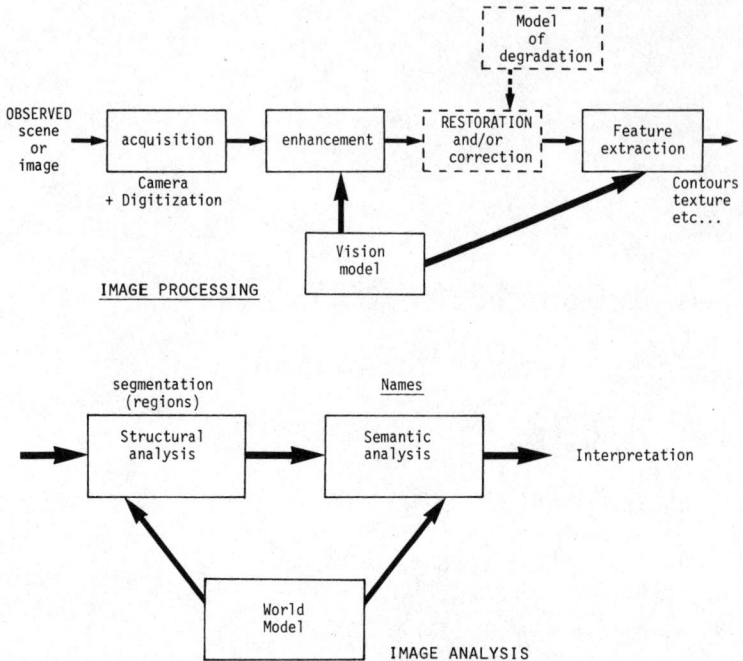

Figure 1

corresponds to those which are aimed to interpret the image automatically, by computer and thus replace human interpreters. Input to image analysis is the output of image processing.

IMAGE PROCESSING

The techniques of digital image processing are divided commonly in four sections as indicated in Fig. 1.

Acquisition

The first one is the acquisition. A scene is observed by an image acquisition system which can be a camera, a telescope or a microscope for example. The analogous image is then digitized and quantized by a A/D converter in order to obtain a computable representation of the image. The acquisition phase commonly converts a rectangular analogous image into a matrix of values coded on 8 to 32 bits per point (called pixel). This matrix is called a digital image (see refs. 1-3 for more precise considerations).

Enhancement

The image obtained through the acquisition system has undergone a certain number of degradations which can be geometrical distortions, blurring and/or noise. The first problem to solve through digital image processing techniques was to "undo" these degradations and have appear details, non visible in the original image, but of importance for the observer.

Two strategies are available, which can be eventually combined, and which depend on the task to be fulfilled. The simplest one is called image enhancement. The quality of the image is heuristically improved in a quick way using ad hoc methods which does not require knowledge of the model of the degradation. Enhancement has to be achieved in regard to the human visual system so that it is highly advisable to use a vision model in order to perform this task.

Point enhancement, local high-pass and low-pass filtering, use of pseudo-color are the most popular techniques used. A very efficient noise cleaning technique preserving edge information has to be specially mentioned: median filtering (see (1)), which is nevertheless limited to the case of elimination of isolated noise points in uniform regions.

Image Restoration and Correction

If the degradation of the acquisition system is important (optical or geometrical distortion) image enhancement techniques are not sufficient to retrieve the information of the "ideal image". It is then necessary to analyze the deformation, obtain a model of it and "undo" the deformation.

All these problems are of image restoration concern. Geometrical distortions have a specific nature and restoration techniques retrieve this specificity. Specialists call these techniques: image correction.

An excellent survey of restoration techniques can be found in (4). The design of restoration filtering is very often heavy computationally. Following S. Attasi's ideas (5) it is possible to reduce this computing time by using recursive approximations of these filters (6,7,8). Such techniques extend Kalman filtering to the case of 2-D signals. I would like to mention and discuss a little bit more extensively the case of adaptive noise filtering where results are very interesting and may be used for astronomical data.

High frequency noise is usually attenuated by the use of low-pass filters. Kalman filters of Wiener filters are optimum for the mean square error between the desired and the obtained images in the case of additive white noise independent of the signal. unfortunately, such low-pass filters also blur the image edges which are very important semantically, so that instead of improving the quality of an image, results are worse than the original image.

The idea of adaptive filtering is to restore the image in a non-homogeneous way: in zones containing no edges, filter the noise out drastically using Wiener filter for example; in zones containing edges, low-pass filter the image only a little or not at all, because we could destroy the edges and because the visual system is not very sensitive to noise around an edge.

The main problem is to determine automatically the existence of an edge which is made difficult due to the presence of noise in the image. Anderson and Netravali (9) proposed the use of a visual masking function:

$$M(i,j) = \sum_{x=i-k}^{x=i+k} \sum_{y=j-k}^{y=j+k} C^D (|\tfrac{\delta f}{\delta x}| + |\tfrac{\delta f}{\delta y}|)$$

$$C = .35$$

$$D = ((x+i)^2 + (y+j)^2)^{\tfrac{1}{2}}$$

where $\tfrac{\delta f}{\delta x}$ and $\tfrac{\delta f}{\delta y}$ are the components of the gradient of the image luminance $f(i,j)$. A great value of M accounts for a great probability of edge. The filter was tuned using the constraint $V(i,j) \cdot M(i,j) =$ cte, where $V(i,j)$ is the amount of the noise reduction brought by the filter. This work has been extended in a more formalized version by Abramatic and Silverman (10).

An example of the results obtained by Abramatic and Silverman is shown in Fig. 2.

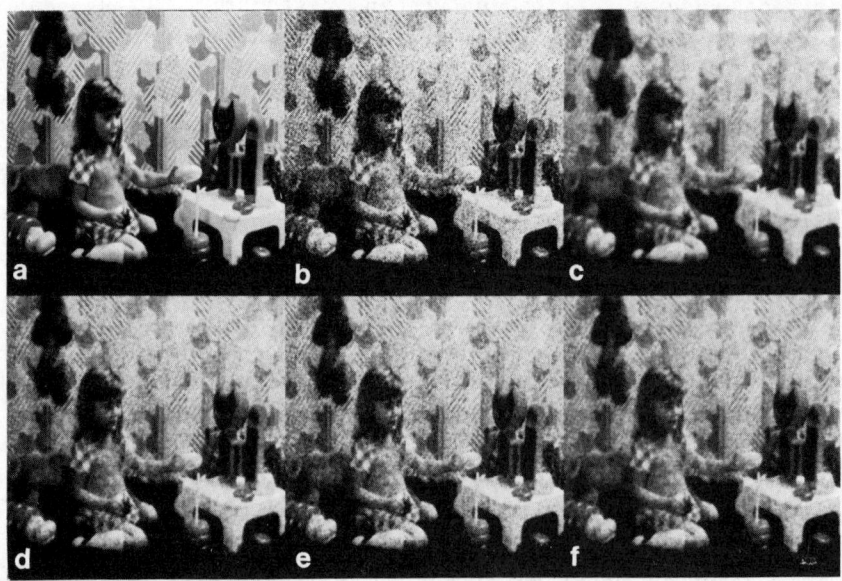

Figure 2: Adaptive filtering results (from Abramatic and Silverman (10)).

- Subpicture a shows an ideal image.
- Subpicture b is the input image. It has been obtained by artificially adding white noise to subpicture a.
- Subpicture c presents the result of Wiener filtering where the blurring effect is noticeable.
- Subpicture d gives the result of adaptive filtering where the masking function has been computed on the ideal image which is theoretically unknown. It gives the best solution attainable using adaptive filtering technique.
- Subpicture e shows the result of adaptive filtering when the masking function is computed on the original image b; noise reduction is small.
- Subpicture f utilized the masking function computed on the output of the Wiener filter c. This result is available without knowing the ideal image and this image is definitely better than the original image b.

Feature Extraction

After the third step, we have obtained and "ideal" image or at least the best image of the observed scene computable from the acquisition system. It is then possible to extract the useful information from the image. This information may be the surfaces reflectance or orientation, edges, textures or motion parameters in the case of successive frames.

Reflectance. Given an object relief map, the position of light sources, the position of the observer and the form of the reflectance function it is possible to extract the reflectance (or albedo) of a surface element. Very often, surfaces are supposed to be Lambertian for simplification, but accurate extraction is difficult for natural surfaces.

Surface orientation. It is also possible to extract the surface orientation if the position of the light source, of the observer and if the surface reflectance are known. Horn's work is the most noteworthy in this domain (see (11)).

Edge detection. Edge detection has been extensively studied (see refs 1-3). We would like simply to mention one type of edge detection which happens to be very powerful for certain kinds of images. This method proposed by Marr and Hildreth (12) and simplified by N. Keskes (13) is called the zero crossing method.

An edge point is localized by a zero value of a second order derivative of the image surface. N. Keskes (13) proposed the use of the difference of two low-pass filters of various sizes as an estimate of this second order derivative.

The image has to be convolved by two convolution masks M_3 and M_7 of size 3 x 3 and 7 x 7, where the value of the convolution mask is constant, and compute the difference between the two filtered images.

$$M_3 = \begin{bmatrix} 1/9 & 1/9 & 1/9 \\ 1/9 & 1/9 & 1/9 \\ 1/9 & 1/9 & 1/9 \end{bmatrix}$$

Similarly, M_7 values are all equal to 1/49. These convolutions may be computed very quickly and are now in the process of being implemented in hardware with VLSI circuits. The zero-crossing points are obtained where the difference image points change their signs. Of course for flat regions, though there is no edge, this difference is also zero so that N. Keskes proposed to add the computation of the gradient values and keep the zero crossings only when this gradient is above a given threshold.

An example of such a technique is described in Fig. 3. Subimage a is an original image showing relay parts lying on a belt and seen from a camera fixed on a robot which has to recognize and grasp them. Subimage b is the result of the zero-crossing algorithm. Subimage c shows the locations of the gradient thresholded at 0.1 and the final result is presented on subimage d which is simply a logical AND between subimages b and c.

<u>Texture extraction</u>. Texture information is spatial (2-D). It is a property of surface on the opposite of edges which describe curves (1-D) on the plane. Texture characterization has been studied extensively since 1960 approximately but until 1982 the concept of texture was not well understood. It is a property of surface perceived by the visual system as homogenous (stationary). We give in (14) (see also (15) a review of all the methodologies studied in the literature and by ourselves and present a texture model obtained after a series of psychovisual experiments, which has proven to be very efficient for texture classification (16) and synthesis.

We have shown in (14) that a good model for homogeneous textures is a set of second order spatial averages $\tilde{p}_\Delta(L_1, L_2)$ where:

$$\tilde{p}_\Delta(L_1, L_2) = \frac{1}{I} \sum_{i=1}^{I} \delta(X_i - L_1)\,\delta(X_{i+\Delta} - L_2), \quad \forall L_1, L_2 \in L$$

for $||\Delta|| \leq 9'$ of solid angle approximately.

- X_i is the luminance of the texture field at location i of the plane.
- L_1, L_2 are possible grey level values belonging to L ($L=\{0,1,..,L-1\}$)
- δ is the Kronecker function.
- $\Delta = (\Delta x, \Delta y)$ is a translation in the texture plane.
- I is the number of pairs of points $(i, i+\Delta)$ in the texture field plane.
- $\tilde{p}_\Delta(L_1, L_2)$ is the number of co-occurences of pairs of luminances L_1, L_2 when their locations differ by Δ.

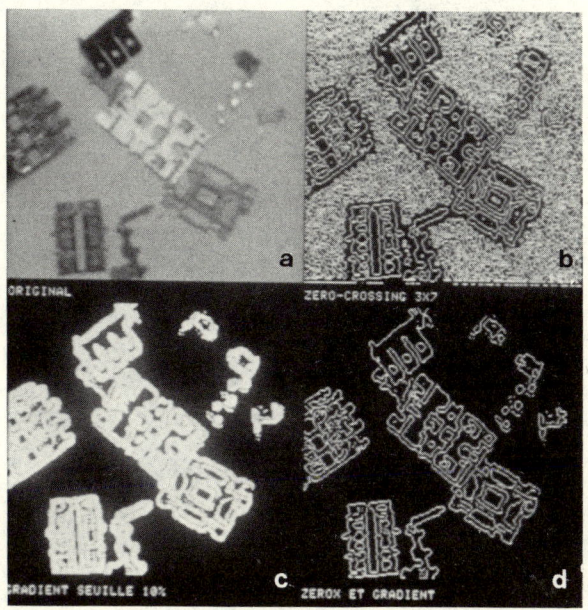

Figure 3: Edge detection of relay parts by a zero crossing method (from N. Keskes (13)).

For normal conditions of view 9' of solid angle corresponds to translations of size 10 to 12 pixels in the x and y directions. We have designed an algorithm which allows the synthesis of a texture image from its model, that is to say its second order spatial averages, and have verified that a homogenous natural texture and its artificial synthesis are almost impossible to discriminate visually (see Fig. 4).

Figure 4 shows synthesis results in the case of natural WOOL. Subimage a presents an image of natural WOOL; subimage b the synthesis of WOOL using a control window of size 4' of solid angle (4 pixels); subimage c is that of a 8' of solid angle control and subimage d the case of the synthesis where the second order spatial averages are controlled for 12' of solid angle. Such experiments prove that the model proposed retrieves all textural information and is thus a good model to use.

Motion characterization. This subject involves the study of successive image frames and has been skipped as we consider that it is beyond the scope of our paper.

IMAGE ANALYSIS

Feature extraction is commonly used as input to image interpretation. Image analysis is commonly subdivided in two parts: structural

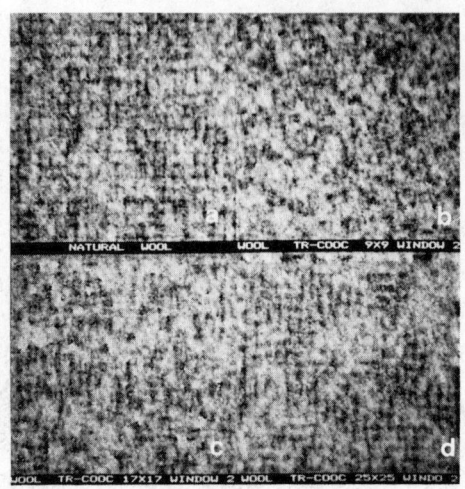

Figure 4: Natural WOOL synthesis (from (14)).

analysis, where regions having similar features are created and semantic analysis, where names are given to these regions. All these techniques have to be driven by a world model (see Fig. 1).

Structural Analysis

We have to assign points to a finite number of classes. Connected points of a given class will then be grouped into homogeneous connex regions. Segmentation of an image is achieved mainly in two ways: region growing or contour closing.

Region growing. The region growing technique starts with randomly chosen points (or small zones) of the image and each region is grown while testing if the frontier points of the growing region may be merged to this region or not. Features of the frontier points and of the growing zones are compared for those purposes.

Contour closing. Contours obtained with edge detection techniques also define limits between zones. These edges present holes which have to be filled up. An easy way to obtain zone boundaries immediately is to use the zero crossing technique because the difference image is made up of positive, negative and zero values and it is always possible to follow the boundaries of the positive (and negative) zones, and delimit zones, this way. A limitation to this procedure is that we obtain luminance edges but up until now, there is no good texture boundary detector so that such a method will work poorly on textured regions.

Mathematical morphology. Mathematical morphology developed by Matheron (17) and Serra (18) is very useful for the study of connectivity for region segmentation.

i. **Connectivity of two points.** On a square grid, two definitions are commonly used: one point M is connected to another one P iff:
. M and P have a common edge (4-connectivity)
. Or M and P have at least a common vertex (8-connectivity).

ii. **Connected path between two points.** A connected path between two points P and Q is an ordered set of points $M_1 = P, M_2 M_3, \ldots, M_n = Q$ such that $\forall i$, M_i and M_{i+1} are connected points.

iii. **A connected region.** One set S of points is connected off $\forall(P,Q)$, a pair of points of S, a connected path between P and Q exists.

iv. **Erosions and expansions.** Let X be a set of points in an image and B another set called structural element. One of the points of B, y_0 is special and called "Center" of B. By is the translated version of B when $y_0 \to y$. Let B' be the symmetrical of B through y_0.
- The expansion of X by B denoted $Y = X \oplus B'$ is the set of points $Y = \{y : By \cap X = 0\}$.
- The erosion of X by B denoted $Y = X \ominus B'$ is the set of points $Y = \{y : By \subset X\}$.
- Erosions and expansions are dual with respect to complementarity: $(X \oplus B')^c = X^c \ominus B'$, which shows that it is the equivalent to erode points of X or expand the background of X.

v. **Opening and closing.** The opening of X by B denoted X_B is defined by:
$X_B = (X \ominus B') \oplus B$
The closing of X by B denoted X^B is symmetrically:
$X^B = (X \oplus B') \ominus B$
Opening is an erosion followed by an expansion. It tends to smooth the contours of X and suppress small irregularities in the image (see (18)). Closing is the dual of opening with respect to complementarity:
$(X_B)^c = (X^c)^B$ and $(X^B)^c = (X^c)_B$
Closing of X tends to smooth the background of X. Mathematical Morphology revealed itself as being a very fruitful image processing technique in a lot of applications of Physics though it was specially designed for granulometry studies. It gave rise to a very popular hardware system called Leitz tecture Analyzer.

Semantic Analysis

After the region segmentation, we have now to interpret the image. For these purposes it is necessary to describe the obtained regions in a condensed way, very often symbolically.

Regions descriptions. It is possible to represent a region by a feature vector whose components measure certain of its quantitative properties.

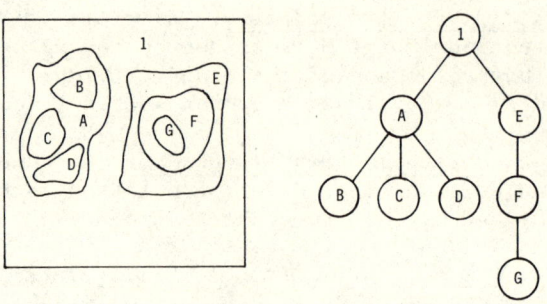

Figure 5: Topological Tree (from (19)).

i. **Topological attributes.** Among classical attributes associated to a region we have the number C of connected components, the number H of holes and the Euler number E = C - T, which are invariant through the algebraic group of affine similarities.

ii. **Metrical attributes.** It is possible to use the perimeter P, the area A, the thinness $TH = P^2/4\pi A$ and the elongation $L = \lambda_1/\lambda_2$ where λ_1 and λ_2 are the principal inertia moments. One can also compute the max. and min. diameters with respect to the center of gravity, the area of the region's convex hull and so on. Generalized moments are also often used:

$$m_{p,q} = \sum_i \sum_j i^p j^q f(i,j)$$

where (i,j) describe X and f(i,j) is the luminance at location (i,j).

Semantic descriptions. It is also necessary to describe the geometrical relationships between various regions. This is done mainly in four possible ways.

i. **Topological trees.** We define a tree whose nodes are the connected regions of the image. Moving down the tree reflects the inclusion of one region in another one. The tree root is the whole image (see Fig. 5).

ii. **Adjacency graphs.** Nodes of the graph are the connected regions of the image and arcs of the graph connect adjacent regions. (See (20) and Fig. 6).

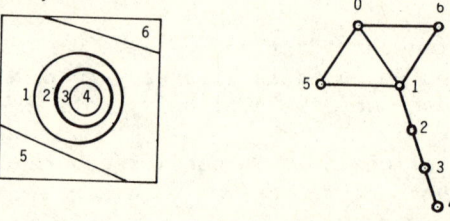

Figure 6: Adjacency graph (from Pavlidis (20)).

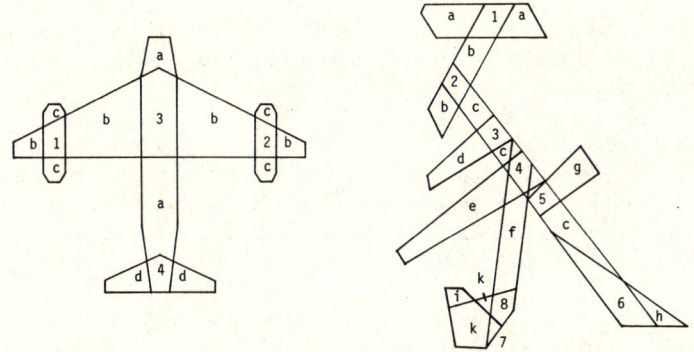

Figure 7: Graph of convex components (from Pavlidis (20)).

iii. Graphs of convex components. It is always possible to represent a connected region by a graph of convex components (see (20) and Fig. 7). This representation is difficult to compute and is not unique.

iv. Semantic graph. Is a graph in which each node represents a connected region with its attributes and the arcs of the graphs express relationships between regions as shown in Fig. 8.

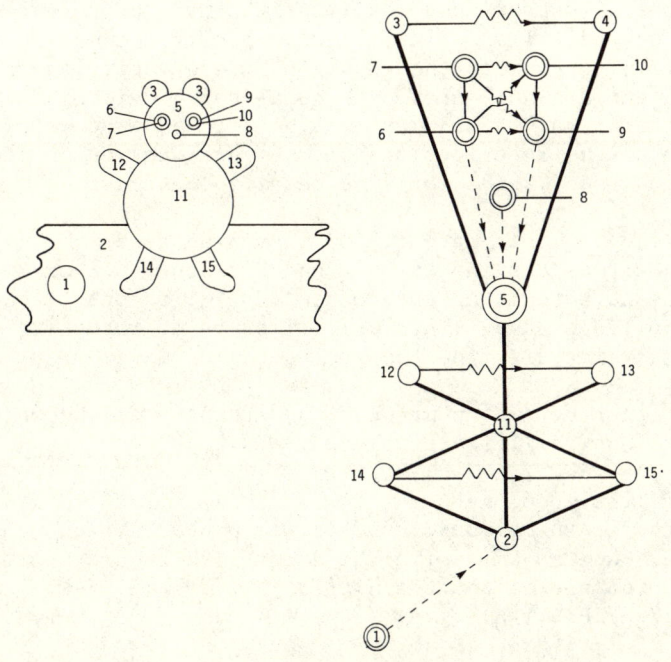

Figure 8: Semantic graph (from (21)). ⊙ Designs circular zones. Arcs denoted by —, - →, ⤳ express respectively adjacency, inclusion and left positioning of regions.

353

It is also possible to use the contour of the regions for the semantic descriptions of the image (see (20,22,23)). This dual approach is skipped for compactedness.

Image interpretation. Some methods use global features of regions (or objects): they can be applied only in the case where all regions (or objects) are isolated on the scene. In the case of occlusions former methods fail and specific techniques have to be invented.

i. Case of isolated objects. In the most general framework we are dealing with images of 3-D objects so that one object has got a certain number of possible silhouettes on the image.

a) Learning. The first step is to create the world model also called knowledge representation. This is done by learning (see (24,25) for more detailed study). To each silhouette S_j of an object S is associated a feature vector X_j. Each couple (S_j^j, X_j) is a model of S. If the objects lie on a plane parallel to the image plane, it is sufficient to compute only one model per stable state of an object as used attributes (see Regions descriptions) are always rotation and translation invariants.

If the object orientation is arbitrary, the number of models becomes theoretically infinite. Practically, we may restrict this set to a finite number (though big) while quantifying the rotations of the object around 2 axes orthogonal to the vision axis, (see (26, 27)). Statistical models may also be defined more generally to take into account the imperfections of the segmentation, noises in the acquisition (see (22,28) for more detailed discussions).

b) Recognition. Each region localized after the structural analysis defines a silhouette S_0 and a feature vector X_0 associated to it. We have to match the couple (S_0, X_0) to one of the models of the knowledge base. This will allow us to recognize and position the object related to the considered region S_0.

Pattern recognition methods, now rather well known, (see (22,28)) are used. We recall the simplest ones.

- Maximum likelihood method (22)
In the learning phase conditional probabilities $p(X/S_j)$ have been evaluated. One may minimize the error risk when associating S_0 to the most probable silhouette S_j if:
$p(S_j/X_0) \geq p(S_i/X_0) \; \forall i$

- Closest neighbor method
We choose the closest feature vector X_j from X_0 in the feature space and the silhouette S_j associated to it:
$||X_j - X_0|| \leq ||X_i - X_0|| \; \forall i$.

Other distances taking into account the variances of feature vectors associated to each model have also been developed (see (29)).

- K-nearest neighbors method (27).

If, for each silhouette S_j, we define a set of representative feature vectors $X_{j,i}$ it is possible to classify S_0 having X_0 as feature vector in the following way: we take the K-nearest neighbors of X_0 in the feature space and take the silhouette S_j containing the greatest number of feature vectors in this set of K vectors.

- Decision tree method

The important drawback of former methods is that they imply the computation of N distances between X_0 and all X_i corresponding to the N models of the knowledge base. An interesting alternative (30) consists of the successive comparison of certain components x_k of the feature vectors, to successive thresholds s_k and reduce after each comparison the number of potential silhouettes S_j to which S_0 may be compared. The learning procedure must be adapted to such a decision tree. It is a binary tree whose nodes contain the name of the component x_k of X to test and the threshold value s_k. The leaves of the tree contain the name of one silhouette S_j. At the level of the root of the tree, we analyze the component x_1 which clusters the best the set of model attributes X_j in two subsets and so on at each node of the tree. This way in the ideal case where half of the silhouettes are rejected at each test, the tree depth is only $Log_2 N$. This method is robust only if each test is robust (very important errors in recognition occur if simply one decision was badly taken).

c) <u>Positioning</u>. As soon as a silhouette S_0 was identified with a model S_i, it is possible to compute the affine similarity which transforms S_i into S_0. This enables the position of the observed object relative to the model to be obtained.

ii. <u>Case of partially visible objects</u>. In the case of occluded objects or if the number of models increases drastically, the global methods described above, which present the advantage of being simple and easy to implement, fail. More powerful methods must thus be used.

a) <u>Hough transform</u>. This method utilizes silhouettes information through their contours. We take an analytical formulation of a family of curves C_α of $\mathbb{R}^2$ of the form $f(z,\alpha)=0$, where $z=(x,y)$ is a point of $\mathbb{R}^2$, and α is a parameters vector. Given a set of contour segments in an image represented numerically by a set of points $z_i=(x_i,y_i)$ we would like to know if one or several z_i belong to a curve C_α whose parameters have also to be determined. The idea is to compute for each z_i the set of parameters α which verify $f(z_i, \alpha)=0$. It is a surface in the parameter space whose equation is $f(z_i,.)$, which corresponds to all curves C_α passing through z_i.

When two points z_i and z_j belong to the same curve C_α, $f(z_i,.)$ and $f(z_j,.)$ have a common intersection at α. To avoid the computation of intersections of surfaces, the parameter space is quantized in an hypercube considered as an accumulator A. For each point z_i of the contours, we increment by one all the points of the hypercube where $f(z_i,.) = 0$ and finally detect the maxima of A. To each local maximum of A corresponds a curve C_{α_0} and the height of $A(\alpha_0)$ corresponds to the number of points z_i belonging to this curve. Duda (31) used this technique to detect line segments, Tsuji (32) could detect ellipses and Ballard (33) showed the way to generalize this technique to general curves.

The Hough transform is interesting because it requires only a low level description of images (contour points plus eventually the orientation of the tangent of the contour points) and because all the partial occurrences of a model are obtained in one pass. These advantages have their counterparts. The memory space and the complexity of the algorithms become enormous rapidly and the choice of a good quantization of the parameter space is cumbersome.

b) <u>Tree search techniques.</u> Two very similar methods are proposed in references (34) and (35) to interpret real images using corners, line segments and circular holes on contour building silhouettes. When a model is selected as a candidate each of its primitives is matched with a primitive of similar nature in the image. Position and orientation constraints are then deduced for the matching of other primitives of the model. A tree search algorithm called A* (see (36)) allows a matching to be obtained which tends to maximize the number of identified primitives while minimizing position errors between matched primitives. These methods seem efficient when silhouettes contain only corners, line segments and circular holes but the number of primitives per model is small (≥ 8) in the proposed examples which avoids the combinatory explosion. Furthermore these methods were not tested in the case of variable homothetic factors which is also a limitation. More general tree search techniques are applied on semantic graphs in reference (37) but all examples are synthetic.

c) <u>Maximal cliques technique.</u> Ambler (38) proposed the following method: initially all local primitives detected in the image (circular holes, corners, line segments, etc.) are matched with model primitives of the same nature. A compatibility graph is built afterwards. Its nodes are made of each of the couples (image primitive/ model primitive) previously formed and its arcs connect compatible nodes. The compatibility of two nodes is determined by the relative positions of the primitives in the image and in the model. An algorithm of maximal cliques research determines the greatest subset of nodes, all compatibles two by two, and leads to the identification (or not) of the model. This method is time consuming because the number of nodes N in the graph grows rapidly with the number of primitives of the model (m) and of the image (n). This upper bound

of N is indeed m*n. Furthermore, the number of arcs of a clique
of size N' is N'!!!.

Bolles (39) solves this problem partially while reducing the set
of primitives of the image to subsets of "similar" primitives containing three to four elements, and applying the former method only to
one of the subsets of "similar primitives" at a time. The potential
positions of the model deduced from thid analysis are confirmed or
not using tests of global superposition of the primitives.

d) <u>Relaxation techniques</u>. Relaxation techniques are iterative
processes to modify a labeling of primitives using local measures on
these primitives and a world model described by a set of local constraints between neighboring primitives (neighborhoods are described
in the wide sense by a graph of neighboring primitives) (see (40,41,
42)). Such techniques were used successfully for the interpretation
of various types of images including aerial photographs described
by semantic graphs (see (41)). Their application to the study of
occluded objects described by the polygonal approximation of their
contours was introduced by Bhanu (43) and developed by Ayache (44).

e) <u>Correlation methods</u>. Perkins (45-47) proposes a representation of edges in a 2-D space (θ,s) where s is the curvilinear abscissa
and $\theta(s)$ is the orientation of the tangent to the edge at s. A first
segmentation of the edges in circle arcs and line segments leads to
a first selection of models. A model is tested while computing the
cross correlation between a part of its contours and a part of the
contours of the image. This correlation is computed in the (θ,s)
space. The maximum of the correlation gives a position of the model
in the scene. This position is confirmed or rejected by a test of
global superposition of the model contours. This method gives good
results on industrial scenes but requires heavy computations and is
not adapted to variations in the scale factor.

Dessimoz (19, 48-50) used a similar method replacing simply the
orientation by the curvature of the contour $K(s) = d\theta(s)/ds$. Mc Kee
(51) used the correlation of Freeman codes of model and images contours, but his method was restricted to the case of one object on
the scene.

f) <u>Recursive methods</u>. These methods refine the estimation of
the position of the model in the scene while more and more parts of
the model are recognized (see (52-55)). The a priori estimation of
the position of the model is given either by an external system
(52,55) either defined in a non-unique way by labeling a subset of
primitives (44,53,54). An a priori estimation of the position of
the model allows a certain number of possible matchings of primitives
to be deduced, which themselves allow new possible matchings (recursivity).

g) <u>Syntactic methods</u>. Each model is given by a description of a set of symbols (representing primitives) and a syntax relating the various symbols. This syntax is described by a list of constructive rules applied to subsets of symbols. In the image analysis process, one looks for models with a syntax compatible to the greatest number of symbols (56). Those syntactic methods have been used for a long time in character or chromosome recognition (57). Their application to the description of more general scenes remains seldom. This is probably due to difficulties occurring in the modeling of complex scenes.

CONCLUSION

A lot of problems are still unsolved in image interpretation and will be very challenging in the near future as it will be a major task to solve for the control of robots or sophisticated scientific or industrial tasks.

REFERENCES

1. W. K. Pratt, "Digital Image Processing", John Wiley & Sons editors, London, (1978).
2. A. Rosenfeld, "Picture Processing by Computer", Academic Press, New York, (1969).
3. R. C. Gonzales and P. Wintz, "Digital Image Processing", Addison-Wesley Publishing Company, Reading MA., (1977).
4. H. C. Andrews and B. R. Hunt, "Digital Image Restoration", Prentice-Hall. Signal Processing Series, Englewood Cliffs, NJ., (1977).
5. S. Attasi, "Modélisation et Traitement des Suites à Deux Indices", Thèse de Doctorat d'Etat, Paris VI, (1975).
6. J. F. Abramatic, F. Germain and Emm Rosencher, "Design of Two-Dimensional Separable Denominator Recursive Filters", IEEE Trans. Acous., Speech and Signal Proc. 27(5), pp.445-453, (1979).
7. R. M. Mersereau, W. F. G. Mecklenbrauker and J. F. Quatieri, "Mc Clellan Transformations for Two-Dimensional Digital Filtering I and II Design", IEEE Trans. Cir. and Syst. 23(7), pp.405-422, (1976).
8. J. W. Woods and V. K. Ingle, "Kalman Filtering in Two Dimensions: Further Results", IEEE Trans. On Accoust. Speech and Signal Proc. Vol. 29, pp.181-197, (1981).
9. G. L. Anderson and A. N. Natravali, "Image Restoration Based on Subjective Criterion", IEEE Trans. Syst. Man and Cybern., 6(12), pp.845-853, (1976).
10. J. F. Abramatic and L. M. Silverman, "Nonlinear Restoration of Noisy Images", IEEE Trans. On Pattern Analysis and Machine Intelligence, Vol. 4(2), pp.141-149, (1982).
11. B. Horn, "The Image Dissector Eyes", MIT Artificial Intelligence Laboratory, Memo 178, (1969).

12. Marr and Hildreth, "Theory of Edge Detection", Report of the Artificial Intelligence Laboratory of the MIT, Memo 618, (1979).
13. N. Keskes, "Applications des Techniques d'Analyse des Images aux Signaux Sismiques", Thèse de Doctorat d'Etat, Paris VI, (1984).
14. A. Gagalowicz, "Vers un Modèle de Texture", Thèse de Doctorat d'Etat, Paris VI, (1983).
15. M. A. Song De, "Synthèse de Textures", Thèse de Doctorat de 3ème cycle, Paris VI, (1983).
16. R. W. Conners and C. A. Harlow, "Some Theoretical Considerations Concerning Texture Analysis of Radiographic Images", Proceedings of IEEE Conference on Decision and Control, (1976).
17. G. Matheron, "Random Sets and Integral Geometry", Wiley & Sons Inc., New York, (1975).
18. J. Serra, "One, Two, Three... Infinity: Quantitative Analysis of Microstructures in Materials Science, Biology and Medecine", J. L. Chernant, Editor, Dr. Riechener Verlag Gmbh, Stuttgart, West Germany, (1978).
19. J. D. Dessimoz, "Traitement des Contours en Reconnaissance de Formes Visuelles. Applications en Robotique", Thèse Ecole Polytechnique Fédérale de Lausanne, (1980).
20. T. Pavlidis, "Structural Pattern Recognition", Springer-Verlag, New York, (1977).
21. H. G. Barrow, A. P. Ambler, and R. M. Burstall, "Some Techniques for Recognizing Structures in Pictures Frontiers of Pattern Recognition", S. Watanabe, Ed., pp. 1-29, (1972).
22. R. O. Duda and P. E. Hart, "Pattern Classification and Scene Analysis", Wiley, (1973).
23. L. S. Davis, "Understanding Shape, I: Angles and Sides", IEEE Trans. On Comp., Vol. C-26, pp. 236-242, (1977).
24. P. H. Winston, "The Psychology of Computer Vision", Mc Graw-Hill, Computer Science Series, (1975).
25. P. H. Winston, "Artificial Intelligence", Addison-Wesley Series, (1979).
26. P. Coiffet and P. Rives, "Reconnaissance par un Robot de l'Orientation d'objets Tridimensionnels un Vue de Tâches de Saisie Automatique", RAIRO Automatique/Systems Analysis and Control, Vol. 14, No. 1, pp. 5-32, (1980).
27. S. A. Dudani, "Aircraft Identification by Moments Invariants", IEEE Trans. On Comp., Vol. C-26, No. 1, (1977).
28. J. C. Tou, and R. C. Gonzales, "Pattern Recognition Principles", Addison-Wesley, (1974),
29. E. Persoon et. al., "Shape Discrimination using Fourier Descriptors", IEEE Trans. On Systems, Man and Cybernetics, Vol. SMC-7, No. 3., (1977).
30. G. J. Agin, and R. O. Duda, "SRI Vision Research for Advanced Industrial Automation", 2nd USA-JAPAN Computer Conference, Tokyo, Japan, (1975).
31. R. O. Duda and P. E. Hart, "Use of the Hough Transform to Detect Lines and Curves in Pictures", Commun. Ass. Comput. Mach., Vol. 15, pp. 11-15, (1972).

32. S. Tsuji, and A. Kakamura, "Recognition of an Object in a Stock of Industrial Parts", Int. Joint Conf. on Art. Intel., Tbilissi, Georgia USSR, pp. 811-818, (1975).
33. D. H. Ballard, "Generalizing the Hough Transform to Detect Arbitrary Shapes", Pattern Recognition, 13, Vol. 2, pp. 11-112, (1981).
34. F. Attneave, "Some Informational Aspects of Visual Perception", Psychology Review, No. 61, pp. 183-193, (1954).
35. P. Rummel, and W Beutel, "A Model-Based Image Analysis System for Workpiece Recognition", Proc. of the 6th Int. Conf. on Pattern Recognition, Munchen (BRD), (1982).
36. Nils. J. Nilsson, "Principles of Artificial Intelligence", Tioga Publishing, Palo Alto, California, (1980).
37. L. G. Shapiro, and R. M. Haralick, "Structural Descriptions on Inexact Matching", IEEE Trans. on Pattern Analysis and Machine Intelligence, Vol. PAMI-3, No. 5, (1981).
38. A. P. Ambler, et. al., "A Versatile Computer Controlled Assembly System", Proc. IJCAI, Stanford, California, pp. 298-307, (1973).
39. R. C. Bolles, and R. A. Cain, "Recognizing and Locating Partially Visible Workpieces", Proc. of Conf. on PRIP, Las Vegas, Nevada, (1982).
40. A. Rosenfeld, R. A. Hummel, and S. W. Zucker, "Scene Labeling by Relaxation Techniques", IEEE Trans. on Syst. Man and Cyb., Vol. SMC-6, No. 6, (1976).
41. O. D. Faugeras, and K. E. Price, "Semantic Description of Aerial Images Using Stochastic Labeling", IEEE Trans. on PAMI, Vol. PAMI-3, No. 6, (1981).
42. O. D. Faugeras, "Decomposition and Decentralization Techniques in Relaxation Labeling", Computer Graphics and Image Processing, No. 16, pp. 341-355, (1981).
43. B. Bhanu, "Shape Matching and Segmentation Using Stochastic Labeling", Ph.D. Thesis, University of Southern California, USA, (1981).
44. N. Ayache, "In Système de Vision Bidimensionnelle en Robotique Industrielle", Thèse de Docteur Ingénieur, Paris Sud, Orsay, (1983).
45. W. A. Perkins, "Model-Based Vision Systems for Scenes containing Multiple Parts", 5th Int. Joint Conf. on A.I., (1977).
46. W. A. Perkins, "A Model-Based Vision System for Industrial Parts", IEEE Trans. on Comput., Vol. C-27, No. 2, (1978).
47. W. A. Perkins, "Simplified Model-Based Part Locator", 5th Int. Conf. on Pattern Recognition, Vol. 1, Miami Beach, Florida, (1980).
48. J. D. Dessimoz et. al., "Recognition and Handling of Overlapping Industrial Parts", Int. Symposium on Computer Vision and Sensor Based Robots, Warren, Michigan, (1978).
49. J. D. Dessimoz, "Visual Identification and Location in a Multi-Object Environment by Contour Tracking and Curvature Description", 8th Symposium on Industrial Robots, Stuttgart, (BRD), (1978).

50. J. D. Dessimoz et. al., "Recognizing and Handling of Overlapping Industrial Parts", 9th Symposium on Industrial Robots, Washington, (1979).
51. J. M. McKee, and J. K. Aggarwal, "Computer Recognition of Partial Views on Curved Objects", IEEE Trans. on Comput., Vol. C-26, No. 8, (1982).
52. C. A. Darmon, "A New Recursive Method to Detect Moving Objects in a Sequence of Images", Proc. of the Conf. on PRIP, Las Vegas, Nevada, (1982).
53. D. B. Genner, "A Feature-based Scene Matcher", Proc. 7th Int. Joint Conf. on A.I., Vancouver B.C., Canada, (1981).
54. C. S. Clark et. al., "Matching of Natural Terrain Scences", Proc. of. Conf. on PRIP, (1980).
55. H. G. Barrow, J. M. Tenenbaum, R. C. Bolks, and H. C. Wolf, "Parametric Correspondence and Chamfer Matching: Two New Techniques for Image Matching", Proc. IJCAI, pp. 659-663, (1977).
56. R. C. Gonzales, and M. G. Thomson, "Syntactic Pattern Recognition", Addison-Wesley, Reading (MA), (1978).
57. K. S. Fu, and P. H. Swain, "On Syntactic Pattern Recognition", Software Engineering, J. Tou Ed., Vol. 2, pp. 155-182, (1969).

2-D PHOTOMETRY

Massimo Capaccioli

Institute of Astronomy
University of Padova
35100 Padova, Italy

INTRODUCTION: BOUNDARY CONDITIONS

Strictly speaking, the projected images of all astronomical objects appear to possess a two-dimensional light distribution. In fact, due to the resolution constraints of the earth atmosphere and of the observational equipments (telescopes as well as detectors), even point-like sources such as distant stars and extremely remote galaxies are spread into <u>blurring disks</u> which sometimes call for a 2-D photometric investigation (see, e.g., Newell 1979, Agnelli <u>et al</u>. 1979, Zou <u>et al</u>. 1981, Kron 1983). However, this review will be focused only to the the scientific and technical aspects of 2-D photometry of classical <u>extended sources</u>, i.e. sources whose angular scale length is considerably larger than the characteristic length of the <u>Point Spread Function</u> (hereafter PSF). Among the class of celestial sources sharing this property (and to which most of the following considerations apply) we will select <u>galaxies</u>. Recent reviews on the same subject were published by Kormendy 1980, 1982, de Vaucouleurs 1983, Capaccioli 1984. Finally, we want to remark that our discussion refers only to studies in the <u>optical</u> <u>waveband</u> (from UV to near IR); observational methodologies and data reduction can differ very much in the other bands (X and gamma rays and radio domain) and thus require a separate analysis.

2-D PHOTOMETRY OF GALAXIES: GENERAL CONSIDERATIONS

With only very few exceptions, mainly the members

of the Local Group, galaxies are never resolvable into individual stars with the presently available ground-based telescopes (but very many will with the high resolution cameras of the Space Telescope). Already at the distance of the Virgo cluster (about 15 Mpc) medium size features such as small clumps of bright young stars or HII regions also become barely discernible against the smooth texture of the overall stellar light distribution and larger structures such as globular clusters acquire a quasi-stellar appearance. In other words most of the galaxies whose projected images can be investigated by a detailed 2-D mapping lack of mini-features; still they exhibit structured subsystems. Among the most obvious examples of the latter we may enumerate the spiral arms in intermediate and late Hubble types and the bars in lenticular and spiral galaxies (and possibly in some ellipticals too), as well as peculiar features such as dust lanes, rings, jets, plumes, bridges etc., which are the rule rather than the exception in galaxies of all morphological types, and which are not easily washed out by distance effects. To start with our discussion we will 'pro tempore' ignore such features and concentrate our interest on the underlying (and usually dominant) smooth stellar content of normal galaxies.

Our task is to review the state of the art of the technology of galaxy surface photometry, i.e. to examine how galaxy images are presently mapped by optical astronomers. While doing that we will try to keep in mind the scientific motivations of such a work in order to illustrate the needs for improvements or changes of route to those who may suggest new ideas and methodologies. This attitude qualifies the interaction between astronomy and science of data analysis; the first one poses the questions, provides the data and asks the second to procure software facilities to obtain the answers.

The first step of galaxy photometry consists of the transformation of the rough data coming from panoramic detectors into digital arrays of mean surface intensity values $I(x,y)$. The original material may be already in digital form and even in a quasi-linear photometric scale (e.g. CCD images), or may require off-line digitization (e.g. electronographic images) and linearization (e.g. conventional photographs). The many aspects of data pre-processing and the linearization techniques will not be commented here, since they differ much from one detector to another. For now we want to make clear that the derivation of the intensity array $I(x,y)$ represents just the starting point of the photometric analysis. Further and qualifying steps of the game are the <u>removal</u> of all the possible disturbances and the <u>resolution restoration</u> of the digital images, and then the <u>extraction of the information</u> needed for <u>astrophysical applications</u> such as the light profiles in one or more colors, the properties of substructures and peculiar features, the values of characteristic lenghts and of isophotal parameters, the total magnitudes, etc.

From the point of view of both the scientific expectations from surface photometry and the technical problems, three main regions may be identified in the optical images of <u>normal galaxies</u>, namely the central peak of light or <u>nucleus</u>, the intermediate range or <u>body</u> and the very faint <u>outskirts</u>. The nucleus is a bright spot whose light distribution is dominated by the atmospheric and instrumental convolution; it may be the site of strong activity and/or of density singularities, fenomena of key importance in the understanding the evolutional history of galaxies (Young et al. 1978, Schweizer 1978, de Vaucouleurs and Capaccioli 1979, Nieto 1984, Begelman et al. 1984). The body carries most of the visible light and is responsible of fixing the morphological (Watanabe et al. 1982), structural (de Vaucouleurs 1959, King 1978, Kormendy 1982, Capaccioli and Rampazzo 1984, Benacchio et al. 1984), dynamical (Capaccioli 1979 and references therein, Lake and Norman 1983, Capaccioli et al. 1984a) and ,in general, physical characteristics of the objects. Finally, the outskirts, which are at the limits of detectability against the sky background light and data noise, are linked to problems such as dynamical evolution of galaxies (for instance, truncated profiles in dwarfs; King 1966, Prendergast and Tomer 1971), background light in clusters and interactions with the environment (Quinn 1982, Thuan and Romanishin 1983, Malin 1983) and cosmological aspects as those of the missing mass, existence of population III, etc. These three regions need to be considered separately.

PHOTOMETRY OF THE NUCLEAR REGIONS OF GALAXIES

In intrinsic absorption-free systems usually the trend of the light profiles in the very inner regions is a steep decline from a bright central value. For instance, the mean B-band surface brightness at the very centers of ellipticals is equivalent to one star of about 14 mag per square second of arc. What is actually measured even with the highest achievable ground-based resolution is instead an intensity at least one order of magnitude fainter. This difference is clearly due to atmospheric blurring and to instrumental problems.

The latter concern sampling (detector and/or digitizing machine resolution), saturation, side-effects as Eberhard's in photographic emulsions (Nieto 1984) or adjacency effects in electronic detectors, binning for noise suppression, telescope guiding (bearing also on geometry of isophotes) and others which are specific to individual equipments. For instance, background removal problems (to be considered in the Section dealing with the faintest regions) may be severe in detectors with limited dynamical range such as photographic emulsions, where saturation may occur before high signal-to-noise ratios overall the image areas are reached to ensure a correct background estimation. Such obstacles are almost eliminated with modern detectors with large dynamical range, although their small fields of view pose other limitations on this matter for objects with large angular sizes (cfr. Leach 1981, Vigroux 1984, Kent 1984).

Another technical difficulty related to non-linear detectors is _calibration_. It is too often neglected that inaccuracies in the linearization function (poorly designed or badly used calibration devices) affect mainly the brightmost regions (de Vaucouleurs 1983). Paradoxically photographic photometry may be easier to do, and more reliable, at the low that at high density values.

Atmospheric and instrumental convolution represents the main barrier against our knowledge of the photometric properties of galaxy nuclei (needed, for instance, to establish whether massive black holes sit at the centers of some, even 'normal' galaxies). If we assume that the intrinsic 2-D light distribution is $J(x,y)$, and that the PSF may be represented by the parametrized function $G(x,y,p_1...p_n)$ (but a digital mask will also work), then the observed intensity map (in arbitrary units until absolutely calibrated) is: $I(x,y)=J(x,y)\star G(x,y,p_1...p_n)$. Before discussing _deconvolution techniques_, a digression about measurements and mathematical representations of the PSF is in order.

The typical trend followed by a symmetric PSF is shown in _Fig. 1_. It shows two separate regimes which are due to different causes and affect different regions of galaxy images. The central part of the PSF, or _seeing-disk_ (hereafter SD), extends typically to a fraction of one arc minute, and is dominated by variable atmospheric refraction, telescopic diffraction and detector resolu

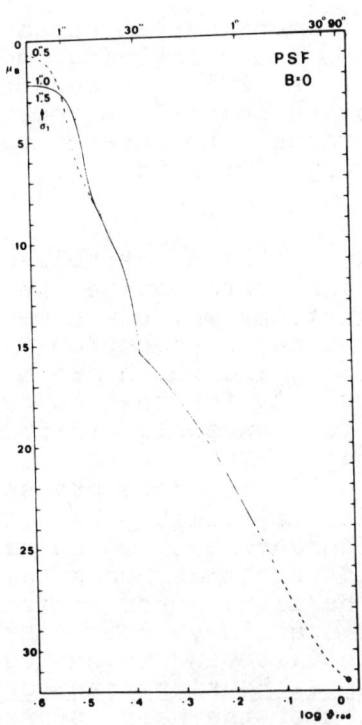

Fig. 1 - PSF profile characteristic of ground-based telescopes. The very inner region (<u>seeing-disk</u>), dominated by the atmospheric conditions and by the intrumental resolution, is highly variable. The extended outer part (<u>aureole</u>), due to scattering in the atmosphere and in the optics, sheds light into the outskirts of galaxies and thus may produce spurious haloes (from de Vaucouleurs and Capaccioli 1979).

tion; thus the seeing disk is time, place and instrument dependent, and must be measured in real time for each image. The astronomical practice is to map the seeing disk by means of stellar images observed simultaneously with the object in study. For a functional representation of the inner PSF (SD) we may assume circular symmetry; this assumption is justified by experience and by the random nature of almost all fenomena involved (excluding systematic guiding drifts and optics aberrations). Also diffraction patterns and the spikes due to secondary mirror supports in reflectors are usually ignored (but this easy practice has to be dismissed with the Space Telescope high resolution images). The favourite functional representation of the SD radial profile is by a sum of <u>Gaussian functions</u> with the number of components and the values of variances fixed by empirical rules (see Capaccioli and de Vaucouleurs 1983, and references therein). However other empirical laws have been proposed which work as well as pure Gaussians; for instance Newell (1979) has suggested to fit the SD flat wings by a <u>Lorenzian function</u>.

It must be stressed that only the variable peak (or SD) of the entire PSF controls the convolution of bright spots as nuclei of galaxies. The remaining faint tail or <u>aureole</u> (caused by scattering in the optics and in the atmosphere and covering 6 orders of magnitude in surface intensity) is rather stable in time and, carrying not more than 10% of the total light of the PSF, represents just a second order effect in smearing point-like sources. However, due to its wide angular range, the aureole can be of crucial importance in the study of the faint outer regions, as it will be seen later.

Once the SD profile (2-D map) has been established, the deconvolution of the digital data may be performed by one of the classical algorithms which are usually available as black boxes in the major reduction facilities (e.g. IHAP and MIDAS at ESO), and which are not always clearly understood by users. As for the sources of such algorithms we may refer to Bracewell (1955), Lucy (1974), Bryan and Schilling (1980), Bendinelli et al. (1981). However direct deconvolution does present some well known difficulties which may limit the results. The first one is the noise enhancement, which has to be controlled by suitable and somewhat uncertain filtering techniques (note that the signal from bright spikes of light may contain high frequency components which mix with noise). One of such filtering techniques that proved satisfactory is a local Fourier transform based on a two-dimensional extension of the Haar transformation (Fritze et al. 1977; for astronomical applications see Capaccioli et al. 1984b). The second problem concerns the uniqueness of the solution. Numerical experiments show that quite different light profiles, such as King's (1966) flat law (or quasi-isothermal) versus de Vaucouleurs's (1948) peaked law (known also as $r^{1/4}$ law), may look very much similar once convolved; this is mainly because of the usually large values of the smearing parameters and the noise of the data (Capaccioli 1984).

The last comment is also a criticism to the procedure of model convolution. It consists in convolving a parametrized model of the intrinsic light distribution of an object and in comparing the result with the observed data; such a technique, practicable only for galaxies with almost circular symmetry, was adopted for instance by de Vaucouleurs and Capaccioli (1979) for the standard galaxy NGC 3379, and worked well because of the rather high quality of the data. A significant methodological improvement to the above scheme, which added reliability to the results, was introduced by Nieto (1984). Performing the comparison of a convolved empirical model of the intrinsic light profile of the same round galaxy NGC 3379 with sets of data at different resolution (Fig. 2),

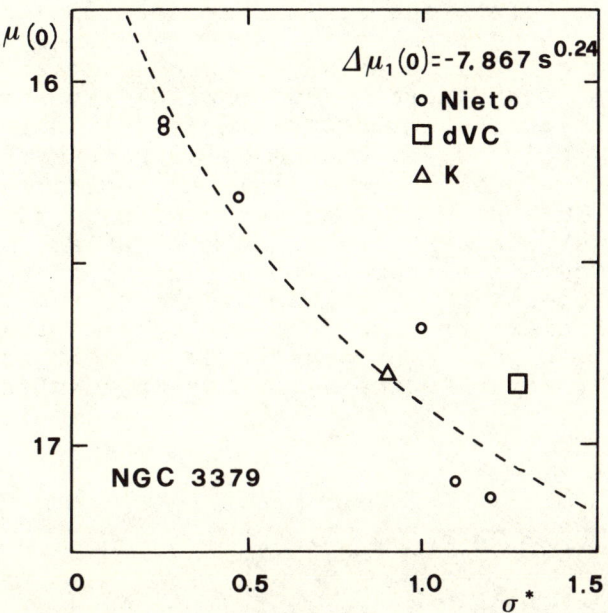

Fig. 2 - Observed peak brightness values for NGC 3379 at various resolutions compared to the convolution correction (dashed line from formula at right top) for the $r^{1/4}$ law fitting the galaxy body (from Nieto 1984).

he demonstrated unquestionably the correctness of the statement formulated by de Vaucouleurs and Capaccioli (1979): <u>NGC 3379 has a central spike of light</u>. In passing we wish to note that this result, combined with the radial profile of the velocity dispersion, claims for the existence of a massive black hole (about 10^6 solar masses) at the center of NGC 3379 on the basis of the same arguments brought by Sargent <u>et al</u>. (1978) for M87.

When nuclei of galaxies are structured (for example multiple as in the case of NGC 6166, patchy, crossed by dust lanes or prolonged into jets), then astronomy may call for some image processing able to enhance the peculiarities. This is conveniently done by a digital version of the <u>unsharp masking technique</u> developed by Malin (1978) for dark room treatment of conventional photographic material. In brief (see also Schweizer 1984) one may apply a cleaning algorithm to the data in order to remove medium frequency features as foreground stars or defects (scratches in photographs and particle events in CCD's, for instance); lacking better ideas a classical median filter may be used. Then the data are passed through a low-pass filter (e.g., gaussian) and subtracted from the original. The result is an image containing only high frequency structures over an almost flat background. This technique is extremely powerful and allows unexpected discoveries (Fig. 3). It would be extremely valuable if statisticians would work out effective and easy-to-understand recipes to help astronomers in this important game which has to become a chapter of quantitative 2-D photometry.

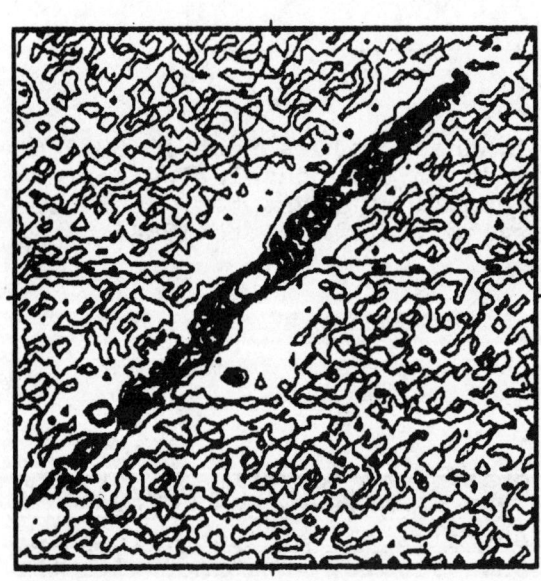

Fig. 3 - Computerized unsharp masking of the sum of five 2^m exposures of the lenticular galaxy NGC 3115, taken in the V-band with the CCD camera of the 2.2-m telescope at La Silla (frame size is 46.4 arcsec). The thin disk structure and a previously unknown central bar-like feature inclined to the disk are clearly seen (Capaccioli et al. 1984c).

THE INTERMEDIATE REGIONS

The intermediate regions of a galaxy image are defined operatively as everything outside the blurred nucleus and still bright enough to be measured with ease; they typically cover a brightness interval of two orders of magnitude. The goals of surface photometry in this luminosity regime are too numerous to be effectively reviewed in a limited space; for this reason we will arbitrarily select some of them only.

The most obvious output of the photometry of intermediate regions is the trend of the surface intensity versus some radial distance from the center, i.e. the so called light profile. In their simplest form light profiles are centered cross sections of the intensity maps (e.g. along major and minor isophotal axes). Such profiles in one or more colors have a very wide range of uses; they may serve to establish the morphology of the light and color distributions in galaxies of different Hubble types (problem of the empirical laws), to flanck kinematical studies (problem of the mass-to-light ratio and of the initial conditions for N-body simulations), to localize the presence and to study the characteristics of substructures as bars or lenses. In addition, if data are of high quality, they may help to disclose peculiar deviations from smooth trends such as the 'waves' found by de Vaucouleurs and Capaccioli (1979) in the E-W light profile of the elliptical galaxy NGC 3379 (Fig. 4); the meaning of such departures is not yet understood.

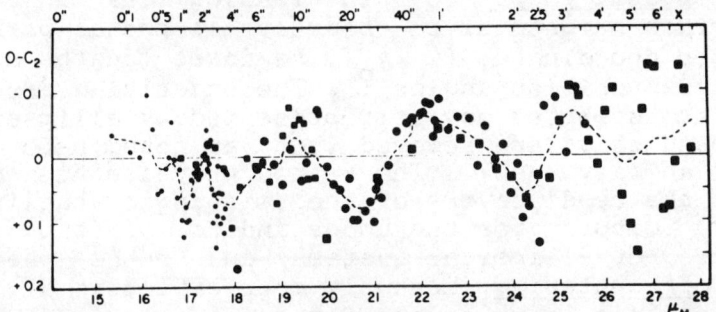

Fig. 4 - Wavy pattern in the residuals of E-W profile of the elliptical galaxy NGC 3379 from a smooth empirical model consisting of the sum of an $r^{1/4}$ law and a gaussian core (from de Vaucouleurs and Capaccioli 1979).

Omitting any comment on calibration deficiencies, which are mostly ascribable to lack of care in the hardware and software linearization pipeline, and postponing the discussion of background removal, a major technical problem in deriving clean light profiles of galaxy bodies regards the removal of as many foreground (Milky Way) stars as possible from the linearized intensity map. To the best of my knowledge this cleaning (to be extended to any other spurious object or defect in the field) is basically performed by some simple filtering technique or by an interactive identification-removal procedure. This one is certainly an area of 2-D photometry where more theoretical work and better application routines are urgently needed.

Another technical problem is related to the <u>optimum use</u> of all of the information stored in one image to achieve the highest possible signal-to-noise ratio everywhere in light profiles. In fact, while the luminosity and S/N decrease with radius, the isophotal area increases providing more and more pixels at the faintest levels; and no advantage of this reserve of information is taken by plain cross-sections. For this reason, and to smear out the irregularities presented by late type galaxies, more than 20 years ago de Vaucouleurs introduced the notion of <u>equivalent profile</u>; this is the profile of an ideal round galaxy whose isophotes encircle the same light as the real galaxy. In this way all of of the 2-D information is used, although some structural features may get lost.

Along this line and in the context of the implementation of the <u>Padova Numerical Mapping Package</u>, Barbon <u>et al</u>. (1976) developed a method for an analytical fit of <u>elliptical isophotes</u>; the leading idea was to link the surface brightness to objective isophotal parameters as center coordinates (x_o, y_o), semiaxes lengths a,b, and major axis position angles ϑ. The underlying assumption was that isophotes may be represented by ellipses once foreground stars are removed (but, as noted before, this is not an easy task). The method proved efficient and lead to the re-discovery of the twisting of the isophotes (a key feature for the undestanding of the spatial structure of elliptical systems) and to the phenomenon of the <u>off-centering</u>. Recently more effective algorithms than the plain <u>least square fit</u> originally used by Barbon <u>et al</u>. have been proposed (cfr. Leach 1981 and Kent 1983; also Cawson, private communication), which may permit to investigate departures from pure elliptical shapes. In spite of the work done so far, isophote fitting methodology and related problems (star identification and removal, noise suppression, shape analysis) are still open subjects for theoretical improvements.

In closing this Section it seems worth noting that, while in principle the mapping of the intermediate regions of galaxies presents only moderate difficulties, still results so far available in the literature exhibit large discrepancies which limit the astrophysical significance of the wide body of data (<u>Fig. 5</u>). In other words it is apparent that in the field of galaxy surface photometry there is room for astronomers with good will to work and for data analysts with good ideas. For instance it is still unresolved the problem of isophotal fitting in multistructured systems such as lenticulars.

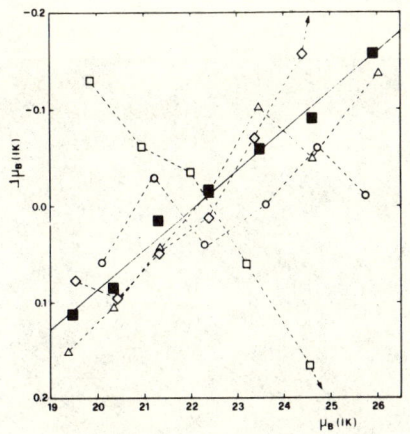

Fig. 5 - Open symbols represent magnitude residuals of from the comparison of the equivalent B-band light profiles of four Virgo galaxies in common to King (1978) and Benacchio et al. (1984). Full squares and solid line (computed ignoring M87 = open squares) indicate systematic (calibration?) differences between the two studies.

THE FAINT OUTER REGIONS

The main problem of the photometry of the outskits of galaxies is the low S/N; for example, at the brightness level of 27 B-mag/arcsec^2 the signal from a galaxy is typically 1/100 of that from the night sky, and all sources of disturbance (telescope vignetting, color effects and detector's sensitivity variations, faint and diffuse galactic nebulosities, foreground stars and background galaxies, overlapping haloes of nearby extended objects) blow up in this luminosity range. Fig. 6 illustrates a number of such disturbances.

The subject of background removal has been recently reviewed by Capaccioli and de Vaucouleurs (1983), who also discussed the aspects of absolute calibration which will not be considered here. The essence of the problem is the following; how can we isolate and eliminate all the signals which are accumulated in the rough data and still do not pertain to the object under study ? It must be noted that not all of the signal components are additive; this is the case, for instance, of sensitivity variations which call for a pixel by pixel calibration. In addition, the main contribution to the background comes from a time-varying source, the night glow. Thus, in principle the mapping of the background light has to be done in clean regions of the same specimen containing the object in study; and this mapping has to be extremely accurate or the results will be increasingly meaningless as the surface brightness decreases. In fact, if the background is underestimated, then the galaxy profile develops a spurious halo; on the contrary, if the background is overplaced, the result mimics the trend of a tydally truncated profile (Fig. 7).

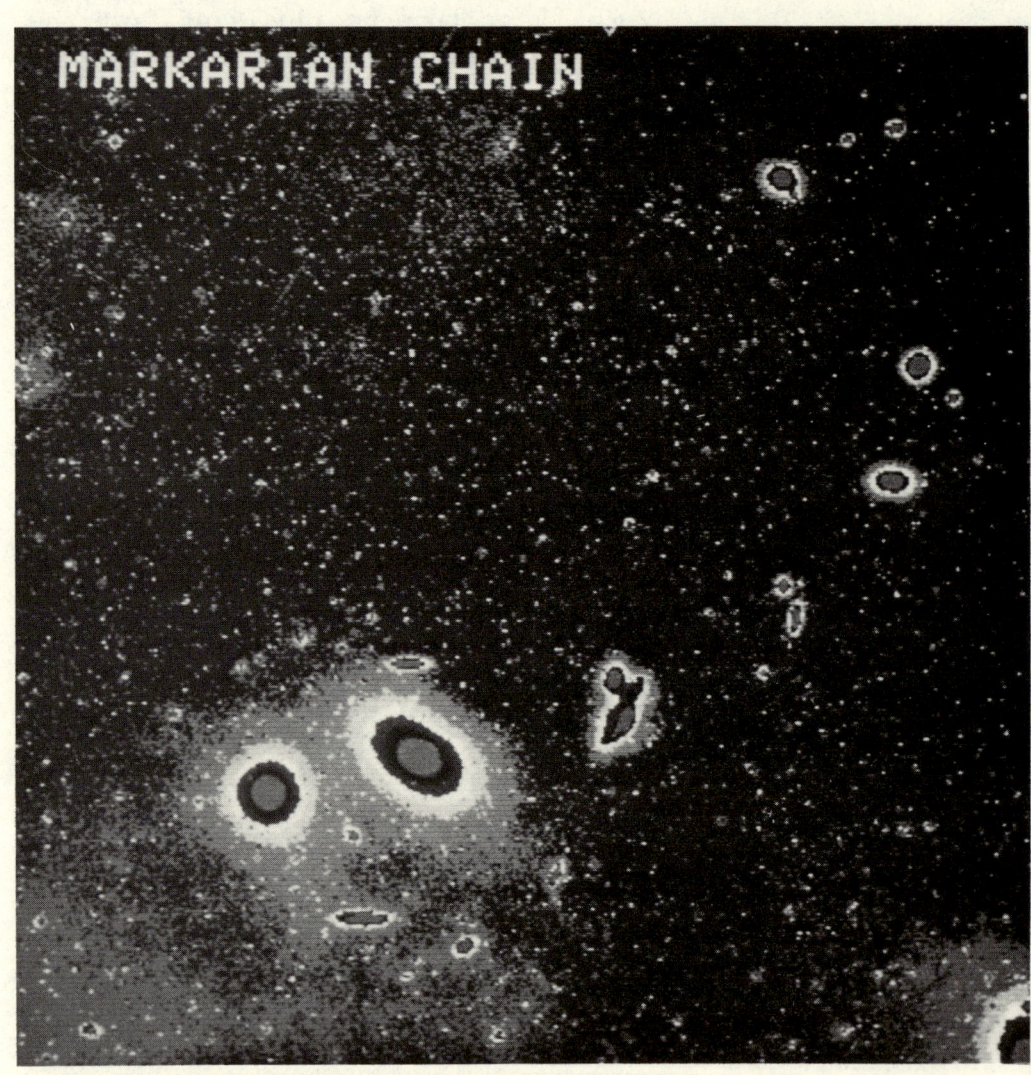

Fig. 6 - The PDS scan of a long exposure UK Schmidt photograph of the Markarian chain in Virgo (reproduced from a map in false colors) exhibits clearly some of the complications coming out in the photometry of the faint outer regions of galaxies (uneven background, foreground stars, background galaxies, nearby companions). The two overlapping giant galaxies are M84 (left) and M86. M87 is visible at the bottom-right corner.

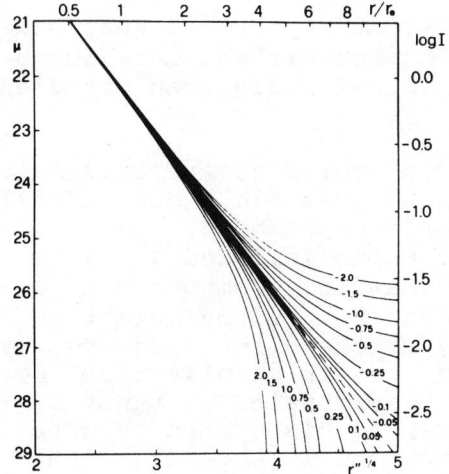

Fig. 7 - Simulations of the effects of background errors in an otherwise $r^{1/4}$ luminosity profile (dashed straight line in the figure). Note the spurious cuts and the extended haloes resulting from setting the background level too bright or too faint respectively (from Capaccioli and de Vaucouleurs 1983).

A digital technique of background low frequency removal, pioneered by Jones et al. (1967) and lately improved by Barbon et al. (1976), consists in fitting a two-dimensional canonical polynomial to all those pixels which are thought to be unaffected by the galaxy light (outer field). Astronomical sources, which are inevitably present in the outer field and produce a systematic skewing of the hystogram of the residuals (only some defects contribute negative flux), are controlled by a clipping technique. To be applicable the procedure requires the galaxy-free portion of the image to be large enough (at least as large as the polynomial wavelength, and this may not always be the case, e.g. in CCD frames). In addition a reasonable assumption must be made about the true size of the object because the outer field need to be the smallest possible but it shall not contain any trace of the galaxy light. Among the limits of this naive approach there are the arbitrary choice of too many parameters, the lack of quantitative verifications, and the systematic tendency to set the background too bright. In addition, with this method high frequency components cannot be taken into account. In any case, however, the latter may be studied only relatively to sensitivity variations (with the grid technique for photographic plates and with the flat fielding for digital detectors).

A further problem arises from the possible presence of light scattered from the nucleus of the galaxy, which builds a false halo about the object. The contribution of scattered light can be evaluated provided that the complete PSF is known (cfr. Capaccioli and de Vaucouleurs 1983) and if the galaxy nucleus has been studied photometrically.

In conclusion, outskirts of galaxies are extremely difficult to map even in the best possible environmental conditions (outside clusters or groups and at high galactic latitude). According to Capaccioli and de Vaucouleurs (1983) it seems unlike that measurements can extend to levels much fainter than about 29 B-mag/arcsec^2; this limit is set principally by the noise produced by faint stars and galaxies and possibly holds also for extra-atmospheric telescopes). Nevertheless, if not measurable, fainter structures can be discovered by appropriate data cooking of the kind used in the dark room by Malin (1983) which led to many discoveries.

CONCLUSIONS

In this paper we have quickly visited some of the scientific objectives and technical problems of galaxy photometry, and listed some of the techniques currently used by astronomers. The purpose was to provide a quick look into a field which might of some interest for data analysts. Clearly not all of the problems we are facing now in this discipline may be solved by improvements in software methodology. A progress in the hardware (in particular detectors and observing conditions) is also needed, and in fact significative steps in this direction have been recently made.

ACKNOWLEDGEMENTS

This review, based mainly on my work in collaboration with Prof.G.de Vaucouleurs, was supported in part by CNR Grant 100601C and NSF Grant INT 8022847 under the US-Itlay Cooperative Science Program.

REFERENCES

Agnelli, G., Nanni, G., Pittella, G., Trevese, D., Vignato, A. 1979, Astron.Astrophys., 77, 45.
Barbon, R., Benacchio, L., Capaccioli, M. 1976, Astron. Astrophys., 51, 25.
Begelman, M.C., Blandford, R.D., Rees, M.J. 1984, Rev. Mod.Phys. 56, 255.
Benacchio, L., Capaccioli, M., De Biase, G., Santin, P., Sedamak, G. 1984, preprint.
Bendinelli, O., Parmeggiani, G., Zavatti, F. 1981, Astr.

Sp.Science, 83, 239.
Bracewell, R.N. 1955, Austr.J.Phys.Ap.Suppl., No. 8, 54.
Brian, R.K., Skilling, J. 1980, M.N.R.A.S., 191, 69.
Capaccioli, M. 1979, in 'Photometry, Kinematics and Dynamics of Galaxies', ed. D.S.Evans, Univ. of Texas, p. 165.
Capaccioli, M. 1984, Mem.S.A.It., 55, 3.
Capaccioli, M., de Vaucouleurs, G. 1983, Ap.J.Suppl., 52, 463.
Capaccioli, M., Fasano, G., Lake, G. 1984a, M.N.R.A.S., 209, 317.
Capaccioli, M., Rampazzo, R. 1984, to be published in 'New Aspects of Galaxy Photometry', Toulouse.
Capaccioli, M., Lorenz, H., Richter, G.M., Ziener, M. 1984b, in preparation.
Capaccioli, M., Held, E., Nieto, J.-L. 1984c, in preparation.
de Vaucouleurs, G. 1948, Ann.d'Ap., 11, 247.
de Vaucouleurs, G. 1959, in 'Handbuch der Physik', Vol. 53, ed. S.Flugge, Springer-Verlag, p. 311.
de Vaucouleurs, G. 1983, in 'Astronomy with Schmidt-type telescopes', ed. M.Capaccioli, Reidel Publ.Co., p. 367.
de Vaucouleurs, G., Capaccioli, M. 1979, Ap.J.Suppl., 40, 699.
Fritze, K., Lange, M., Mosze, G., Oleak, A., Richter, G. 1977, Astron.Nach., 298, 189.
Jones, W.B., Obitts, D.L., Gallet, R.M., de Vaucouleurs, G. 1967, Publ. Astron. Dept., Univ. of Texas at Austin, Ser. II, Vol. I, No. 8.
Kent, S. 1983, Ap.J., 266, 562.
Kent, S. 1984, preprint.
Kormendy, J. 1980, in 'ESO Workshop on Two-dimensional Photometry, eds. P.Crane and K.Kjar, ESO: Geneva, p. 191.
Kormendy, J. 1982, in 'Morphology and Dynamics of Galaxies', XII Advanced Course of the Swiss Society of Astron. and Astropys., eds. L.Martinet and M.Mayor, Geneva Obs., p. 115.
King, I. 1966, Astron.J., 71, 64.
King, I. 1978, Ap.J., 222, 1.
Kron, R. 1983, in 'Astronomy with Schmidt-type telescopes', ed. M.Capaccioli, Reidel Publ.Co., p. 315.
Lake, G., Norman, C. 1983, Ap.J., 270, 51.
Leach, R.W. 1981, Ap.J., 248, 485.
Lucy, J. 1974, Astron.J., 79, 745.
Malin, D.F. 1978, Nature, 276, 591.
Malin, D.F. 1983, in 'Astronomy with Schmidt-type telescopes', ed. M.Capaccioli, Reidel Publ.Co., p. 57.
Nieto, J.-L. 1984, Dissertation, University of Paris.

Newell, B. 1979, in 'International Workshop on Image Processing in Astronomy', eds. G.Sedmak, M.Capaccioli, R.Allen, Trieste Astro. Obs., p. 100.
Olson, D.W., de Vaucouleurs, G. 1981, Ap.J., 249, 68.
Prendergast, K.H., Tomer, N. 1970, Astron.J., 75, 674.
Quinn, P. 1982, Dissertation, Australian National University.
Sargent, W.L.W., Young, P.J., Boksenberg, A., Shortridge, K., Kynds, C.R., Hartwick, F.D.A. 1978, Ap.J., 221, 731.
Schweizer, F. 1978, Ap.J., 220, 98.
Schweizer, F. 1984, to be published in 'New Aspects of Galaxy Photometry', Toulouse.
Thuan, T.X., Romanishin, W. 1981, Ap.J., 248, 439.
Vigroux, L. 1984, preprint.
Watanabe, M., Kodaira, K., Okamura, S. 1982, Ap.J.Suppl. 50, 1.
Young, P.J., Westphal, J.A., Kristian, J., Wilson, P.C., Landauer, F.P. 1978, Ap.J., 221, 721.
Zou, Z., Chen, J., Peterson, B.A. 1981, Chinese Astron. Astrophys., 5, 316.

SPECTROMETRY

R. A. E. Fosbury

Royal Greenwich Observatory
Herstmonceux Castle
Hailsham
East Sussex, BN27 1RP
England

ABSTRACT

In common with other observational techniques, our concern is with the three processes of data acquisition, reduction and analysis. This paper addresses all of these stages and stresses, in particular, the advantages of a coordinated approach. As illustrations of modern rather complex systems, I describe two examples of three-dimensional spectroscopic techniques: TAURUS, an imaging Fabry-Perot interferometer and ASPECT: a slit scanning mode for a long-slit spectrograph.

INTRODUCTION

Spectrometry is a technique with a history stretching back to the contemplation of the action of glass prisms by Isaac Newton in 1666. It is fair to say that the physical sciences have been built upon the application of the spectrometer; from the development of quantum mechanics to the discovery of the expansion of the Universe.

The field is vast. My task is to pick a path throught the field of astronomical spectroscopy in the hope of making one or two remarks of relevance to the subject of this meeting. I shall not be discussing the spectrometrists' requirements from image processing; my emphasis, in fact, will be rather on the need to put more effort into the systems around the telescopes themselves in order to ease the tasks of reduction and data analysis and also to make available new and more powerful observational techniques. Due partly to my own experience, but also due to the wide variety of techniques, my examples will represent the optical band.

ACQUISITION

Spectrometric data necessarily occupy one dimension for wavelength; or equivalently frequency, energy or radial velocity. Other dimensions which typically are associated with the data set are space (distance on the sky, one or two dimensions), time and polarisation. Instrumental systems which use various combinations of these dimensions are in use at observatories around the world.

The acquisition of data at the telescope is a process which requires, in support, an array of appropriate calibrations and a description of the data which is adequate to allow its subsequent reduction and analysis. This carries implications for the design of the data structure and I propose an approach which considers the instrument and the telescope as peripherals under the command of the detector/data reduction system. This allows rather sophisticated interactions between the components of the telescope/instrument/detector system which can then be used, eg for area scanning applications and automatic, real-time assement of data quality.

One of the tedious complications of the reduction process is that the calibrations in each of the qualitatively different dimensions have to be treated rather differently. What can be done at the telescope to ease these reduction tasks? The first and most obvious response is to design the hardware to obviate the need for as many of the calibrations as possible. An example of this is the problem of differental refraction in the atmosphere as a function of wavelength: atmospheric dispersion. Attempts to perform accurate spectrophotometry using spectrographs with small entrance apertures are usually beset by this problem of not accepting all wavelengths with equal weight into the slit. The problem is particularly severe with modern, wide bandwidth, faint object spectrographs where there is a requirement for small apertures to keep sky contamination to a minimum. It is compounded still further when a multi-object mode is used since this generally removes the freedom to rotate the slit to lie along the parallactic angle so that the atmosphere disperses along its length. It is now possible to design and build Atmospheric Dispersion Correctors (ADC) for large optical telescopes which will correct this dispersion over all reasonable zenith distances from about 350 to 1000nm in wavelength. Such a device has been designed (Wynne 1984) and is being constructed for the f/15 Cassegrain focus of the 2.5m Isaac Newton telescope on La Palma. This ADC is seen as an essential component of the instrumental system for the study of galaxy evolution using spectrophotometry of galaxies at high redshift.

The second response is to perform the necessary calibrations properly, describe them adequately and collect them into an intelligently designed data structure. This will allow the reduction software to perform many or all of its tasks automatically. This might sound trivial for some simple type of observation but the spectroscopist is often confronted with an instrument which can be configured in a very large number of different ways for which it is actually quite difficult to write general software. One approach, which is quite compatible with the FITS data transport system is to carry, as part of the header, look-up tables and algorithms which enable the reduction programs to interpret instrument encoder readings as relevant starting points for calibrating the data. An obvious example is the conversion of a grating description and angle into a central wavelength and range which can be used to identify directly the lines in the wavelength calibration source.

REDUCTION

As an example of the kind of reduction processes which are required, let us consider the rather common example of spatially resolved spectrophotometry using a 2-dimensional digital detector. The necessary calibrations might be the following:

1) Non-uniformity of response of the detector (flat-fielding). Possibly also a calibration of intensity non-linearity.

2) Instrument/detector geometrical correction. This is often performed separately in the wavelength and spatial dimensions but one can argue that a simultaneous process is superior. The end point of this should be the association of an observed wavelength scale and a position on the sky, in some well-defined coordinate system, for each spatial pixel.

3) The correction of atmospheric extinction. This includes both the smooth component and the molecular absorptions which are often saturated and therefore not subject to a linear calibration.

4) Calibration of the energy flux scale. This is usually achieved using observations of standard stars interleaved with the object observations. With photon counting detectors, there is usually the problem of finding calibration stars which are sufficiently faint to avoid saturation difficulties.

5) Subtraction of the night sky spectrum. This becomes progressively more serious as one moves into the infrared part of the spectrum. CCD's used for faint object spectroscopy in the near infrared are particularly

susceptible to the problems of imprecise subtraction.

6) For objects which are only a small fraction of the brightness of the sky, the extraction of the 1-D spectra from a 2-D format in an optimum way needs careful thought. This is particularly true for curved echelle formats.

For the optical astronomers, at least, it has been traditional to separate the observing and reduction processes. Initially this was due to lack of computing power at the telescope and one of its results was inevitably to strip the raw data of much of its valuable descriptor information during transfer to the "foreign" data reduction computer. This situation has improved greatly since the introduction and widespread acceptance of FITS but the way forward is surely to do the work at the telescope where the expertise resides. Computing power will not remain a problem for long.

What then should be the end result of this observing/reduction process? It should be a properly calibrated AND PROPERLY DESCRIBED data set ready for astronomical analysis. This, of course, necessitates "reducing" the descriptor information as well as the data arrays themselves.

As an example of an environment which is designed to allow the full integration of instrument control and running of application programs, I should like to mention the Astronomical Data Acquisition Monitor (ADAM) system currently in use at Herstmonceux and La Palma.

The system allows control of both device or D-tasks and application or A-tasks and thus gives the observer opportunities to coordinate the instrument control with the results of on-line data reduction and analysis. ADAM is running currently on Perkin-Elmer 3220 machines but is being extended with the help of the Royal Observatory Edinburgh to run on VAX. A description of ADAM can be found in the ADAM manual(RGO 1984).

EXAMPLES

As examples of the power of coordinating at least the control of detector, instrument and telescope, I shall describe briefly two applications which produce 3-dimensional data arrays covering wavelength and two spatial dimensions. These are TAURUS (Atherton et al. 1982), an imaging, scanning Fabry-Perot interferometer and ASPECT (Clark et al. 1984), a slit scanning mode for the RGO spectrograph on the Anglo-Australian telescope. Both of these techniques rely on the noise-free photon counting performance of the IPCS and, it is true, were developed before the ADAM

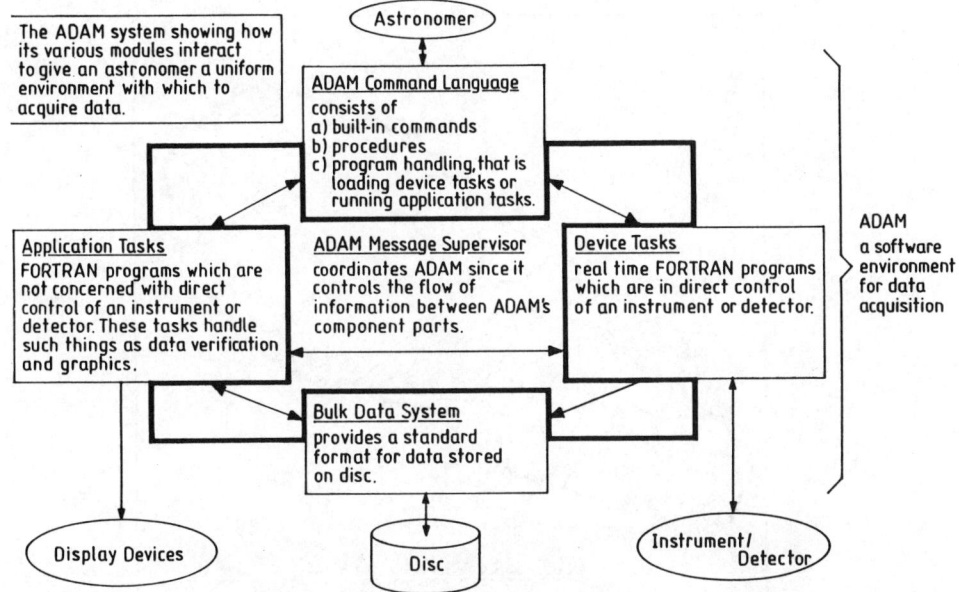

Fig. 1 A representation of the structure of the ADAM environment developed over the last few years at the RGO.

environment. It is clear, nonetheless, how naturally the next implementation will be achieved using ADAM.

TAURUS uses an extremely precise capacitively controlled, piezoelectrically scanned Fabry-Perot etalon to produce, on the IPCS, a sequence of 2-D images of emission line sources at different spacings. This 3-D array can then be geometrically corrected to produce a true spatial-spatial-velocity image of an object. Figure 2 is an example of a velocity map made from such a data set.

The reduction and analysis of such data have benefited greatly from the radio astronomical experience of spectral line mapping. The observational requirements of the technique are for the detector to synchronise the scanning of the FP with its own data acquisition activities and for the telescope to maintain stable pointing throughout the observation. It also requires several different

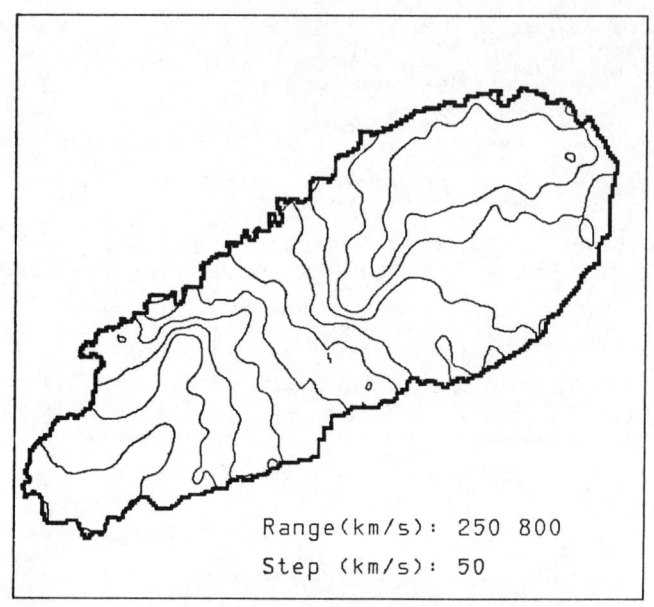

Fig. 2 An emission line velocity contour map of NGC 5128 (Cen A) from observations made with TAURUS on the AAT.

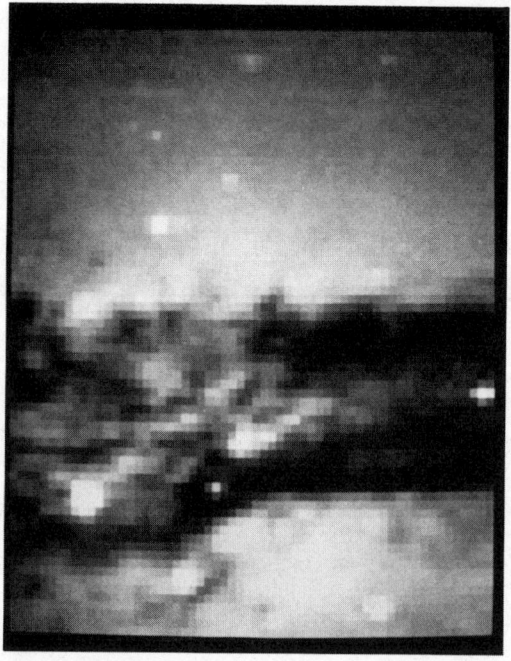

Fig. 3 The inner 2 by 3 arcminutes of NGC 5128 in the yellow part of the spectrum obtained with ASPECT on the AAT.

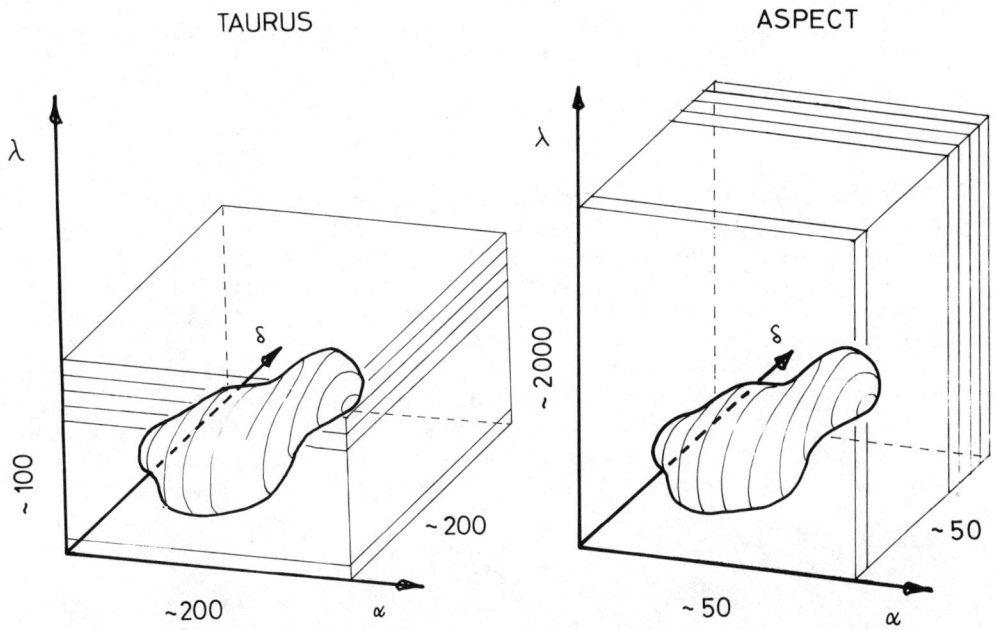

Fig. 4 The data formats for the 3-dimensional techniques TAURUS and ASPECT.

types of calibration arrays of different dimension which have to be recognised by the reduction software.

ASPECT is an entirely software based technique which uses a conventional long-slit spectrograph. The reduced data have a form similar to those produced by TAURUS but, because of the different performance characteristics of the interferometer and dispersing spectrometer, it has a complementary rather than competitive range of applications: in particular for wide wavelength range spectrophotometry and for absorption line work. ASPECT exploits the precise pointing and scanning capability of the AAT to scan the image of an object across the slit of the spectrograph, taking a 2-D slice of data at each position on the sky. Because of the characteristics of the IPCS, the scanning can be done relatively quickly and repeated many times during a particular observation.

Figure 3 shows an example of the application to absorption line work on an elliptical galaxy: the original data set was 5 X 14Mby in size! One advantage of this system is that, although large, the data set is readily handled using conventional 2-D reduction software since the calibrations for the second spatial dimension are all identical.

Diagramatic representations of the data formats for these two techniques are given in Figure 4.

ANALYSIS

The type of astrophysical analysis to which the reduced data are subjected depends, of course, on the nature of the observing programme and the class of spectrometric instrument used. There are, however, some very common requirements for line and continuum measurements. The ability now, in several different wavebands, to make complete line and continuum maps of objects is producing some exciting new results. The application of global analysis techniques such as Fourier and cross-correlation methods is now becoming routine. With the increasing power of computers at the telescopes, astronomers will demand the availability of the reduction and the analysis software during their observing run.

CONCLUSIONS

Spectrometry is a technique which can exploit complex interactions between detector, instrument and telescope systems. In the future, the integrated design of these subsystems and their softwate will allow much more uniform assessment of data quality and fully automatic data reduction. This, in turn, will allow more efficient telescope scheduling and remove the operational barriers to remote observing.

ACKNOWLEDMENTS

I thank Keith Taylor and Joss Bland for allowing me to discuss TAURUS and reproduce Figure 3. I also thank the La Palma computing group at RGO for the use of Figure 1.

REFERENCES

Atherton,P.D., Taylor,K., Pike,C.D., Harmer,C.F.W., Parker,N.M. & Hook,R.N., 1982, Mon. Not, R. astr. Soc., 201, 661.
Clark,D., Wallace,P.T., Fosbury,R.A.E., & Wood,R., 1984, Q. Jl. R. astr. Soc., 25 No.2, 114.
The RGO La Palma computer group, 1984, The ADAM manual(an internal publication).
Wynne,C.G., 1984, The Observatory, 104 No.1060, 140.

TIME DEPENDENT ANALYSIS

Richard G. Kron

University of Chicago
Yerkes Observatory
Williams Bay, WI 53191   USA

1. IMAGE PROCESSING AND TIME SERIES

Analysis of a time series provides information about variation in intensity and position, and astromomers have traditionally used variability as an important physical diagnostic. Most sources that are detected to be variable are unresolved, exceptions being solar system objects and nova shells and the like. Therefore image processing is relevant to variability studies not for the information contained in the image structure, but rather because imaging techniques allow new variables to be discovered in an efficient manner. Pictures are appropriate for measurement of intensity changes if there are at least a few variables per picture, or if a variable is confused by a neighboring image. The astrometric application of time-dependent analysis is qualitatively different only in the sense that the signal is contained in the two angular coordinates. Astrometry is dealt with rather cursorily in the following for the sake of brevity. The main point is that many of the general image processing issues of time series analysis apply to both astrometry and to photometry.

Couched in these terms, the central problem of the automatic analysis of pictures for variability is to isolate the variables from everything else. This is a classic application of pattern recognition, or rather the recognition of a change in the pattern. Subsequently it may be required to classify the candidate variables -- this may be done either directly from the light curve characteristics, or by introducing new data into the analysis.

Section 2 reviews the scientific motivation for establishing
complete samples of variable objects, using supernovae and quasars
as illustrative examples. Section 3 expounds upon certain problems
specific to the detection of small changes in images. Section 4
briefly discusses a few recent applications.

2. SCIENTIFIC MOTIVATION

Intensity Variations

Supernovae. Supernovae are usually discovered via deliberate
efforts in the comparison of new pictures with old. The surveys can
be divided into those for relatively bright supernovae, and those
for quite faint ones, as each group has a distinct use. To the
extent that one might want to follow up the discovery of a supernova
as quickly as possible with other instrumentation, this is an extreme example of the need for efficient picture analysis. An added
dimension to the problem is that of the reliability of the announced
supernova: large telescope time is valuable (especially when pre-
empted!), and one does not want to chase miscellaneous faint objects
for nothing.

Bright supernovae can be studied in detail and are therefore
important for understanding the physics of the events. There is a
need, especially, for good spectrophotometry during the ascending
phase of the light curve, which means that such observations must be
made soon after time zero. Better statistics are needed to determine the dispersion in the energetics, which will in turn help to
constrain the characteristics of the supernova progenitors. The
supernova rate depends on the stellar population content of the
parent galaxy. Better statistics are needed here too in order to
refine the details of this connection.

Faint supernovae cannot be studied in spectrophotometric detail, not only because of low signal-to-noise ratio, but because of
confusion with light from the underlying galaxy, at least for
ground-based observations (Wagoner 1980). Still, faint supernovae
are important for what they can tell us about distant galaxies.
Specifically, the Type II supernova rate depends on the rate at
which massive stars are born, on details of the shape of the mass
function, on the distribution of absorbing material in the parent
galaxy with respect to these massive stars, and so on. Any systematic change in the supernova rate with redshift would be a valuable
diagnostic of changes in the stellar population.

The other great hope is that supernovae can be used as cosmological probes (Colgate 1979; Wagoner 1980). It is tantalizing that

H_o could be found from just a single object at a distance that is physically determinable (say, v ~ 5000 km/sec), and likewise for q_o (say, z ~ 0.5), thus obviating the need for a distance ladder. What is required is a physical model for the expansion of the photosphere that adequately describes the emergent flux, an unassailable determination of the total absorption between us and the supernova, and good data.

Anything that varies in a predictable way, and which can be seen at high redshift, should display cosmological time dilation. The light curves for distant supernovae of known redshift would be valuable in this regard.

Quasars. There are a number of motivations for studying variability in quasars. An obvious one is that the characteristics of the variability may provide information about the energy source or its immediate environment. At the most basic level, the timescale for the variability puts causal constraints on the dimensions of the source. Beyond this, the physical interpretation of quasar light curves is not far advanced, but the fact that quasar light curves look like some kind of noise does carry physical information. Fahlman (1977; see also Fahlman and Ulrych 1975) has for instance shown that a shot-noise model for quasar activity appears reasonable, from which can be derived the shot amplitude and frequency. However, this mathematical model may not apply to all quasars, and there may be other models that are as satisfactory.

Another motivation for quasar variability studies is to reduce the scatter in the magnitude-redshift relation by adopting some appropriate measure of the apparent brightness. The resulting relation can then be used to argue for the hypothesis that quasar redshifts are cosmological (Bahcall and Hills 1973, Usher 1978, Pica and Smith 1983). A related issue is the evaluation of the effect of variability on deductions from the $<V/V_{max}>$ test (Bahcall 1980). Also, the distribution of points in the magnitude-redshift plane with respect to other objects, like N galaxies, can be used to argue for generic relationships between classes.

Finally, quasar variability surveys can be used simply as a discovery technique. Complete samples of quasars are needed for the determination of statistical distribution functions, e.g. the luminosity function, and how these distribution functions depend on redshift. Thus it is important to establish unbiased samples, particularly with respect to redshift. A variability criterion for quasar discovery no doubt introduces a bias of some sort, but the important thing is that this selection technique is independent of the others.

Many issues of quasar variability remain unresolved. What is the incidence of radio-quiet BL Lac objects? Do optically-selected quasars otherwise have identical variability characteristics to radio-loud quasars? Does variability depend on redshift, or on luminosity? Do there exist "genuine" quasars that are not variable at levels that are practically detectable? Is it possible to classify quasars according to the characteristics of their light curves, and, if so, does this classification correlate with other observables? Is the effect of time dilation evident in the data? Are Optically Violently Variable Quasars a physically distinct class, or part of a continuum? All of these questions could be better addressed with improved statistical samples.

Position Variations

Ground-based astrometric work is growing in sophistication, and measurement precision is increasing accordingly. The theoretical limit for IIIa plates appears to be between 0.2 and 0.3 microns (Lee and van Altena 1983), or about 5 milliarcsec at a scale of 50 microns/arcsec. This theoretical limit can be approached in practice (Lee and van Altena 1983). Independent measurements can improve the result, of course. CCD detectors have achieved a precision in the measurement of parallaxes of 2 milliarcsec (Monet and Dahn 1983). Photoelectric devices with the image output modulated by a Ronchi grating may be limited at about 1 milliarcsec. These developments are exciting for parallax studies of individual objects, but as an example of an image processing application, it is more relevant to consider wide-field proper motion surveys. For instance, an error of 3 milliarcsec and a baseline of 5 years would allow giants in the Galactic bulge to have transverse velocities measured to a precision of 25 km/sec. Combined with radial velocities, such a survey would allow the velocity dispersion of the bulge and the anisotropy of the velocity distribution to be determined. Furthermore, a distance to the Galactic center could be derived by the method of statistical parallax. With such high precision, essentially every star in any field, no matter how faint, would have a detectable motion. This raises the possibility that new classes of problems in Galactic dynamics could be addressed with large-volume kinematic data that would be supplied by a good astrometric time series.

3. PROBLEMS

The key problem in the comparison of two pictures is to distinguish between real celestial events and artifacts of the detection process. Aside from relatively straightforward differences like changes in the background level or gain, the most common difference between pictures is in the image profile shapes due to seeing, guiding, focus, optics, or atmospheric dispersion. For linear detectors a star profile characteristic of one picture can be transformed into another by a linear operation, and this aspect should not be the cause of any confusion with real variability. A potentially major problem however is in the treatment of galaxy images. The observed image is the intrinsic image convolved by the instrumental point spread function, plus noise. Thus without actual deconvolution and its associated computational expense, it is not easy to compare two galaxy images obtained with different point spread functions. The problem in practice pertains to supernova surveys, as for any $z > 0.1$, a supernova can be expected to be no more than a few arcsec away from the galaxy nucleus. In this case one is testing for a slight perturbation on top of the galaxy profile, especially if the supernova is far from maximum brightness. An example of the difficulty of measuring a variable point source in a galaxy has been given by Green, Huchra, and Bond (1977). Their particular application was iris photometry, but the problem is the same in principle for any technique.

It may be that a universal linear transformation applied to galaxy images obtained at one epoch can make the images look sufficiently like the images at another epoch that the difference between the pictures still can reveal the small perturbation of a supernova. A definitive answer to this question can probably only come from experimentation.

The alternative might be to consider not changes in individual pixels, but rather changes in the integrated light of each galaxy in a picture, as the integrated light is independent of seeing effects. This approach has the further advantage that once the galaxy positions have been catalogued, the likely regions of the picture within which to find supernovae are known, and less computational effort needs to be spent in evaluating the other pixels. The two flaws with this alternative are the lower signal-to-noise ratio achievable by comparison of integrated magnitudes, and, if the area of the search is restricted, one is throwing away data that are of value for surveys of other types of variable objects.

Faint galaxy images outnumber stars at high latitude at magnitudes that are routinely measured (B ~ 21). At some low value of the signal-to-noise ratio, errors in image classification are inevitable. This means that even if one is restricting attention to nominally stellar images, there may well be contamination from galaxies that are both erroneously classified and erroneously determined to be variable for the above reasons.

If two pictures are obtained at different air masses, there will be second-order effects in the photometry of an object accord-

ing to its color with respect to the colors of the comparison
objects (this is of course a well-known problem in broad-band photometry). Thus pictures need to be obtained not only at different
times but also in more than one band in order to evaluate and
correct for this effect.

Another major class of problems concerns discrimination against
picture defects. In a given picture of, say, 10^7 pixels, a faint
supernova may detectably affect one of them. This depends on the
supernova rate and on the fraction of the sky that is covered by
galaxy images, which itself depends on the survey limit. For
Schmidt plates, the ratio may be more like one pixel out of 10^8. [A
good analysis of the supernova detection rate in the presence of
noise has been presented by McGlynn and Turner (1980).] Every
picture defect must be at least provisionally flagged for further
processing. Ideally the incidence of picture defects should be
small compared to one part in 10^7. Realistically the alternative is
to design efficient software filters for the recognition and rejection of picture defects (Grosbøl 1980; Ratnatunga and Newell 1984).

A real, unresolved astronomical source is constrained to have
an intrinsic profile like that of the point spread function. Therefore, a variety of defects can be rejected by applying a test
whereby one asks whether the observed image is consistent, within
the statistical noise, of being a realization of the point spread
function. A less direct but still effective (and faster) method is
to compute various image profile parameters for stars, determine the
dispersion of these parameters, and then reject anything outside an
allowable volume in the parameter space. Common types of photographic emulsion defects can also be distinguished from galaxies by
the same principle. Something along these lines is done by the Bell
Labs FOCAS software (Jarvis and Tyson 1981; Valdes 1982). In its
simplest form this approach may have difficulties when applied to
variables projected against galaxy images. Another technique to
screen out picture defects is to require a match between every
object between two pictures obtained at essentially the same epoch
(e.g. Kron 1980; Hawkins 1983).

A possible source of confusion would result from a non-Gaussian
wing on the error distribution, say in the measurement of integrated
image brightness. The contributors to measurement errors are many
and varied, and in concert usually lead to a net distribution that
looks like a Gaussian. This distribution may still have a wing
extending to large errors, not necesarily attributable to image
defects. If a large number of objects per picture are evaluated,
some will lie in the wing of the error distribution even if the
amplitude of the wing is low. These objects may be flagged as
candidate variables, and must be identified as spurious by some
technique that recognizes the non-Gaussian nature of the statistics.

Table 1. Hawkins' Survey for Variables

amplitude range (mag)	number
A > 0.6	4
0.6 > A > 0.5	4
0.5 > A > 0.4	27
0.4 > A > 0.3	42

As an example of this point, Table 1 gives the amplitude distribution, as a function of amplitude, for Hawkins' (1983) list of 77 faint variables. A number of plates in each series were obtained at nearly the same time, which allows Hawkins to compute the distribution of internal errors. This peaks near 0.1 mag in the interval $20 < B < 21$. Out of a total of 10^5 objects, as many as 1000 could be regarded as being in the wing of the error distribution. The objects with large errors compared to their amplitudes can be screened out, but one thereby suffers incompleteness in the cataloguing of variables at small amplitudes. From Table 1, which includes only those objects that have passed such a screening process, it is clear that the number of variables (in this case mainly quasars) is a very steep function of amplitude (see also Usher, Warnock, and Green 1983). A less conservative criterion would have detected many more variables, but would have also picked up many false ones, the numbers for which must also be increasing rapidly with decreasing amplitude. It is therefore of great importance to do everything possible to reduce the size of the internal errors, both for scientific reasons (finding low-amplitude variables) and for practical reasons (fewer objects need to be reevaluated on a second pass).

The same kind of survey that Hawkins has been doing with Schmidt plates can of course also be done with wide-field large telescopes, with the resulting increase in limiting magnitude. Figure 1 illustrates what can be done. This gives photometry for all objects that passed a classifier for stellar images on Mayall plates of Selected Area 57. Two of these plates, MPF 1561 and MPF 1562, were obtained on the same night, and the magnitude differences as a function of magnitude are shown at the top. This defines the error distribution. Figure 1 also shows MPF 1562 with respect to MPF 1053, a similar plate obtained about a year earlier (Table 2). The variables over this particular baseline are evident. The time series is by now relatively extensive, and the existing set of matched plates (same telescope, correcting lens, filter, and emulsion) is given in Table 2. More telescope time has been granted for the next two years, and in the meantime I hope to make progress in identifying and classifying the variables, with the help of collaborators D. Koo, A. Vignato, D. Trevese, and D. Nanni. Koo has in

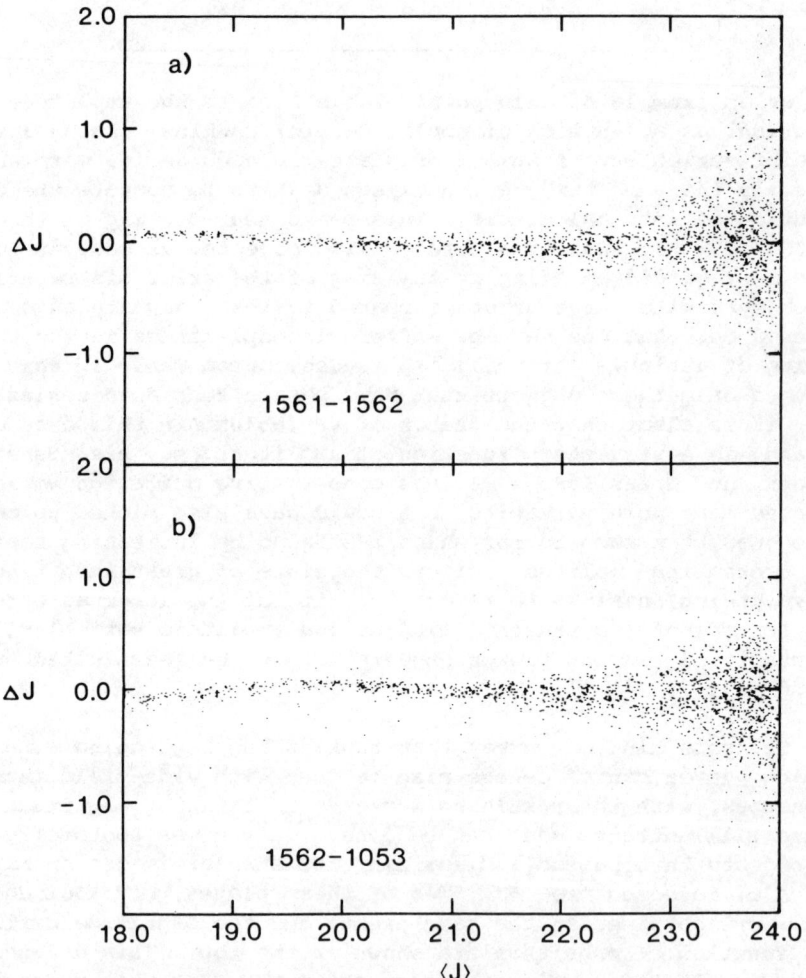

Fig. 1. Detection of variables to very faint limits by comparison of stellar photometry at different epochs (bottom). The "J" nomenclature refers to a photographic band that is similar to B.

Table 2. Mayall IIIa-J + GG 385 Plates of Selected Area 57

MPF	epoch (UT)	exp (min)	comments
1053	21 05 74	45	
1561	04 04 75	45	
1562	04 04 75	45	
2176	01 12 76	45	large air mass
3313	17 05 80	20	some twilight
3622	31 01 82	55	
3919	05 04 84	50	some clouds
3921	05 04 84	70	

fact accidentally discovered a supernova on the MPF 1561/1562 plates during a visual scan. The spectroscopic redshift for the galaxy is $z = 0.12$, and the supernova was observed 3.5 arcsec from the nucleus at $B = 21$. The following night two IIIa-F + GG 495 plates were obtained, and we should be able to derive a color.

The image processing problems discussed above have the general nature that the signal (variable objects) is buried in the noise (all of the things that go wrong with real pictures). The task of designing software to recognize the difference is more or less difficult, depending on how important completeness of the survey is, and how important it is that the final list be uncontaminated by false identifications. Variable objects have traditionally been identified on photographs by eye. This has as much to do with the capability of the eye to reject noise as it does with the capability of the eye to "parallel process". Sometimes it is said that machine identification of variables is superior to that of the eye because machines are objective and the results are reproducible. However it is more to the point to say that the real advantage of machine measurement is in analysis of data volumes that are too large for reasonable human consumption. The machine can be regarded as a filter which produces a very much reduced body of data according to explicit search criteria. The initial design of pattern recognition algorithms that identify certain classes of images, and reject others, always needs to be done interactively -- the machine and the programmer together learn about the various possible manifestations of noise. There will come a point of diminishing returns, when a circumstance is so rare that it does not pay to anticipate it in the software.

Photometry from digitized pictures have been discussed in many contexts (e.g., Newell and O'Neil 1977; Stetson 1979; Newell 1979; King 1983; and Lee and van Altena 1983). Here it suffices to re-emphasize a few general issues.

1) The greater the number of pixels across the seeing disk, the more games can be played in terms of optimizing the integration of the light under the image. On the other hand, the more pixels per seeing disk, the more processing has to be done.

2) From the image profile $f(r)$ one can compute the associated function $g(r) = 2\pi r\, f(r)$. When integrated over r this is of course the total brightness. The differential form $g(r)$ tells how the light is distributed, and in particular one might say that the annulus within which $g(r)$ peaks is the most important. Although less of the total light is contained at smaller radii, the light at the center is after all seen at maximum signal-to-noise ratio. All of this should be taken into account: an <u>unweighted</u> fit of some analytic form to the data for purposes of integration is in general not optimal because it does not recognize the character of the angular distribution of information.

3) Many investigators have successfully improved the internal precision by making local differential measurements. This involves comparing the measured brightness of a candidate variable with those of stars in the angular vicinity. This technique removes residuals due to variations of the point spread function from place to place, and variations in plate fog and sensitivity, to the extent that these variations are on larger angular scales. Effort should be made to understand the nature of these residuals in order to be sure that all of the information that can be used has been used.

4) Not all of the error is inherent in the image. The measuring machines can introduce a non-negligible error depending on all sorts of factors, for example lag in the amplifier, scattered light, quantization in both intensity and position, and internal noise. The size of this problem can be easily determined by repeat scans, say with the plate rotated by $90°$. Also, software can introduce a kind of "noise" according to the specifics of exactly which pixels are used, how they are weighted, how the background is determined, etc. This can be tested by giving the same data to different programs. J.A. Tyson and I have tried a variation on this: the plate was the same but the particular scan was not, nor was the software. The rms deviation for the same star images was about 0.1 mag, indicating room for improvement.

4. APPLICATIONS

For the present application of image processing we want to cover as large an area of sky as possible, as often as possible, to a faint limit with high precision; to suffer few image defects; and to discover new variables no later than, say, 24 hours after the data are obtained. There are clear scientific benefits to be realized according to the degree to which each of these desiderata can be achieved, as discussed in section 2.

Any given system must involve compromises and will be tailored for some particular task. At the telescope there is a trade-off between flux depth and area surveyed per unit time. In practice the faster a measuring machine, the less precise are the positions and magnitudes that it produces. In view of the complications mentioned in section 3, there is some minimal processing time required per pixel or per image in order to screen out adequately the noise. This implies an upper limit to the data that can be processed within the nominal 24-hour interval.

Since the job of finding supernovae quickly is perhaps the most demanding, the various specific techniques that have been considered (see review by Trimble 1983) are instructive in general.

McGraw's (McGraw, Angel, and Sargent 1980; McGraw et al. 1984) solution illustrates a novel approach with numerous advantages. The system consists of a dedicated 1.8 meter transit telescope with a CCD detector clocked out at the sidereal rate. Each row of pixels as it is read out is immediately available for comparison with archival data for the same patch of sky. More than one band can be observed by having more than one CCD in the focal plane. The limiting magnitude depends on the signal accumulated during the minute or so that it takes for the field to cross the CCD array. [Since charge is transferred along columns while integrating, many problems of flat-fielding are circumvented -- see Boroson, Thompson, and Shectman (1983) for a discussion of this technique.] The limiting magnitude is expected to be around $V = 22$ for a precision of 10%.

It is interesting to compare what the CCD transit instrument, or CTI, can do with respect to a conventional technique such as Schmidt photography followed by fast machine measurement. In one night of operation, the CTI samples about 8×10^7 spatially-independent pixels, each 1.6 arcsec on a side, for a total area of 17 $\deg^2$. A Schmidt telescope can survey more area per night, but different bands must be obtained sequentially, rather than simultaneously, as in the CTI. Roughly speaking the data rates and angular resolution are similar. There are a number of fast machines that can scan a large Schmidt plate and process the information in less than a day. An interesting new contribution to the field is David Monet's PDS/IIS Measuring Engine, currently operating at the National Optical Astronomy Observatories in Tucson. The principle here is the use of a commercial CCD television camera to digitize successive areas of the plate; the PDS optics are not used at all -- the PDS is merely a device for moving the plate in a precise way under the CCD camera. The CCD frames are processed in a VAX as the scanning proceeds, and with current software a Schmidt plate takes about six hours to process (this does not count the special analysis required for identifying variable sources). Monet's machine therefore fills a niche in data acquisition hardware: in this instance, photographic techniques are used to survey a large area of the sky rapidly, and a CCD is used to digitize the information rapidly. An important aspect of Monet's system is that the hardware, not including the PDS itself, is very inexpensive.

The general advantage of the CTI is that it is dedicated to the particular task. Every clear night the same strip of sky is surveyed, which means that confirmatory observations of any suspected variable are automatic. The time series on every variable in the strip will be unprecidentally extensive -- one observation in at least two bands per clear night of the dark run, during the period of the year when the object is visible. Supernovae at moderate redshift should be frequently detected. The supernova rate in different types of galaxies should be determined to much higher accuracy than is now known. (The method is however not efficient at finding relatively nearby, bright supernovae, and this is where other techniques have a special value.) The CTI will also of course detect numerous quasars and Galactic variable stars, and the scientific returns from this approach are likely to be enormous.

I am grateful to David Monet, David Koo, and John McGraw for their assistance.

REFERENCES

Bahcall, J.N. 1980 Ap.J. 240, 377.
_____ and Hills, R.E. 1973 Ap.J. 179, 699.
Boroson, T.A., Thompson, I.A. and Shectman, S.A. 1983 A.J. 88, 1707.
Colgate, S.A. 1979 Ap.J. 232, 404.
Fahlman, G.G. 1977 Ap.J. 211, 649.
_____ and Ulrych, R.R. 1975 Ap.J. 201, 277.
Green, R.F., Huchra, J.P. and Bond, H.E. 1977 Pub.A.S.P. 89, 255.
Grosbøl, P.J. 1980, in Applications of Digital Image Processing in Astronomy, ed. D.A. Elliott (Proc. SPIE 264), p. 118.
Hawkins, M.R.S. 1983 M.N.R.A.S. 202, 571.
Jarvis, J.F. and Tyson, J.A. 1981 A.J. 86, 476.
King, I.R. 1983 Pub.A.S.P. 95, 163.
Kron, R.G. 1980 Ap.J. Suppl. 43, 305.
Lee, J.-F. and van Altena, W. 1983 A.J. 88, 1683.
McGlynn, T.A. and Turner, E.L. 1980, in Applications of Digital Image Processing in Astronomy, ed. D.A. Elliott (Proc. SPIE 264), p. 188.

McGraw, J.T., Stockman, H.S., Angel, J.R.P., Epps, H. and Williams, J.T. 1984 Optical Engineering 23, 210.
_____, Angel, J.R.P., and Sargent, T.A. 1980, in Applications of Digital Image Processing to Astronomy, ed. D.A. Elliott (Proc. SPIE 264), p. 20.
Monet, D.G. and Dahn, C.C. 1983 A.J. 88, 1489.
Newell E.B. 1979, in Image Processing in Astronomy, eds. G. Sedmak, M. Capaccioli, and R.J. Allen (Osservatorio Astronomico di Trieste), p. 100.
_____ and O'Neil, E.J. 1977 Pub.A.S.P. 89, 925.
Pica, A.J. and Smith, A.G. 1983 Ap.J. 272, 11.
Ratnatunga, K.U. and Newell, E.B. 1984 A.J. 89, 176.
Stetson, P.B. 1979 A.J. 84, 1056.
Trimble, V. 1983 Rev. Mod. Physics 55, 511.
Usher, P.D. 1978 Ap.J. 222, 40.
_____, Warnock, A. and Green, R.F. 1983 Ap.J. 269, 73.
Valdes, F. 1982, Proc. Soc. Phot. Instrum. Eng. 465.
Wagoner, R.V. 1980, in Physical Cosmology, eds. R. Balian, J. Audouze, and D.N. Schramm (North-Holland:Amsterdam), p.179.

MORPHOLOGICAL ANALYSIS OF EXTENDED OBJECTS

Fabio Pasian and Paolo Santin

Osservatorio Astronomico
Trieste, Italy

INTRODUCTION

   Extended objects in the field of Astronomy show a great variety of shapes, ranging from the simplest (almost circular or elliptical) to the the most complex morphologically, composed by a number of elementary shapes. Furthermore, it is quite common for astronomical images to be severely degraded by noise; this fact induces distortions on the final descriptions of the extracted shapes. It is therefore difficult to describe analytically, or by means of models, morphologically complex shapes, and also the simplest shapes, if they are affected by distortions.
   The shape analysis techniques being at the moment the most popular in Astronomy make use of best fit algorithms; model shapes are compared to the observed ones and the best fit is computed (Benacchio et al.,1979; Cawson,1981). Such techniques, although being quite efficient, in the case of generical shape analysis have two major drawbacks:
(i) the unknown shape is always compared to a "known" shape and therefore a limited number of "known" simple shapes is generally used;
(ii) the error in the fit, if given, is a spatially integrated RMS error, therefore hiding local deviations.

The goal of the proposed approach is to obtain a complete description and analysis of the shapes of extended objects, obeying to the following requirements:
(i) all of the information contained in the shape of the object should be extracted, in order to obtain good description results also in the case of complex morphologies; furthermore, the problem of shape description and discrimination may in this way be approached without any prior knowledge or prejudice on an analytical form or, conversely, the analytical form may be recognized but also significant deviations may be put in evidence;
(ii) the obtained description should univocally define the shape of the object, so to allow comparison with other descriptions, both of real objects and/or of theoretical models.

THE METHOD

In the following, solutions to the above-mentioned problems are proposed and an integrated system for shape analysis of extended astronomical objects is presented.
(i) in the preliminary phase, when the shape of the object is to be extracted, every possible information must be gathered, even related to distortions due to noise. A following step will establish a hierarchy among all of the shape components. When the shape to be extracted has a low signal to noise ratio or represents a complex morphology, some preprocessing steps may be needed.
(ii) a set of parameters, describing as univocally as possible the shape of the object being analyzed, is looked for. A correspondence between the shape and a set of N parameters is established, therefore allowing representation of the object as a point in an N-dimensional space. In the N-dimensional space, comparisons between different extracted shapes or between an extracted shape and a theoretical model can be performed.

The proposed solution is to make use of techniques and algorithms already experimented in the field of pattern recognition, combined in an integrated system which is an original approach to the study of complex morphologies in astronomy (Pasian and Santin, 1984).

Fourier Descriptors

The selected approach to the problem of shape analysis makes use of the Fourier Descriptors (FDs), as defined by Zahn and Roskies (1972). Such a method allows description of any closed

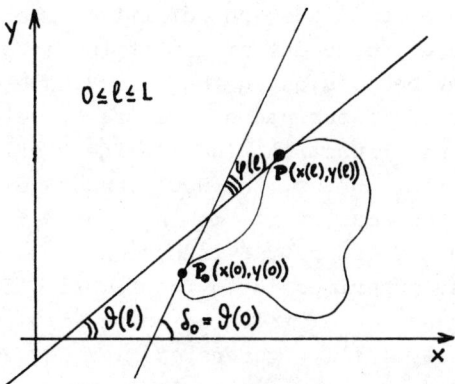

Fig. 1. Fourier descriptors for plane closed curves: definitions.

curve, no matter how complex, with a good degree of univocity, allowing comparison between the extracted descriptions. The shape of the curve describing the object is represented by a point laying in the N-dimensional Fourier descriptors space. If a curve is particularly complex, the description precision may be improved incrementing the dimensionality factor N. The usefulness of such a representation is that the positions of the points in the N-dimensional space may be compared among each other and reference can be made to a grid of positions of theoretical models, thus allowing comparison and, possibly, classification.

With reference to Fig. 1, for a closed curve with parametric representation $x(l),y(l)$ with $0 \leq l \leq L$, a normalized cumulative curvature function (NCCF), defined as:

$$\varphi^*(t) = \varphi\left(\frac{Lt}{2\pi}\right) + t \qquad t = 2\pi \frac{l}{L}$$

can be defined. The Fourier descriptors for the closed curves are the harmonic amplitudes and the phase angles of the $\varphi^*(t)$ function, expanded as a Fourier series and converted in polar form.

Fourier descriptors have some interesting properties for curves frequently dealed with in Astronomy. For circles, all of the harmonic amplitudes are null and consequently, all circles are represented in the N-dimensional Fourier descriptors space by the same point (the origin). As for ellipses, the harmonic amplitudes are zero for all odd indices.

The concept of zero is worth further discussion since, when

403

dealing with shape analysis on digitized real data, truncation errors and noise modify the definition of the zero "functions".

The circle may be considered the most "shapeless" curve since all of its Fourier descriptors are zero, being the normalized cumulative curvature function (NCCF) identically zero. When dealing with digital curves, the NCCF cannot be identically zero, due to truncation errors; it makes sense to define a zero normalized cumulative curvature function (ZNCCF), which may be considered as noise superimposed to the identically zero NCCF. As a test, random gaussian distributed noise with varying amplitudes was added to an analytically generated X-Y-Z gaussian: circles were extracted from this artificial image to analyze the behaviour of the Fourier Descriptors in the case of noisy curves. A number of different realizations of random noise was implemented for each signal to noise ratio. The noise we are dealing with is the noise induced radially on the curve by the intensity noise affecting the image. The behaviour of the ZNCCF was studied: it has been verified (Fig. 2a) that its sigma reaches a well defined saturation level when increasing the noise to signal (N/S) ratio. Two regions may be identified on the plot: in the leftmost, increasing the intensity N/S ratio, the induced ZNCCF sigma increases also, thus increasing the net amplitude of the deviations from the ideal curve. Let's call this region the "micro distortions region". When furtherly increasing the intensity N/S ratio, the induced effect on the curve is to create "macro distortions", generating spurious curve branches which break shape simmetry. The ZNCCF sigma reaches a plateau, but analysis on the curve branches may lead to spurious (erroneous) information.

In order to avoid this artificial effects, care should be taken in order to consider curves extracted from images at reasonable N/S ratios; a system allowing shape analysis with FDs at higher N/S levels is proposed in the following.

Fourier descriptors were computed for circles extracted at various N/S ratios: in the region before saturation the range of their variation from the ideal zero level is narrow, while increasing abruptly in the "macro distortion" region (Fig. 2b and 2c as examples). When choosing a particular N/S ratio, it can be seen that FD values have a narrow range of variations, independently from the various intensity noise realizations. The average FD value at the selected N/S ratio can be identified as a working zero reference level. The uncertainty on this zero reference level may be chosen accordingly to the specific application. We can also take this level as a normalization level,

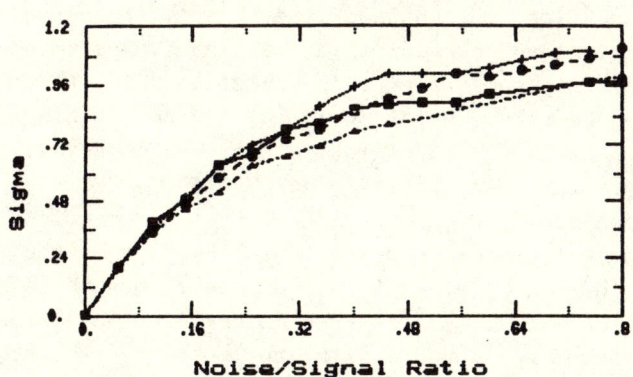

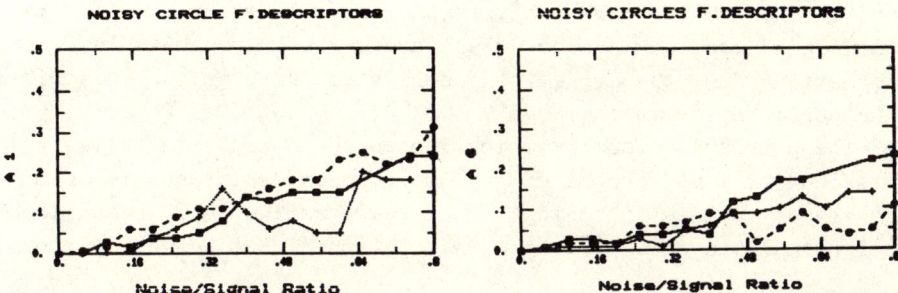

Fig. 2. Fourier descriptors for noisy circles : (a) Sigma of the Zero Normalized Cumulative Curvature Function vs. noise-to signal ratio for different noise realizations ; (b) A1 Fourier descriptor vs. noise-to-signal ratio for different noise realizations ; (c) A6 Fourier descriptor vs. noise to-signal ratio for different noise realizations.

to which all the extracted FDs may be referred to. The normalized FD values therefore will express the number of times each FD is greater with respect to the mean power contained in the zero level FDs.

Curve tracking

The use of the Fourier descriptors leads to the need of describing curves in terms of their parametric equations. The method has been fully described in Pasian and Santin (1983,1984).

The first problem is extracting from the original image a curve representing the shape of the object. The used approach is to follow an isolevel curve with a tracking algorithm and define it by means of its parametrical representation $x(l), y(l)$. The curves extracted may be very complex, both due to the complexity of the object being analized, and/or to the presence of noise.

The algorithm used (Pasian and Santin, 1983) exploits the classical pattern recognition scheme of omnidirectional tracking. Any curve, no matter how complex, is followed and the parametric description of each elementary component branch is stored in a tree-structured array. All of the information contained in the curve is therefore saved, with no losses: anyway, a further identification and segmentation phase is needed and performed in order to obtain the closed curve on which the Fourier descriptor analysis is to be done. A distinction is made between branches forming the curve to be analized and branches to be analized in a following phase or to be disregarded as possibly spurious information due to noise. The segmented curve is then described using parametrized functions.

The particular form in which the curve is represented is well suitable to be processed by many different one-dimensional algorithms, such as filtering, shape registration and comparison, Fourier analysis and, of course, extraction of Fourier descriptors.

Preprocessing

In the case of a low signal-to-noise ratio, some processing before the curve extraction phase might be necessary.

First of all (Pasian and Santin, 1983), the object of interest is extracted from the original image by means of a segmentation operation; different segmentation techniques are available and adaptable to the particular situations to be possibly solved. Then, a curve representing the object of interest is extracted by means of a thinning technique, in order to be followed by the above-mentioned tracking algorithm.

For extended objects, a further preprocessing phase may be needed to help overcoming problems such as background normalization or noise rejection.

APPLICATIONS

In order to study the performance of this method of

morphological analysis, curves extracted from both synthetically generated and real images were compared with a theoretical model. As a first application, the simplest model of astronomical interest has been chosen, the elliptical one. It must be noted that the purpose of these applications is not to perform an ellipticity analysis, but to give a quantitative measure of the similarity between the extracted shape and the elliptical model.

A set of synthetical objects has been generated in order to test the performance of the procedure: the objects were 2-dimensional gaussians having different X and Y axes and with gaussian noise superimposed. From these test images, isolevel curves were extracted, therefore representing ellipses corrupted by topological noise. Any closed curve is represented in the N-dimensional space of Fourier descriptors by a single point and two curves may be compared by computing the distance of the two points representing them in the N-dimensional metrics. In particular, the distances between each extracted curve and the theoretical noise-free ellipses were computed.

For the first test, a set of elliptical shapes with axial ratio ranging from 0.2 and 1.0 were extracted from synthetically generated objects at noise levels of 10% and 20% of the signal level. In Fig. 3 are depicted the distances from elliptical models of the extracted curves vs. the axes ratios of the elliptical models themselves. The white and black dots refer to curves

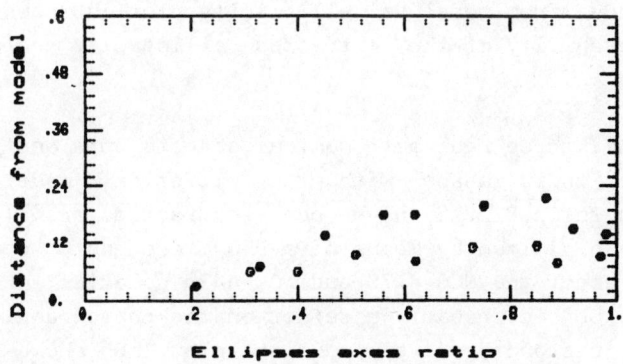

Fig. 3. Distance from elliptical model vs. ellipses axes ratio for curves extracted from synthetically generated images at noise-to-signal ratio of 10% (white dots) and 20% (black dots).

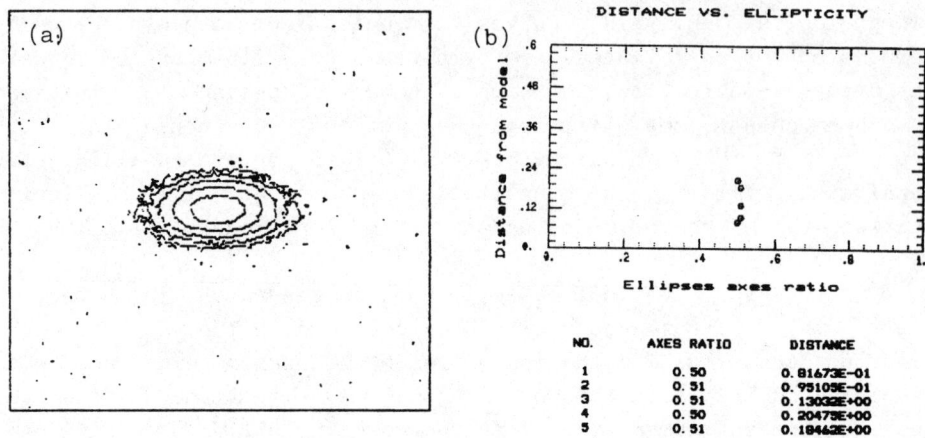

Fig. 4. Synthetically generated image : (a) five isolevel curves; (b) position on the distance from elliptical model vs. ellipses axes ratio plane of the points referring to the isolevel curves of Fig. 4a.

extracted from the test images at noise/signal ratios of 10% and 20% respectively. It is evident that the distance increases with higher noise/signal ratios, raising the uncertainty of classifying the curve as "elliptical". Furthermore, the uncertainty increases also with the axes ratio.

The second test refers to the analysis of curves extracted from a single image with 0.5 axes ratio at noise levels ranging from 3% to 50%. The five isolevel curves are shown in Fig. 4a, and the distance from model vs. ellipticity plot is shown in Fig. 4b, where the stability of the extracted ellipticity values may be appreciated.

Finally, three real astronomical objects were analyzed so as to compare their shape with the elliptical model. The three objects, shown in Fig. 5, have been extracted from photographic plates and digitized by means of a PDS microdensitometer. Fig. 5a and 5b represent the NGC 4473 and NGC 4438 galaxies, respectively, while Fig. 5c is probably a defect on the photographic plate. In Fig. 5d the distance vs. ellipticity plot is shown for curves extracted at different levels from the three preceding images. In this plot, the presence of three clusters of points is evident. The cluster featuring smaller distance from the elliptical model refers to Fig. 5a. The cluster on the left-upper side of the

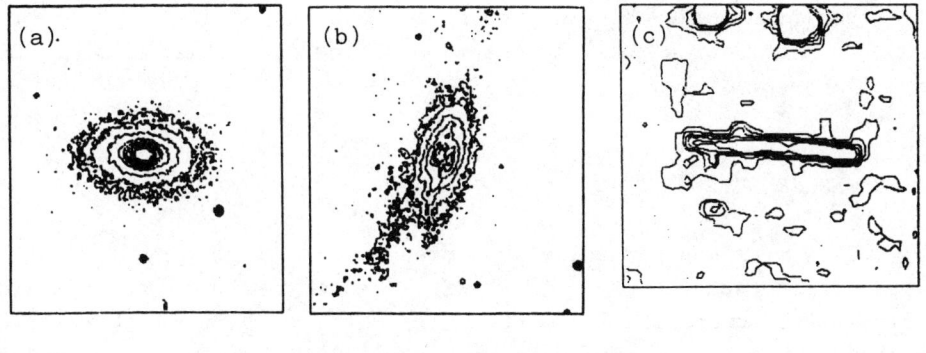

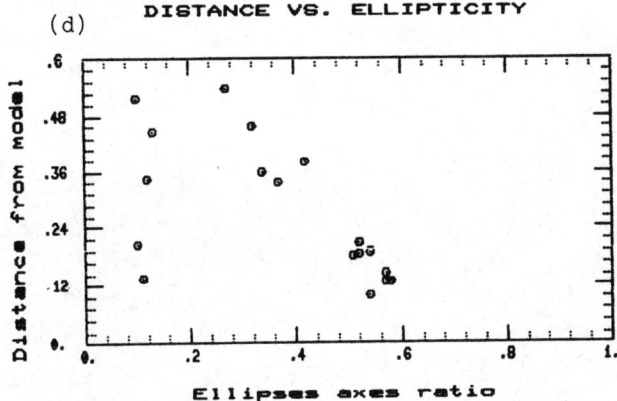

Fig. 5. (a) isolevel representation of NGC 4473 ; (b) isolevel representation of NGC 4438; (c) isolevel representation of unknown object; (d) position on the distance from elliptical model vs. ellipses axes ratio plane of the points referring to the isolevel curves of Figs. 5a, 5b and 5c.

preceding one refers to Fig. 5b, and all of the points show significant distance from the elliptical model. Finally, the leftmost cluster refers to Fig. 5c: it can be seen that the distance from the elliptical model increases at lower levels, where the object clearly shows morphological distortions from the elliptical shape.

The separation of the clusters in the distance vs. ellipticity plot is shown just to give an idea of the performance of the methodology. The analysis of the distribution and clustering of points in the N-dimensional Fourier descriptors space and the comparison between the positions of the objects under analysis and the positions of objects chosen as training set give a much more powerful tool for a possible solution of the morphological analysis problem.

ACKNOWLEDGEMENTS

This work was partially supported by a CNR-GNA grant. Data were processed at the ASTRONET Trieste center, Osservatorio Astronomico, using the local graphic library. The authors are grateful to Prof. G.Sedmak and to Dr. M.Pucillo for useful discussions.

REFERENCES

1. Benacchio L., Capaccioli M., De Biase G.A., Santin P., Sedmak G., 1979, Surface photometry of extended sources by an interactive procedure, in: "Image Processing in Astronomy", G.Sedmak, M.Capaccioli, R.J.Allen eds., Oss. Astr. Trieste, 196:210.

2. Cawson M., 1983, Fast shape analysis of galaxies, in: "Astronomical Measuring Machines", Stobie, McInnes eds., Edimburgh, 175:184.

3. Pasian F., Santin P., 1983, Shape information extraction in noisy environments, Pattern Recognition Letters, 2,2, 109:116.

4. Pasian F., Santin P., 1984, A proposal for shape analysis in Astronomy, Mem. S.A.It. (in press).

5. Zahn C.T., Roskies R.Z., 1972, Fourier descriptors for plane closed curves, IEEE Transactions on Computers, C-21,3, 269:281.

SOLAR IMAGE PROCESSING WITH THE CLARK LAKE RADIOHELIOGRAPH

T.E. Gergely and M.J. Mahoney

Astronomy Program
University of Maryland
College Park, MD., USA

The Clark Lake Teepee Tee radio telescope is located near Borrego Springs, California and is owned by the University of Maryland. The telescope has been operating since 1981 in different stages of software and hardware development, and is used for both sidereal and solar studies. In the latter capacity its ultimate objective is to obtain two-dimensional images of the Sun at high time resolution ($\sim$ 10 msec) in real time. The telescope is continuously tunable between 15 and 125 MHz, but radio frequency interference does limit observations to a number of interference-free bands with detection bandwidths up to 3.0 MHz.

The telescope itself consists of 720 conical log-spiral antennas, grouped in banks of 15 elements. A bank of elements can be electronically steered by imposing a linear phase gradient across it. Thirty-two banks are laid out on a 3000m east-west baseline and 16 banks form the 1800m south arm of the Tee. A detailed description of the telescope has been given by Erickson, Mahoney and Erb (1982); its use as a solar instrument has been discussed by Kundu et al (1983).

A digital correlator permits the simultaneous measurement of the sine and cosine components of the visibility function on 512 baselines. Telescope pointing and data acquisition are controlled by a Perkin-Elmer 7/16 minicomputer. Data processing may be accomplished on site with a Perkin-Elmer 7/32, or off-site using AIPS or other programs.

Because the sun can vary in brightness by factors of 10^4-10^6 on time scales of a few seconds or less, solar data processing presents difficult problems over and above the ones usually associated with astronomical data sets. Three of the major problems are the data volume, flux calibration and positional errors. We will now discuss each of these briefly.

Data Volume: At short wavelengths solar sources sometimes exhibit temporal variations on a millisecond time scale (Slottje, 1978). Very fast changes have also been observed at longer wavelengths by the Nançay Radioheliograph group (1983). In the wavelength range observed by the Clark Lake Radioheliograph some bursts are also known to occur on time scales of a second or less (McConnell, 1982). In order to study such bursts, observations of the highest possible temporal resolution are needed. Unfortunately this causes a commensurate increase in the volume of data which must be processed.

The output of the data reduction process is a 64 x 64 pixel map. The reduction process involves essentially four steps: fringe stopping, the application of gain and phase corrections, a two dimensional FFT, and a coordinate transformation. At the present time this takes about 6 seconds to accomplish, with the last two steps taking most of the time. The required computing times for each step are shown in table 1.

Table 1.

Fringe stopping + Gain and phase correction	1s
Fast Fourier Transform (2D)	1s
Coordinate Transformation	4s
Total Time	6s

Data is currently being acquired every 0.67 seconds, and eventually this will be reduced to every 11 milliseconds. Real time processing is possible only if an integration time of 6 seconds or greater is used. Considering 4 hours of observing, the current output of the telescope calls for processing a volume of data $\sim 10^9$ bits day^{-1}. This figure is to be compared with the output of "conventional" aperture synthesis telescopes, estimated to be the range $5-10 \times 10^6$ bits day^{-1} (Disney, 1984).

In order to achieve real time processing, it is necessary to reduce the current volume of data by a factor of about ten. This is accomplished by "stacking" the high time resolution data, thus simulating data acquired with large time constants. By looking at the data at high compression rates (e.g. effective time constants of 10 or 20 seconds) the observer can usually (but not always) determine if processing the data (or selected portions of it) at high time resolution would be of interest.

Flux Calibration: The large and rapid variability of the sun's flux makes it particularly difficult to calibrate solar observations. Invariably the data will be smoothed over even the shortest of integration times, and in addition the very large signal-to-noise ratio forces the data into the non-linear region of a correlated power measurement. As a practical matter, little can be done to overcome the data smoothing, other than using even shorter integration times with a proportional increase in the volume of data acquired.

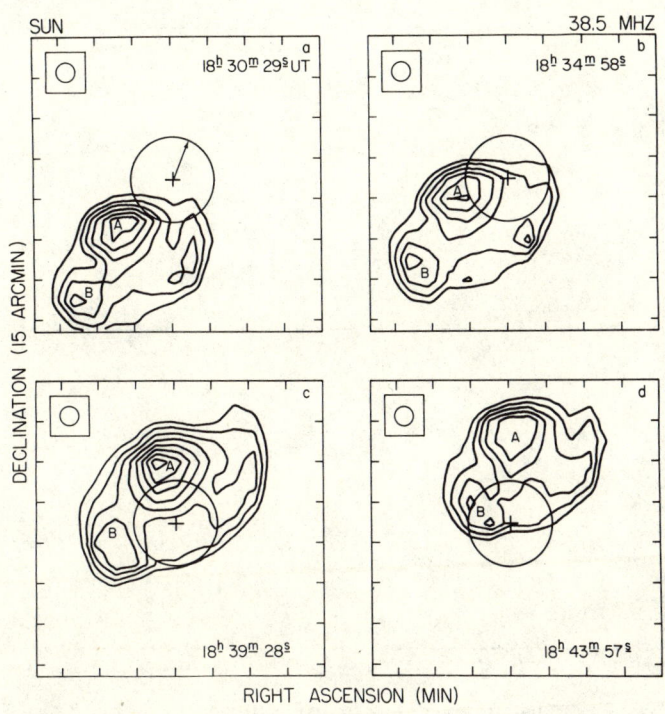

Figure 1 a-d. Maps of the Sun obtained on November 16, 1983 at 38.5 MHz. Each map was obtained by integrating 99 individual frames. The integration time for each frame was 0.67 seconds. The mean time of the observation is noted on each map.

If an ALC circuit is used, the response time of the circuit must be taken into consideration as well. ALC circuits of course are essential if more than 1 bit digitization is performed, and self-correlators can be used to help correct for the response time of the ALC circuit. If however 1 bit digitization is performed the ALC circuit makes little difference because self-correlators provide no useful information. So in either case we are faced with the dilema of unraveling the correlated from the uncorrelated signal. This ultimately must done in one of two ways: either by reducing the S/N ratio or else by applying corrections based on ALC levels,

or very frequent receiver gain calibration. We will be able to do
the former by disabling parts of the array in real-time. The second
approach will be used as well, since we plan to digitize the ALC
levels. In addition the very rapid steerability of the antenna will
allow frequent gain calibration. Astronomical sources must be used
instead of a noise source since we have no means of injecting noise
at the individual antennas forming the array.

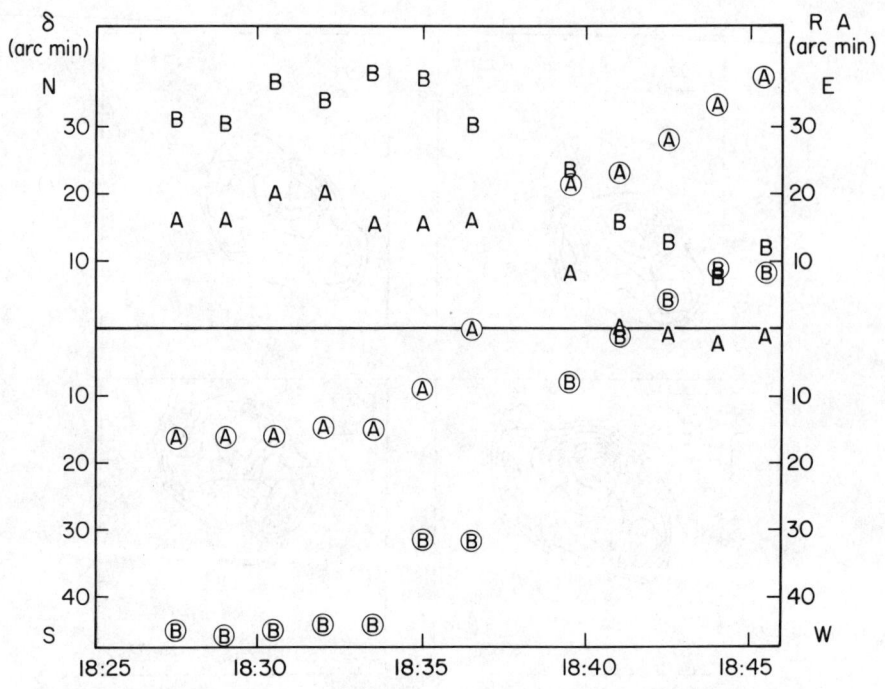

Figure 2. The displacement of sources A and B in Right
Ascension vs. time is shown by the letters A and
B. Displacement of the same sources in Declination
vs. time is indicated by the letters enclosed in
circles.

Positional Errors: At the meter-decameter wavelengths observed by
us, the ionosphere presents serious problems. In addition to the
quasi-steady displacement due to the spherical and wedge components,
the ionosphere often causes a periodic oscillating displacement of
the apparent position of the Sun. The period of this displacement
is usually about 20 minutes, its amplitude is frequency dependent
and may be larger than the solar radius itself. Figures 1 a-d show
one example, obtained on November 16, 1983 at 38.5 MHz. The four
maps of the Sun were obtained during an interval of ~ 15 minutes.
The integration time used during the observation was 0.67 seconds,
but the maps were processed by compressing 99 individual frames

spanning ~ 90 second intervals. Right Ascension (RA) and declination (δ) displacements with respect to the center of the field (the center of the optical disk) vs. time for the sources marked A and B on the maps are shown in Figure 2. An angular displacement of ~ 52" in declination occurred in ~ 12 minutes. The displacement in RA over the same period amounted only to ~ 20 arcminutes. At the frequency of observation the displacement of the image is approximately one beamwidth every minute and a half. The average direction of the motion is indicated by the arrow on Figure 1. This motion results in a "fuzzyness" or stretching out of the image in the apparent direction of the motion. Since relative positions can be determined with an accuracy of ~ 1" with the telescope (Erickson, 1984), the effect becomes important if integration times or compression of frames over a period longer than ~ 30 seconds are required. Since the beamwidth and the displacement due to the ionosphere increase with f^{-1} and f^{-2} respectively, the "smearing" will become more pronounced at lower frequencies. The problem is compounded by the fact that the source may undergo intrinsic changes during the time of the observation. On the map of Figure 1d, for example, source B has become double.

We have mentioned only a few outstanding problems encountered in observing a rapidly varying radio source at long radio wavelengths. Work is currently being done on these problems.

REFERENCES

Disney, M., 1984, This volume.
Erickson, W. C., Mahoney, M. J. and Erb, K., 1982, The Clark Lake Teepee Tee Telescope, Astrophys. J. Suppl. Ser, 50:403.
Kundu, M. R., Erickson, W. C., Gergely, T. E., Mahoney, M. J. and Turner, P. J., 1983, First Results from the Clark Lake Multifrequency Radioheliograph, Solar Phys., 83:385.
McConnell, D., 1982, Spectral Characteristics of Solar S Bursts, Solar Phys, 78:253.
Slottje, C., 1978, Millisecond Microwave Spikes in a Solar Flare, Nature, 257:520.
The Radioheliograph Group, 1983, The Mark III Nançay Radioheliograph, Solar Phys., 88:383.

DISPLAY OF THREE-DIMENSIONAL DATA IN RADIO ASTRONOMY

Arnold H. Rots

National Radio Astronomy Observatory
Socorro, N.M. 87801
U.S.A.

INTRODUCTION

The basic objective of image display is to provide the user with a tool that enables him to get as complete a grasp of the contents of the data as possible with a relatively small effort. With the advent of aperture synthesis techniques maps in (radio) astronomy have grown to sizes where digestion of their contents has become impossible by just staring at large collections of numbers. The problem is especially acute in the case of aperture synthesis spectral line observations where the astronomer has to deal with intensities in three dimensions: two spatial coordinates and velocity; we shall refer to this as "the cube". Historically, spectral line observers used to spread out large numbers of profiles (plots of intensity as a function of velocity at various positions in the sky) on any available surface. Aside from the fact that this becomes highly impractical for large spatial grids, it also does not enable the astronomer to build up a mental three-dimensional picture very easily. This is also true for the alternative approach of leafing through stacks of two-dimensional maps (spatial maps at different velocities and so-called "Position-Velocity maps").

So the question is: how can we provide such cube displays to the user that the three-dimensional structure of the data becomes immediately clear to him? Four methods have been investigated: time sequences of two-dimensional images, intensity-hue images, stereo pairs, and three-dimensional solids. I want to emphasize that the main purpose of these "fancy" displays is to aid the astronomer in interpreting his observations; as such they constitute invaluable tools even though the pictures may not (easily) be publishable.

A good reference for some of the techniques described in this paper is Foley and Van Dam (1982).

TIME SEQUENCE OF TWO-DIMENSIONAL IMAGES

Displaying a time sequence of two-dimensional cross-sections of the cube is, in a sense, a more sophisticated version of looking at a multitude of two-dimensional maps. This movie-like display does make it easier to detect continuity along the third axis. It is, therefore, a very powerful tool that usually can take advantage of functions present in, or connected with, the display device (e.g., video disk or zoom-and-pan) without putting an undue burden on the host computer. Still, it does not provide an "instantaneous" view of the cube that can be taken to an office and stared at in a peaceful atmosphere.

Two methods for generating time sequences are in use at the present moment. The GIPSY system in Groningen has a video disk attached to its IIS image computer. Individual frames are put on the TV monitor and transferred to the video disk; the sequence can then be played back from the disk while a special purpose control panel provides control over length, direction and speed of the playback. A similar system, employing digital Winchester-type disks, is under development at the N.R.A.O. for use with its IIS machines. The other method that is in use in AIPS puts a mosaic of two-dimensional maps in each refresh plane and employs the zoom-and-pan feature of the display computer to show one map at a time and quickly step through the frames. Trackball and buttons provide speed, direction, and dynamic still frame control; a maximum frame rate of about 10 Hz can be achieved. There is obviously a trade-off between the displayed image size and the number of frames in the sequence which is ultimately limited by the number of available refresh planes.

In addition, it is of course possible to transfer images to video tape. However, for this application one needs an NTSC (or PAL) encoder, and a video tape recorder with editing capability; one has to be aware that this will most likely change the aspect ratio of the image, as well as (in the case of NTSC) chop off the bottom 32 lines. The advantage of producing a video tape is obviously that it enables the user to take the tape and view it at his leisure on a very inexpensive work station, rather than tying up more expensive resources.

INTENSITY-HUE IMAGES

Intensity-hue images have been in use for a long time now and are usually deployed to catch the contents of the cube in one colorful picture. In most cases the profile integral (zeroth moment) is coded as intensity and the velocity (first moment) as color - generally fully saturated spectral colors conforming with the redshift-blueshift terminology. Their use is limited, however, since the common color sequence does not provide enough resolution to distinguish subtle changes in velocity and since these images do not carry any information on the shapes of the profiles (e.g., width or multiple peaks). The first disadvantage can be partly overcome by subtracting large-scale motions (such as galactic rotation) and allowing the user control of the color tranfer function. The second problem can sometimes be solved by constructing the intensity-hue image from the entire cube, rather than just the moments, and/or introducing saturation, but there are still severe limitations.

Another application of intensity-hue images is the display of two-dimensional maps of complex quantities (e.g., visibility functions). Here amplitude may be represented as intensity and phase in a cyclic color scheme (yellow-red-magenta-blue-cyan-green-yellow).

To be truly useful, these displays should be complemented by a good color hard-copy device.

STEREO PAIRS

Early attempts at producing three-dimensional images concentrated on making stereo pairs (or anaglyphs) of stacked two-dimensional gray scale images. On the whole they have not been too successful because of the foreground-background confusion in these "transparent" images, but more experimentation is needed. Stereo pairs of three-dimensional "wire frame" plots are usually easier to interpret, but become much clearer when rotated in space (e.g., with an Evans and Sutherland-type machine). In a sense, these plots are very much like the images discussed in the next section. A problem that one should be aware of when working on stereo pairs is that stereoptic perception is extremely subjective and that some people do not have stereopsis.

In general, use of stereo pairs in a stereoscope is to be preferred (some three-dimensional vector display computers actually can provide a device for this purpose), but the use of anaglyphs (the red-green images) is simpler in combination with color TV monitors and for audience presentations. The phosphors of the monitors are usually well matched to the filters in common anaglyph spectacles, as are the dyes in Kodak Ektachrome film when exposed through the proper filters. Another consideration here is whether one wants to display the anaglyph image in positive or negative; most people seem to prefer positive images (bright lines on dark background) for the wire frame plots.

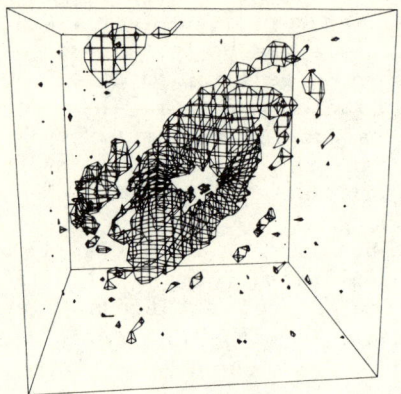
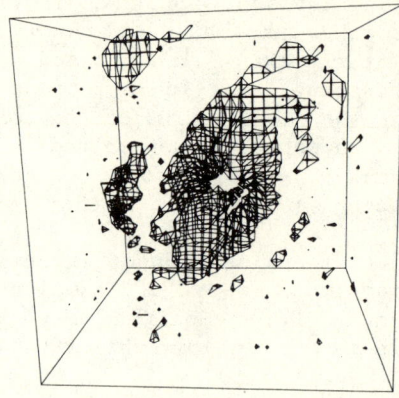
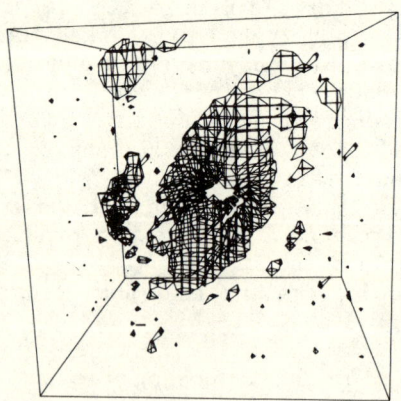
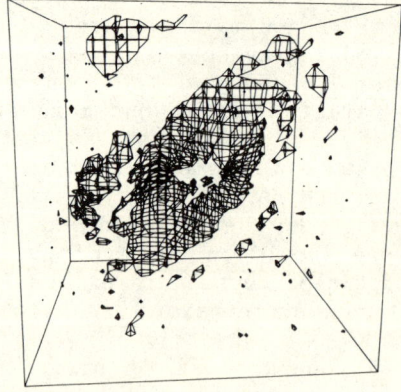

Fig. 1. Three-dimensional wire frame plots of HI spectral line observations of M81; depth is Doppler velocity.
1a. (upper panels) A stereo pair intended for parallel viewing.
1b. (lower panels) A stereo pair intended for cross-eyed viewing.
(See text for instructions).

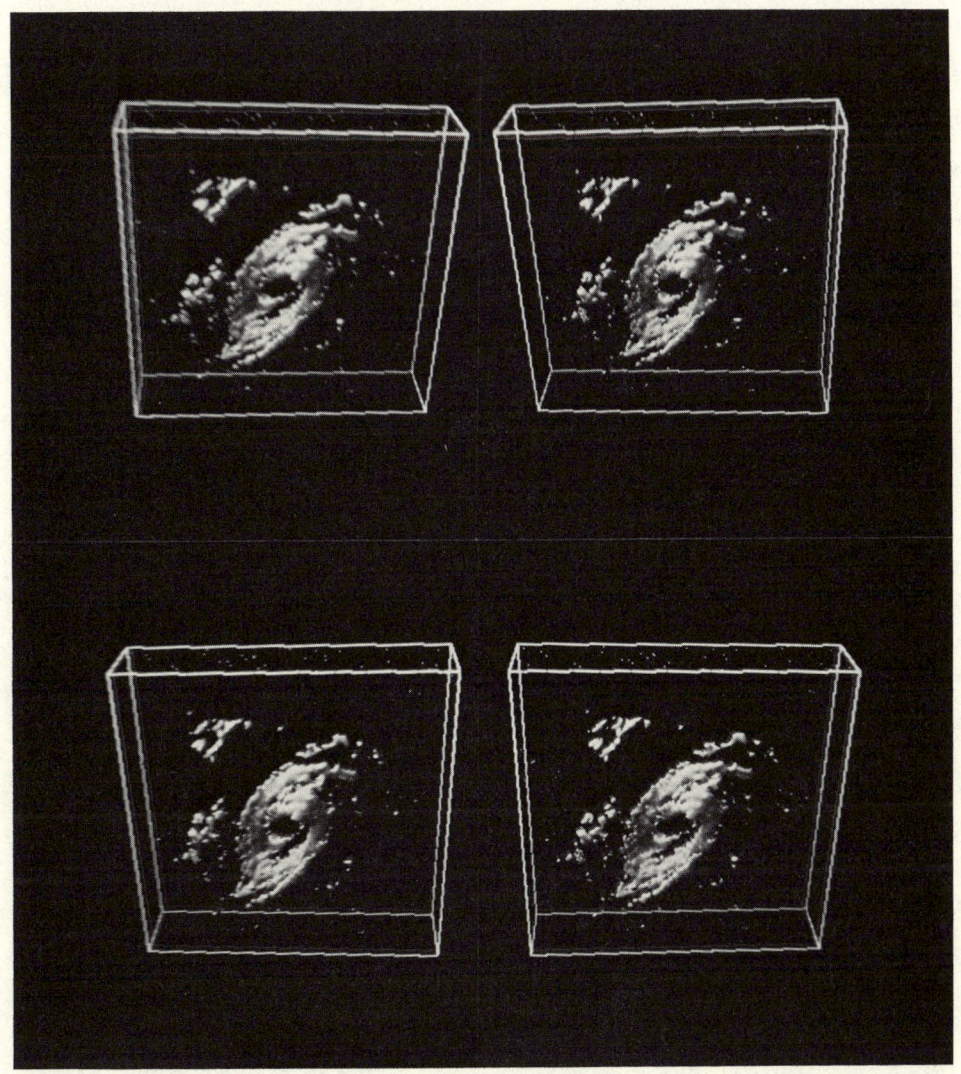

Fig. 2. Three-dimensional solid body representation of HI spectral line observations of M81; depth is Doppler velocity.
2a. (upper panels) A stereo pair intended for parallel viewing.
2b. (lower panels) A stereo pair intended for cross-eyed viewing.
(See text for instructions).

Stereo pairs can also be viewed without any special devices, but this usually requires some training. For this purpose one puts the two images of a stereo pair (each approximately 6 to 7 cm) next to each other. One can either put the image for the right eye on the right hand side and the left image on the left side, or one can interchange them. In the former case the viewer has to look over the images to a distant subject, shift his eyes to the images such that he sees three of them (the center one being the overlapped image from right and left eye), and gradually focus on this center image while making sure that his eyes remain parallel. In the latter case the viewer has to cross his eyes (it might help to focus on a finger held half-way between the eyes and the images) such that the left image in the right eye overlaps with the right image in the left eye, and focus on this overlapped image. Which way works best is very subjective. Fig. 1 shows a stereo pair of a wire frame plot of HI observations in M81; the depth represents Doppler velocity. Fig. 1a presents a pair for parallel viewing (also suitable for viewing through a stereoscope), Fig. 1b a pair for cross-eyed viewing. The observations were made by Butler Hine at the VLA.

THREE-DIMENSIONAL SOLIDS

The most sophisticated three-dimensional display is the 3-D solid surface. A three-dimensional surface is wrapped around the object(s) in the cube at a threshold intensity level and displayed in space by adding a viewing direction, perspective, depth cueing, ambient light, and direct light. A colored backdrop is to be recommended for such images to distinguish between true holes in the object and heavily shaded parts. The object can be shown in gray scale or can be intensity color-coded; this last feature is useful for objects that are "cut open". With the proper lighting and positioning 3-D solids are capable of displaying full spatial and kinematic detail, including velocity width and subtle changes in velocity.

Fig. 2 presents such a 3-D solid display of the same M81 data. The imaging algorithm used is rather simple and imperfect. It uses first-order partial derivatives for calculating the shading, a Z-buffer hidden surface removal method, and it images single voxels into single pixels. It therefore does not perform any anti-aliasing or dejagging, nor does it make any attempt at ray tracing or calculating shadows. Producing one frame from a 256x256x60 cube takes about 2.5 minutes of CPU time on a VAX-11/780 under AIPS.

Obviously, these images are eminently suitable for producing stereo pairs and anaglyphs. As a matter of fact, since astronomical objects in phase space are not as familiar to most people as common household items, it is not quite as easy to mentally visualize the three-dimensional structure from 3-D solid projections as one might expect. Stereo pairs are therefore an important adjunct. Fig. 2a shows a stereo pair for parallel viewing, Fig. 2b one for cross-eyed viewing.

The full potential of 3-D solids becomes apparent when the object is rotated. This is very effective for showing the full and detailed three-dimensional structure of the data. Animation is fairly easy, but real-time rotation is only the privilege of super-computer users. I have animated some rotations by exposing single frames on 16 mm film; it is a laborious process, but the results are very rewarding. As an alternative to film, the techniques mentioned in the second section of this paper can, of course, be used. It should be stressed again that the main purpose of these fancier displays (and this includes the animations) is to provide the astronomer with tools that make the interpretation of his data easier.

Finally, one can, of course, combine the options and make anaglyph animations (3-D movies) of 3-D solids, as well as of wire frame plots.

REFERENCE

Foley, J. D., and Van Dam, A., 1982, "Fundamentals of Interactive Computer Graphics", Addison-Wesley Publishing Company, Reading, MA.

DATA COMPRESSION TECHNIQUES IN IMAGE PROCESSING FOR ASTRONOMY

C. Cafforio*, I. De Lotto**, F. Rocca* and M. Savini**

(*) Dipartimento di Elettronica, Politecnico, Milano
(**) Dipartimento di Informatica e Sistemistica e unità
     del GNCB, Pavia

The classical way to digitally encode an image consists in sampling it on the nodes of a two dimensional grid and assigning to each pixel (picture element) a numerical value which expresses, to a desired accuracy, the luminance value on that node. A digital image can, therefore, be thought of as a matrix. Fig. 1 shows a plot of the luminance values of a bright star. Several effects are evident; namely saturation, irregular sampling and noise. Sampling irregularities are the most complex problem indeed. They are clearly evident in highly structured image areas, but they are obviously present all

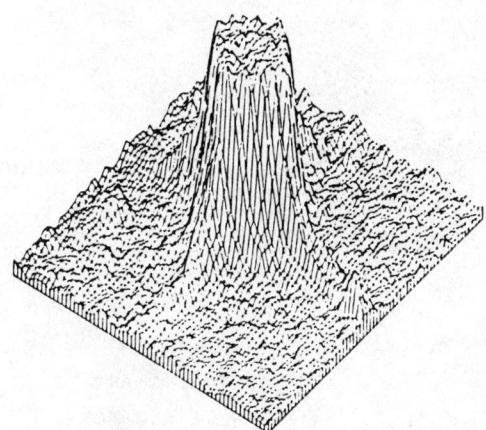
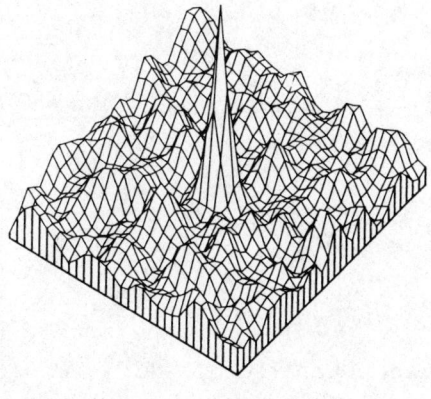

Fig. 1. Luminance plot of a bright star.

Fig. 2. Autocovariance of a background region.

over the image. They might not be as evident within the noisy (close
to purely random, see fig. 2) background, but this could depend on
aliasing noise, due to the small size of the spot of the scanner.
In other words the exploring beam does not act as a prefilter of the
data prior to sampling; thus it does not smooth completely the geometrical irregularities of the scanner which show up in the digitized
image. (Notice that deterministic irregularities of the scanning
system can be corrected by a suitable program once they are precisely
determined.) According to sampling theory, signal prefiltering is
mandatory in order to avoid aliasing noise. The only way to two-
dimensionally prefilter an image is through the use of an adequately
wide spot beam. As a consequence of the necessary filtering digital
samples are correlated. This means that some redundance has to be
present within the data and coding can be useful.

DPCM AND BIT-PLANE ENCODING

Among the compression techniques which do not alter the content
of information of the digitized pictures, or, in other words, that
permit an exact reconstruction of the original image from the coded
one, two have been chosen and tested: DPCM (Differential Pulse Code
Modulation) and bit-plane encoding process. Picture used (fig. 3)
has been scanned line by line and compression process is reinitiali-
zed each time the rigth edge of the image is reached.

DPCM algorithm[1] is applied on multilevel coded images; it enco-
des the prediction error pixel by pixel and requires a previous estimation of the average gray level of all the picture and the normalized correlation coefficient of adjacent pixels on the same line.

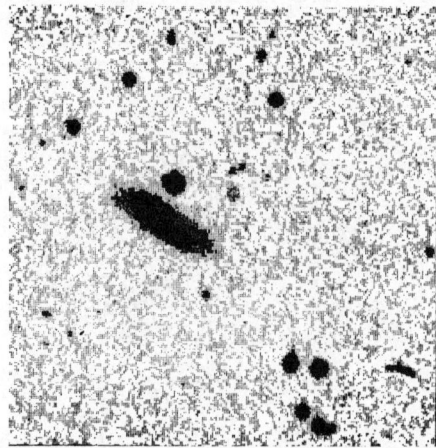

Fig. 3. Test image (ESO11.STD)

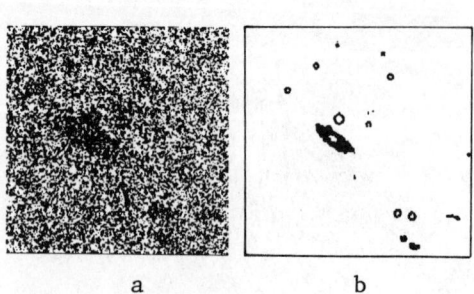

Fig. 4. Bit-planes 2 (a) and 6 (b) of test image of 256x256 pixels and 8 bits/pixel (0 to 7).

Tab. 1. Compression factors obtained with bit-planes method and the Gray code for three vector lengths (in pixels). Notice that taking care of transversal correlation is particularly useful for high level bit-planes. EOL means the end of line symbol used to stop propagation of transmission errors to following lines. As a reference, compression factors obtained with DPCM for the same image are 1.83 with EOL and 1.88 without EOL.

Bit-planes	with EOL Vector Dimension			without EOL Vector Dimension		
	1	2	4	1	2	4
0	0.802	0.8	0.944	0.853	0.831	0.968
1	0.801	0.797	0.944	0.86	0.83	0.966
2	0.896	0.9	1.04	0.974	0.941	1.069
3	1.743	2.236	2.556	2.018	2.486	2.718
4	0.98	0.968	1.09	1.07	1.012	1.122
5	5.784	7.914	10.747	9.392	11.455	13.974
6	8.15	13.063	20.023	19.09	25.352	32.883
7	10.277	16.8	28.519	32.363	44.613	60.569
Global Comp. Fact.	1.61	1.7	1.8	1.73	1.76	1.85

Bit-plane encoding splits the image of 2^n possible shades of gray into n bilevel images. Two gray codes have been tested: natural binary code and Gray code, the latter one having the property of only one bit change between any two adjacent code words. Here we present only the results obtained with the Gray code because of its slightly better performance. Moreover the transcoding process from natural binary code to Gray code and viceversa is very simple and low cost. Bit-plane encoding process operates on each two level image independently. Among the existing methods to code bilevel images we chose to test runlength[2] (RL) coding which consists in looking for consecutive occurrences, on the same line, of pixels of the same colour (runs) and in coding their length. In order to evaluate the effects of keeping into account also transversal correlation, vector runlength coding on each bit-plane has been tested for vector lengths of 2 and 4 pixels. This means that two or four lines have been coded together instead of the single scanning line, so looking for runs of vectors with the same configuration of black and white pixels perpendicular to the scanning direction.

Huffman code[3] has been used to code both the prediction errors in DPCM and the runlengths in the RL method. Table 1 summarizes the results obtained with the bit-planes method. Global compression factors are clearly limited by the least significant bit-planes that are not very much compressible, as it can be seen comparing a low significance bit-plane with a high significance one (fig. 4).

TRANSFORM CODING

An alternate description of an image considers it as a weighted sum of appropriate "base" images. The amplitude of each base image, instead of the luminance value on every sampling point, is to be given. The base set must be so chosen that each representation may be obtained from an other one through an orthogonal transformation. The two representations are entirely equivalent and need the same amount of information to describe the same image. However, if the transformation is properly chosen, an energy compaction results: the number of large coefficients is smaller. This means that it is possible to delete some of them if a small error in the data may be tolerated. The best choice would be the Karhunen-Loeve transform, but ease of implementation usually dictates other transforms[4]. The cosine transform is today the most widely used in television signal coding[5]. Its base functions are two-dimensional trigonometric functions and the coded image is therefore a two-dimensional trigonometric polinomial fit to the original data. As base images of higher order are needed to describe highly detailed areas and as human observer finds image degradations in such areas less objectionable, more transform coefficients may be deleted or described with less accuracy than energy compaction alone would allow. Besides, to assign an approximate luminance value to one pixel produces a luminance "dot" on the reproduced image, while errors on transform coefficients give rise to errors which are spatially distributed and therefore much less visible. Coding of astronomical images puts other more stringent needes. In fact images are not only to be visually inspected, they may be used to get instrumental measures. Fig. 5 shows root mean square and maximum absolute

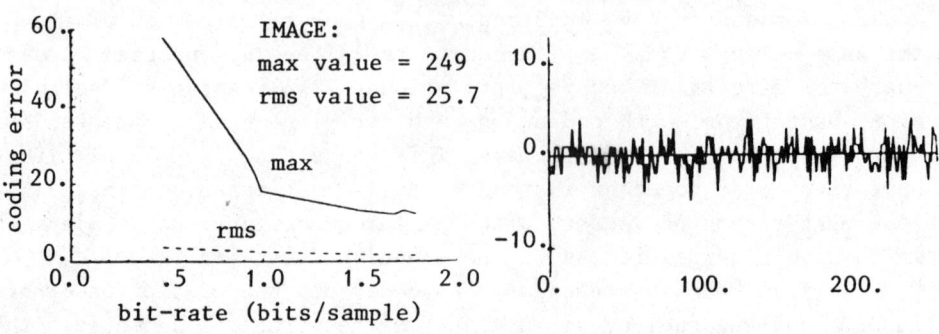

Fig. 5.   Fig. 6.

errors versus the number of bits per pixel used to code the test image
shown in fig. 3. Fig. 6 shows the error between the original image
and the same image coded with 1.77 bits/pixel as measured along an
image section. Transform coding cannot distinguish between different
features within an image (background and luminous bodies). As a consequence when transform coefficients are deleted luminous objects may
give rise to false structures (artifacts). It would be much better to
code luminous bodies and background separately, but then an algorithm
is required to identify them. A model of the source and of the acquisition system is needed. Stars are so far that it seems reasonable to
consider them as point sources. The more or less large "blobs" on
plates may be considered as due to atmosphere-telescope system. If
this system may be accurately modeled as linear, it must be possible
to infer the point spread function (p.s.f.) from the data. If the
system p.s.f. could be considered as space-invariant simple linear
deconvolution techniques could be used. The basic idea of the proposed two-component coding is to obtain a very compact but accurate
description of the position and intensity of the luminous bodies.
Based on this description, a syntetic image can be generated and subtracted from the original one. The residual image would contain no
longer bright areas and therefore the problem of artifacts would be
removed. One such technique has been tested with very poor results.
The reason of this failure is the strong limiting effect evident in
fig. 1. This non-linearity within the acquisition system, a problem
well known to the astronomers[6], clearly accounts for the failure of
any linear technique. The goal of this work was to code the original
data introducing the minimum amount of error. Therefore Gaussian fitting has been discarded because it would leave structured information
in the residual image. A simplistic approach has been used just to
show the potentiality of two-component coding. The bright areas have
been identified and coded at full 8 bits accuracy as they represent

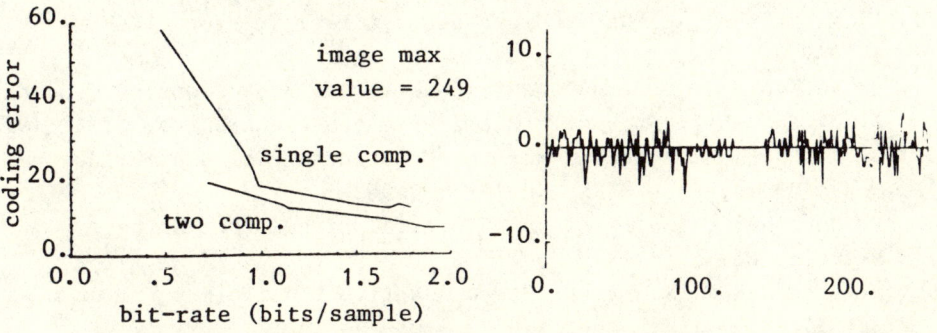

Fig. 7. Maximum absolute error.   Fig. 8.

a small part of the whole image. The residual image (background) has been coded with the already described DCT processing. Fig. 7 shows the obtained results. The gain in maximum absolute error obtained through the use of two-component coding is evident. Changes in r.m.s. error are, of course, much less sensitive to coding scheme, as large errors in small image areas result in small increments. Fig. 8 shows a plot of coding error along a scan line obtained with two-component coding. It is high frequnecy error and therefore it is not clear wether coding has resulted in real coding error or in a useful filtering of noise due to various sources. Should a good modeling of the imaging system be available, it could be very useful to obtain a more efficient coding. The model of the acquisition system should include a limiter as a last block. This certainly increases the difficulty of identifying the p.s.f. of the linear block, given the output and the limiter characteristic. Simple spectral techniques will not work. However the needed information is there in the data. In fact the brightest stars can be used to get the "tails" of the p.s.f., while the smaller stars (the onces whose maximum luminance value is below the saturation level) can give accurate estimations of the "central" area.

CONCLUSIONS

This work must be considered as a first attempt to deal with astronomical imaging problems made by a group of communications and computer science technicians who have worked for years in the field of digital image processing. We have just given a sample of our experience. We leave to the astronomers to judge if they may do any use of it.

REFERENCES

1. A. K. Jain, Image Data Compression: A Review, Proc. IEEE, 69:349, (1981).
2. U. Rothgordt, G. Aaron and G. Renelt, One-dimensional Coding of Black and White Facsimile Pictures, Acta Electronica, 1:21 (1978).
3. D. A. Huffman, A Method for the Construction of Minimum Redundancy Codes, Proc. IRE, 40:1098, (1962).
4. V. Oppenheim, Applications of Digital Signal Processing, Prentice-Hall, (1978).
5. W. Chen and W. K. Pratt, Scene Adaptive Coder, IEEE Tr. on Comm., 32:225, (1984).
6. Proc. of the 5th Colloquium on Astrophysics, Trieste 6/4-8/1979.

FAST DIGITAL IMAGE PROCESSING ALGORITHMS
AND TECHNIQUES FOR OBJECT RECOGNITION AND DECOMPOSITION

V. Cappellini, A. Del Bimbo and A. Mecocci

Dipartimento di Ingegneria Elettronica
Florence University and IROE - C.N.R.
Florence, Italy

INTRODUCTION

Fast digital algorithms are required in many applications of digital image processing, especially when many images of large size are involved (as in astronomy data analysis) or real-time implementation is needed.

Making reference to digital images given, for instance, by a TV camera with a digitizing interface, typical problems encountered in the applications are the following ones: noise reduction, image segmentation, object recognition, tracking of moving objects, complex object decomposition. The last problem, indeed not easy to be solved, is of high interest in astronomy to analyse complex celestial bodies or patterns.

In this paper several fast digital image processing algorithms and techniques are presented to solve the above problems. In particular two complete processing systems are described, based on these algorithms and techniques: the first system for object recognition and classification; the second system for decomposition of complex objects and classification of the extracted subparts.

A PROCESSING SYSTEM FOR OBJECT RECOGNITION AND CLASSIFICATION

This system can perform recognition and classification of the objects in the analysed digital images in a fast way, due to the special algorithms for noise reduction, segmentation, recognition and classification presented in the following.

Noise-Reduction Algorithm

The use of spatial masks to consider the spatial correlation of each pixel is widely covered in the literature (Huang, 1981; Gonzales, 1982). Anyway, its processing requirements often pose objective limits to its implementation on on-line systems.

A five-order moving average, moving along a preferential direction, line by line, is here proposed according to the following recursive algorithm

$$g_{i+1} = g_i + (f_{i+3} - f_{i-2})/5 \qquad i \in [3,253]$$
$$g_1 = f_1 \qquad g_2 = (f_1 + f_2)/2$$
$$g_3 = (f_1 + f_2 + f_3 + f_4 + f_5)/5 \qquad (1)$$
$$g_{255} = (f_{255} + f_{256})/2 \qquad g_{256} = f_{256}$$

where f_i is the original grey level in the ith position (ith pixel) and g_i is the new one.

The above procedure introduces a preferential direction, which could give a non symmetrical deformation: however, experimental tests outlined that this drawback is not too relevant for moving averages of low order (3-7).

Segmentation-Recognition-Classification

Segmentation involves object-background separation and single--object detection. The mean value, for a fixed neighbourhood of the absolute minimum (to eliminate residual noise spikes), is evaluated on the pixels for each line of the image. A symmetrical band is established over the mean value of the line (its width can be in practice chosen according to a look-up table experimentally defined), to separate grey levels belonging to the background from the object's ones. Single object identification is hence obtained: in the first line the object pixels are labelled with different identifiers if non-adjacent; in subsequent lines adjacent pixels are labelled by the maximum identifier of the adjacent pixels in the previous line. A final processing is performed to solve possible ambiguities and to connect with the same label the pixels belonging to the same object.

Recognition and classification are performed, based on "inertial invariants" and a matching procedure with an appropriate "object models space": this, differently from other algorithms, allows some variations in the actual objects against the models, like scaling, rotation etc., without introducing errors in the recognition phase. In particular the area, the centroid position, the maximum elongation axis and the "elongatedness" are used to build a suitable set of state variables for each object under analysis. The elongatedness E is defined as

$$E = (e_2 - e_1)/(e_1 + e_2) \qquad (2)$$

where e_2 e_1 are the eigenvalues of

$$S = \begin{bmatrix} m_{20} & m_{12} \\ m_{12} & m_{02} \end{bmatrix} \qquad (3)$$

and m_{pq} is the central moment of order pq.

Experimental Results

Experimental tests have proved that the above algorithms work well and are suitable to be used as a complete software for on-line processing. An example of experimental test, performed by means of a PDP 11/34 minicomputer with a TV camera and a digitizing interface, is shown in Fig. 1 (digitized image) and Fig. 2 (final recognition-classification), related to four different mechanical objects.

Fig. 1. Digitized image of original external image.

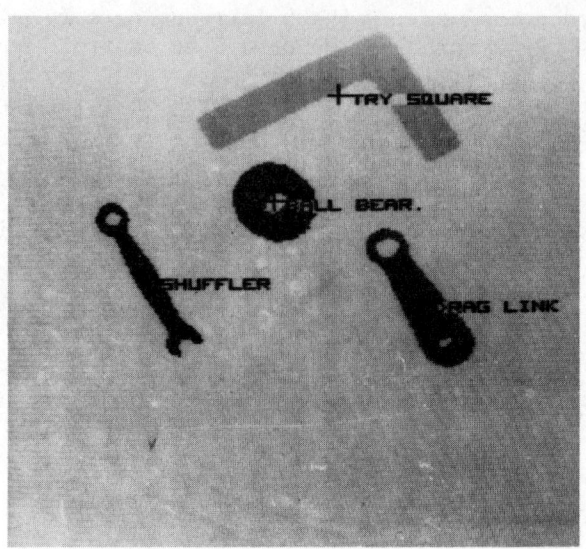

Fig. 2. Final recognition-classification of four objects (after the segmentation step): the different objects are identified by different colours, here appearing as different grey levels).

A PROCESSING SYSTEM FOR DECOMPOSITION OF COMPLEX OBJECTS

This system can perform decomposition of complex objects and classification of the extracted subparts. Three main steps can be identified: image acquisition and filtering, object decomposition in subparts, definition of a descriptive syntax.

Image Acquisition and Filtering

Monochromatic acquisition is currently used in analysis and processing of the global shapes of objects in single or multiple frames. Several experiments have shown that monochromatic acquisition does not give enough information to distinguish the parts composing the single object. Therefore additional processing is required. Three different acquisitions of the external scene (containing the objects) in red, green and blue colour bands are performed to add colour information (Robertson, 1973).

Further, to make the subsequent steps easier, a suitable filtering is applied on each acquired image. The filter uses the algorithm proposed by Moring and Pietikäinen (1983), suitably modified. The 3x3 A-neighbourhood of each pixel is analysed: the pixel satisfying the three following conditions in the grey-level histogram is selected to replace the central pixel

$$p(f_i) - p(f_c) > 0$$

$$\frac{p(f_i) - p(f_c)}{\alpha_\ell |f_i - f_c|} < 1 \qquad \alpha_{\ell+1} = \alpha_\ell/2, \quad \ell \in N \qquad (4)$$

$$\alpha_0 = k = 10$$

$$p(f_i) - p(f_c) = \max_A \left[p(f_i) - p(f_c) \right]$$

where $p(f_c)$ represents the probability of the grey level of the central pixel, while $p(f_i)$ is the probability of the ith grey level for each pixel in the A-neighbourhood; α_ℓ is a sequence of positive numbers that decreases at each iteration (N is the set of integers). The histogram is used as an approximation of the grey-level probabilities. Iterative application of this methodology leads to a smoothed image and to a certain degree of segmentation very useful for subsequent processing steps.

Object Decomposition in Subparts

To achieve automatic object decomposition, it is necessary to derive the significant subsets of a complex object. An initial step is performed, corresponding to a generalized boundary extraction. The following symmetrical algorithm is applied

$$E_m(i_0,j_0) = (1/8) \sum_{i,j} |I(i_0,j_0) - I(i,j)| \qquad i,j \in A(i_0,j_0)$$

$$E(i_0,j_0) = (1/3) \sum_{m=1}^{3} E_m(i_0,j_0) \qquad (5)$$

$$A(i_0,j_0) = \{i,j: |i - i_0| < 1 \text{ or } |j - j_0| < 1\}$$

$$\forall\, i,j \in [1,M]$$

where M is the image size, E_m is the contour map in the mth colour band, E is the final contour map in the image, $I(i,j)$ is the grey level at the (i,j) pixel, and (i_0,j_0) denotes each central pixel of the 3x3 pixel block (A-neighbourhood). Thus the contour information is obtained, by using the three different colour bands.

A second step is performed to obtain homogeneous different subregions of the analysed object. To get a good decomposition of a complex object, grey-level clustering is not enough: spatial relations and proximity information have also to be used (Zahn, 1971; Horowitz, 1976). A decomposition threshold, unique to a single class of objects, is chosen according also to experimental tests.

The used procedure starts from the pixel on the extreme lower left-hand corner and proceeds from left to the right connecting the neighbouring pixels and avoiding closed loops: the final result is the identification of a unique root of an "oriented tree" for each homogeneous region. In this way, each region is identified unambiguously by the tree which connects its pixels.

To test the reliability of the above procedure, it was compared with an "interactive" procedure, in which a human operator gives information about object components: very satisfactory results were obtained.

Definition of a Descriptive Syntax

At this point, some features of the various subsets of the object(s) present in the acquired scene are extracted (as centroid position, area, inertial moments, four FU's invariants) to allow a geometrical description of the subsets (Reddi, 1981).

Mutual spatial relations among the various subsets are evaluated. These relations are based on a scheme, in which the positions are suitably coded (as above, above and left, left, below and left, below, below and right, right, above and right). In addition three relations - inside, outside, partially surrounding - are defined, which describe relative positions.

Information provided by the relative positions of the centroids of the various subsets is seldom good enough to specify spatial relations. Therefore, in evaluating relative positions, this procedure is followed: the angle of sight by which a particular subset is viewed by a "hypothetical observer" from the centroid of another subset is considered. Suitable angular functions are hence evaluated.

Experimental Results

Experimental tests have been performed, according to the above processing system. An example of experimental test, by using the same minicomputer and digitizing unit as previously described, is shown in the following figures.

Fig. 3 shows the original digitized image, representing a circuit board: great acquisition noise is here present (to test the good performance of the proposed system).

Fig. 4 shows the object decomposition into subparts (after boundary extraction (5)): the board and four components are identified and isolated.

Fig. 3. Original digitized image: a circuit board (digital acquisition with great noise).

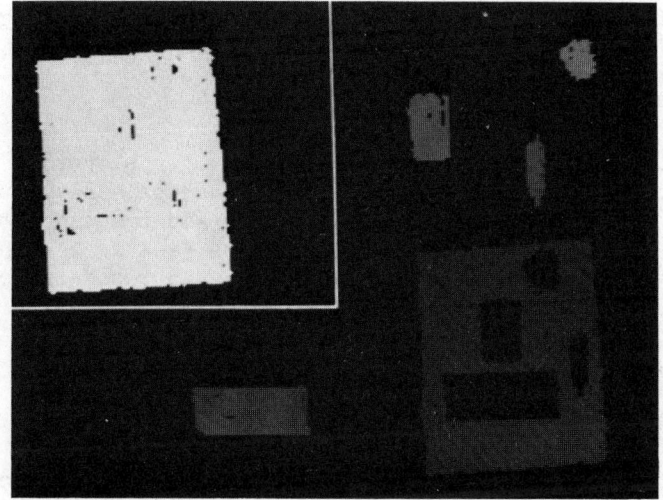

Fig. 4. Object decomposition into subparts: board and four components (colours are appearing as different grey-levels).

Fig. 5 shows the automatic identification of mutual spatial relations among the different components: a component 2 is identified, corresponding to a capacitor, which is inside part 1 (the board), below and to the left of part 3 (a resistance), below part 4 (an integrated circuit), below and to the left of part 5 (a trimmer).

CONCLUSIONS

The two above described processing systems can be very useful

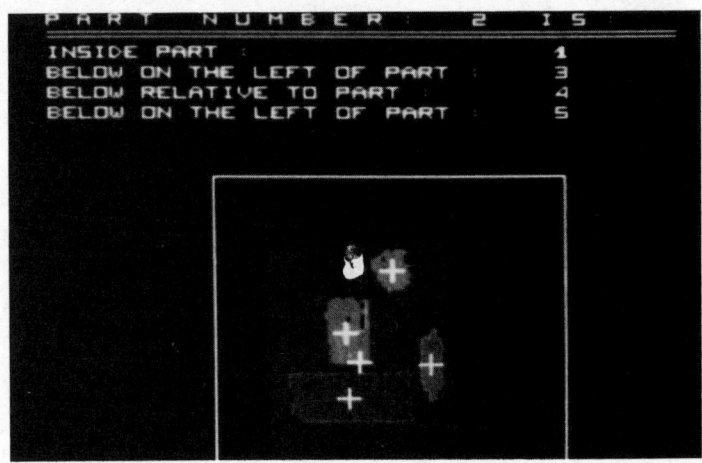

Fig. 5. Identification of mutual spatial relations among the different components: a component 2 (a capacitor) is here identified.

for object recognition and classification (first system) and for decomposition of complex objects and syntactical description of the extracted subparts (second system). The presented experimental tests confirm their good efficiency. These fast processing systems can be applied to astronomy data analysis, to recognize celestial bodies and to decompose complex structures with description of the different subparts.

REFERENCES

Gonzales, R. C., 1982, "Digital Image Processing", Addison-Wesley, London-Sydney.
Horowitz, S. L., 1976, Picture segmentation by a tree transversal algorithm, Journal Assoc. Comput. Mach.13:368.
Huang, T. S., 1981, "Image Sequence Analysis", Springer-Verlag, Berlin.
Moring, I., and Pietikäinen, M., 1983, Experiments with histogram guided image smoothing, in: "Proceedings of the Third Scandinavian Conference on Image Analysis", P. Johansen, and P. W. Becker, eds., Chartwell-Bratt Ltd, Lund.
Robertson, T. V., 1973, "Multispectral Image Partitioning", School of Electrical Engineering Purdue University, West Lafayette.
Reddi, S. S., 1981, Radial and angular moment invariants for image identification, IEEE Trans. Pattern Anal. Mach. Intell., 240.
Zahn, C. T., 1971, Graph theoretical methods for detecting and describing Gestalt clusters, IEEE Trans. Comput., 20:68.

AUTOMATIC ANALYSIS OF CROWDED FIELDS

M.J. Irwin

Institute of Astronomy
The Observatories
Madingley Road
Cambridge CB3 OHA

1. Introduction

There are many important 2D data reduction problems in Astronomy that are not amenable to conventional automatic analysis. Typically these problem areas arise in fields where the number density of images is high and where the local sky background may vary rapidly. Examples of such crowded fields are to be found in globular clusters, nearby resolved galaxies or even in deep frames of data well away from obvious concentrations of objects. Within such regions the image number density becomes so high that the majority of images overlap, even at relatively high isophotes, and simple image analysis algorithms become confused. The confusion can be so bad that even an apparently simple objective like number counting will fail (Irwin & Trimble 1984).

Many authors have suggested ways of dealing with moderately crowded fields. However, the majority of these approaches are semi-interactive and require considerable human interpretation in deciding approximate positions and numbers of images, and many will only work providing not more than a few images overlap at any point. Our goal has been to examine the potential for a fully automatic method that is both robust and efficient in terms of computer requirements, capable of dealing with complex multiple overlaps and able to generate the optimum estimates of image parameters. Further considerations are that the system must be flexible (i.e. capable of dealing with say galaxy/star overlaps) and eventually be capable of being integrated into existing machine systems.

To illustrate the realm of number densities where crowding effects become noticeable consider the number density corrections shown in Fig. 1. The corrections are for three types of analysis:

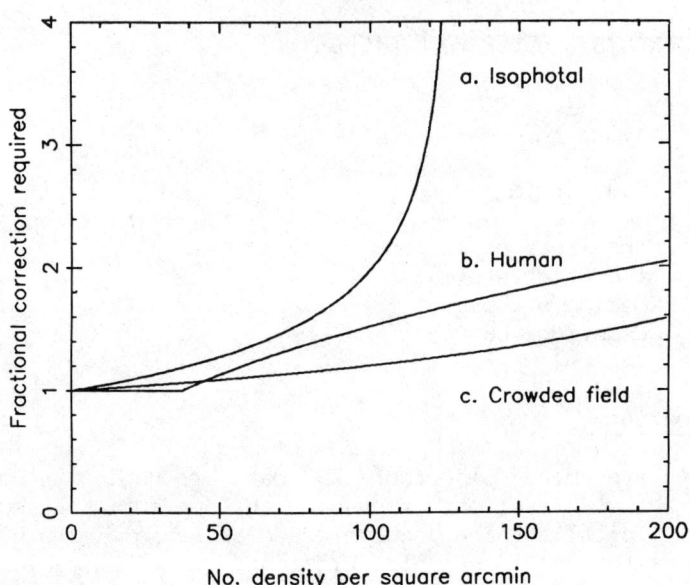

Fig. 1. A comparison of the crowding corrections from King et. al. (1968) with those appropriate for machine measurement, Irwin and Trimble (1984).
  a. Crowding correction for isophotal analysis with average isophotal area roughly 3 times seeing disk area.
  b. Crowding correction for human observer from King et al, (1968) algorithm with image area equal to seeing disk area.
  c. Crowding correction for crowded field algorithm with image area equal to seeing disk area otherwise same formula as a. (see Irwin & Trimble 1984).

The curves are derived for moderate seeing conditions(1.5") and the value of the ordinate gives the correction to be applied to the observed number density. It is straighforward to estimate the corrections for other seeing conditions since the scale of the abscissa varies as the inverse square of the seeing.

human interpretation of a photographic plate, an isophotal method and an idealised crowded field algorithm, the realisation of which will be discussed later, where it is assumed possible to distinguish between independent images at separations greater than their seeing disk size. The corrections to first order depend almost entirely on the product of average local number density with average image area. This area is the isophotal analysis area for the simple machine methods and to a reasonable approximation the seeing disk area for the other methods. From the diagram it is clear that:
- a.) Simple machine estimates of number density (and hence other parameters) always require non-zero adjustments to be made, even for relatively low count levels. In contrast, a human observer, by making use of all the image profile information, is rarely confused in these regions and neither is the idealised crowded field algorithm.
- b.) For moderately high number densities (100/sq.min) simple isophotal analysis fails completely because the majority of images are joined together at the threshold isophote. A human observer can still extract fairly reliable information in such fields and so too can the idealised crowded field method.
- c.) At high number densities (200/sq.min) it should be possible to do even better than a human observer examining a plate by eye. This is essentially due to the difficulty experienced in deconvolving a complex blended image by eye. Deconvolution is an operation that the brain is not very good at, whilst in principle at least, deconvolution by model fitting is well known to produce excellent results.

2. Overall Analysis Strategy

In deciding on a rationale for automatic analysis natural questions to consider are: how does a human observer interpret pixel intensity maps, with particular regard to image detection and deciphering blended images; how accurate are the image parameters likely to be and closely related to this which methods of analysis will give the best results at reasonable computational cost? By reducing the problem of image description to one of estimating certain image parameters (position, intensity and shape) it is then pertinent to use statistical theory to inquire about optimum methods for estimating these parameters. In particular since the method of Maximum Likelihood (Fisher 1958) provides the foundation for an answer to these questions how can we use this knowledge to aid in designing analysis schemes. The Maximum Likelihood method also defines the minimum error it is possible to obtain for each of the parameters. Any new approach, or short cut in calculation, can therefore easily be assessed on an absolute scale in terms of how closely they attain these minimum parameter errors.

In order to break down the overall analysis scheme into manageable portions the following main tasks have been assigned to the automatic crowded field algorithm developed here:

a. Estimate the local sky background over the field.

b. Detect images/image blends and keep a list of pixels belonging to each blend for further analysis.

c. Analyse each image blend detected for multiple images, generate initial parameter estimates, particularly number of images and approximate positions.

d. Refine the parameters for each separate image within a blend, particularly the intensity and revise the local sky estimate if necessary. Decide if more images are required and eliminate spurious images from step c.

This basic strategy has been designed with a view toward incorporation into existing systems, since steps a. and b. are present in conventional isophotal analysis methods. The core of the problem is segmenting a blended image into its component parts. By only concentrating on those pixels belonging to the blend above a certain isophote considerable savings in computation result, with little loss of accuracy. A consequence of this is that more complex refinement schemes may be used.

3. A Crowded Field Algorithm

A flow diagram for a general purpose crowded field algorithm based on the analysis strategy outlined in section 2 is presented in Fig. 2. Several of the steps necessary to implement such a process are common to many analysis systems and will only be briefly commented upon here.

a.) The core of the method is to find an initial global sky estimate for the entire field against which a combination of a detection filter and isophotal connectivity constraint suffice to reliably locate all resolved features. The output from this stage is a list of pixel positions and original pixel intensities for each image. This list contains all the necessary information regarding numbers of images in the blend etc. without recourse to extraneous data.

b.) An obvious method for detecting overlapped images is to re-examine this pixel list at multiple isophotes and decide if distinct images separate out. This is basically one of the ways the eye interprets contour maps. Two images that

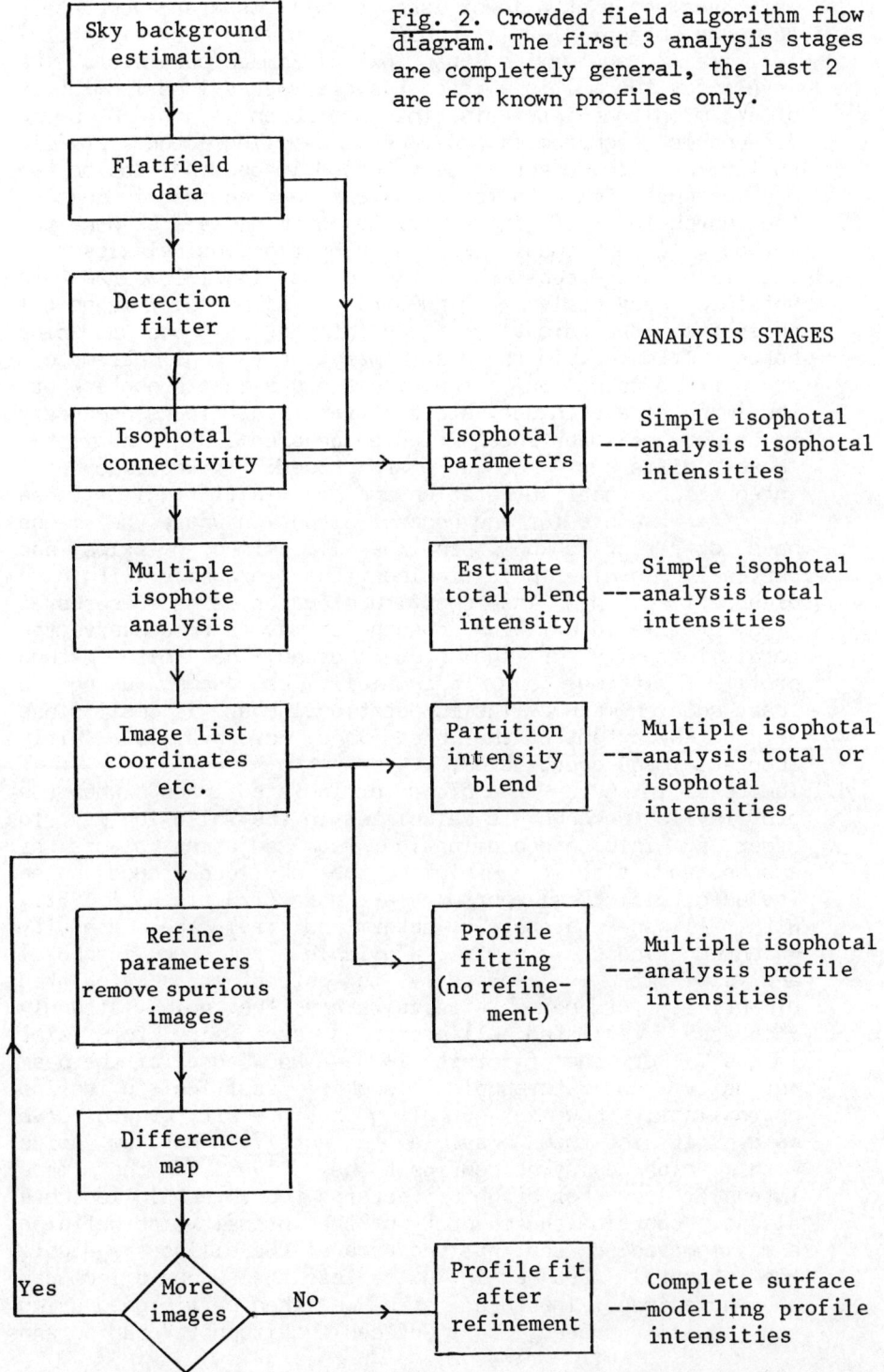

Fig. 2. Crowded field algorithm flow diagram. The first 3 analysis stages are completely general, the last 2 are for known profiles only.

were overlapped at a lower isophote will separate completely at some higher level enabling a simple new isophotal analysis to determine the initial image parameters. The levels for the multiple isophotes are spaced logarithmically at 1/4 magnitude intervals. This has been found to be a reasonable compromise between sampling the profile sufficiently to detect closely spaced images and wasting too much computer time. So far we have impose no restrictions on the morphology of the overlapping images and as such the method will work equally well with overlapping galaxies.

c.) The final stage consists of taking the initial parameters obtained previously - number of images, isophotal intensities, coordinates, shape information - and producing better estimates. In the preliminary analysis all parameters are derived using only the pixels above the isophote where the image separated out. At the very least it is necessary to adjust the image intensities to be a consistent estimate. Three different methods are used: fixed isophotal integration, total integration and profile fitting. Of these the first two are for the general situation where the images have different unknown profiles (i.e. mix of galaxies) and basically involve partitioning the intensity within a blended image into its constituent parts using measures of image profile information in such a way as to conserve the total intensity of the overall group. The third method profile fitting can essentially be done using a least-squares method. The computational cost is small since we know the extent of the blend and have available fairly accurate image coordinates.

d.) The total intensity of a blend, or isolated image, when the profile is unknown is calculated in the following way. In order to minimise contamination from adjacent blends or images and also to minimise the sky background noise included, elliptical apertures are used (Irwin & Hall 1982), with ellipse parameters determined from the intensity weighted second moments available from the isophotal analysis. (It is necessary to correct the calculated ellipticity for noise effects since the noise strongly increases estimated ellipticities especially for faint images). For the faintest images knowledge of the peak intensity and threshold isophote suffice to define approximately how far out along the intensity growth curve we are. It is then possible to specify a maximum radius within which we may be confident that 99% of the image intensity lies. For brighter features the threshold isophote already contains the majority of the intensity, so defining a maximum radius such that the area of the ellipse is double the isophotal area again guarantees that we have included the whole image. The total area is then partitioned using ten equally spaced radii between the isophotal radius and the maximum radius and the integrated intensity interior to these radii is calculated. An automatic examination of this sequence of intensities is made to determine the most likely value for the total intensity. We are in effect looking at the intensity growth curve of each feature, so a suitable cutoff point can be chosen to include the whole image, the minimum amount of sky and to avoid any other images entering the aperture.

e.) All refinement and profile fitting steps involve an independent coordinate and intensity stage. After the initial coordinates are obtained from the multiple isophote stage we can either stop after a single stage of linear least-square profile fitting or enter the full surface modelling stage. The advantage of the single pass though the data lies in computing speed. Typically the multiple isophote analysis with either single profile fitting or intensity partitioning takes roughly twice as long as a conventional isophotal analysis scheme. The only real disadvantage when profile fitting lies in failing to resolve closely spaced images. Even if the profile is not known accurately this will only introduce a constant offset in magnitude. The intensities estimates are also fairly resilient to coordinate errors. This in itself anticipate an obvious refinement scheme: calculate image coordinates (maximum likelihood method), profile fit (least-squares method) and repeat until convergence.

f.) During refinement for full surface modelling spurious images can usually be easily detected and removed. By studying the difference map between data and model function within each blend previously unresolved images can also be detected and included in the refinement. With caution this procedure can be repeated (usually one extra cycle is sufficient) until the difference map is featureless, implying that all resolved images have been accounted for. Although this procedure is similar to that used in CLEAN in radio astronomy (Hogbom, 1974) it differs in one vital respect. Here we are interested in using the absolute <u>minimum</u> number of images to model the surface there is no such restriction in CLEAN.

g.) The sky background over the blend may also be refined during profile fitting stages. However it is only worthwhile doing so if the profiles are known to be <u>accurate</u> and <u>all</u> images in the blend have been accounted <u>for</u>. Otherwise it is possible to end up with a worse sky estimate than before.

h.) When analysing multiple frames of the same area of sky once a master coordinate list has been obtained and frame-to-frame coordinate transformations ascertained several short cuts are possible. For example: for complete surface modelling once the data has been flat fielded it is better to go straight to the final profile fitting stage. This will ensure that different frames are analysed in exactly the same way. Instead of using the data to define threshold isophotes the model can be used in an analogous fashion. Only those images within a certain distance need appear in any one blend and only pixels above a specified threshold or within a fixed radius need be used in refinement. (A similar approach can also be used for dealing with an external coordinate list, with or without coordinate refinement etc.).

4. Discussion

For a 512 x 512 frame of data with number densities up to 200/sq.min. The full surface modelling takes roughly 30 min. on a VAX 11/780, whilst the simple multiple isophotal analysis with profile fitting takes around 5 min. In many cases the latter approach will produce acceptable results and in terms of ability to detect the presence of images in blends performs as well as the idealised algorithm mentioned previously. Since this stage could easily be added on to existing isophotal techniques it also seems a very promising candidate for integrating into machine measuring systems. The full surface modelling, which it should be stressed is completely automatic, seems to perform close to the theoretical limits. At present it probably is too expensive computationally for large scale photographic studies, but it is certainly useable for CCD data. The main difference between the profile methods is that the full surface modelling results in virtually no spurious magnitudes due to the failure to detect unresolved images. There are still a few of these with the simple multiple isophotal method.

The complete crowded field algorithm has various options which enable it to be used as a general analysis tool in addition to its ability to operate reliably in severely crowded fields ($£$ 200 images/sq. min). It therefore should open up a large volume of astronomical data to automatic analysis and provide useful results into the centre of nearby resolved galaxies and globular clusters in addition to its potential uses in deep galaxy photometry.

Acknowledgements

This work was supported financially by the UK Science and Engineering Research Council through the auspices of the APM facility at Cambridge. I would also like to thank the rest of the APM group Ed Kibblewhite, Mick Bridgeland and Peter Bunclark for many invaluable discussions and ideas.

References

Fisher, R.A., 1958. Statistical Methods for Research Workers (Oliver & Boyd).
Hogbom, J.A., 1974. Astr. Astrophys. Supl., 15, 417.
Irwin, M.J. & Hall, P., 1982. Occ. Reports of ROE Edinburgh, 10, 111.
Irwin, M.J. & Trimble, V., 1984. Astron. Journal, 89, 83.
King, I.R., Hedemann, E.Jr., Hodge, S.M. & White, R.E., 1968. Astron. Journal, 73, 456.

DWARF: THE DWINGELOO-WESTERBORK ASTRONOMICAL REDUCTION FACILITY

J.P. Hamaker, R.H. Harten, G.N.J. van Diepen and K. Kombrink

Radio Observatory, 7990 AA Dwingeloo, Netherlands

SUMMARY

After an introduction summarizing the reasons for developing DWARF, those aspects which in our view distinguish it from other data processing environments are presented. Subjects covered are:
- The rationale for committing ourselves to VAX/VMS.
- The program parameter interface as seen by the user; defaulting mechanisms; generation of multiple functionally distinct versions of a single program.
- Messages, logging and file history.
- The tree-structured data file catalogue.
- Virtual vs programmed I/O.
- Processing of files from different sources with different layouts though one table-driven interface.
- The major benefits of using dynamic core for control arrays.

INTRODUCTION

Modern procedures for the reduction of data from the Westerbork telescope call for a wide variety of operations, ranging from pure batch to highly interactive, on large data sets (100.000 points and up). As a service centre for Dutch astronomy, our institute also sees increasing demand for processing of all sorts of imported data.

DWARF is to provide a common base for all this processing, replacing several mini-environments that evolved over the preceding decade. Of course, it has features in common with other systems, either accidentally or because they were deliberately copied. In this paper we concentrate on those in which we feel DWARF is distinct from its peers. We also present the rationale for some of our basic choices.

MOTIVATION FOR DWARF'S DEVELOPMENT

A good environment should minimize the efforts required on the parts of the user and of the programmer to work with it. It should provide in the simplest possible way for all operations except those specific to individual applications. Such operations include:
1. Program-user interaction: The interface must cater for all modes of operation ranging from straight batch to highly interactive. It must free the user from all redundant typing. As much as possible, user input should be checked immediately for validity.
2. Creation and administration of data files.
3. I/O on a variety of data files, including those created by other systems.
4. Error reporting.
5. Record-keeping should minimize the need for manual administration.
6. Simultaneous execution from one terminal of more than one program, either interactively or in batch mode.
7. Programming: The interface for application programs should be as simple as possible, in terms of number of subroutine calls and number of arguments per call, as well as in terms of special declarations (e.g. of control arrays) to be made on behalf of the system. Not only does all this make programming per se easier; as a consequence it also promotes easy exchange of programs with other systems.

We found all environments existing or planned a few years ago unsatisfactory in one or several of the above areas. In particular the user, data I/O and program interfaces fell short of what we wanted and judged to be feasible. Given the constraints on the length of this paper, we shall concentrate our discussion on these three aspects.

PORTABILITY VS SOPHISTICATION: THE RATIONALE FOR A COMMITMENT TO VAX/VMS

Much has been said in this conference about the advantages of a system that is portable in its almost-entirety. DWARF blatantly went the other way: We chose to exploit the VAX's excellent operating system rather than to develop a portable layer on top of it that would inevitably be much more primitive. Clearly, this implies a short- and medium-term commitment to VAX/VMS. By insisting, on the other hand, on the use of Fortran where possible, we have taken the possible later need of a migration into account.

DWARF does not use a keyboard monitor of its own, but receives its input through the VMS command interface; its command language is an extension of that of the VAX (DCL) employing the same basic syntax. While working with DWARF, a user retains unrestricted immediate access to all DCL's facilities such as
- interactive symbol and logical name manipulations;
- command procedure creation and execution;
- program development tools;
- multiple-user support with proper protection against mutual interference;
- new features to appear in future system releases.

One disadvantage is that DCL does not allow for any preprocessing of terminal I/O, such as saving commands for later use in the way MIDAS does (see elsewhere in this volume). In this area we depend on future developments outside our control (and in fact some are already underway in VMS version 4).

THE USER-PROGRAM INTERFACE

Like several of its peers, DWARF incorporates a table-driven interface that controls all aspects of parameter communication between program and user.

Parameter values may be set before program execution. The scope of such predefined "defaults" may be either program-specific or global. The program will only prompt for parameters that have not been predefined. The DCL symbol table is used to store the defaults.

We shall demonstrate the most important aspects of the interface by means of an exemplaric "structured walk-through" (Allen, this volume). In the following, the screen dialogue is shown left-justified, with user input capitalized. Indented text represents our comment. The '$' sign is the standard DCL prompt.

$ LET PI=3.14159
    The LET command is used to define DCL symbols, whose values may contain floating-point expressions.
$ SPECIFY PLOT/MENU
    The SPECIFY command is used to pre-define parameter defaults; it may be used in menu mode (the default) or to specify selected parameters through inputs in the form "keyword=value".
Node: BIANCO;ROSSO
    Program PLOT runs in a loop whose execution is controlled by the NODE parameter. We intend to execute the loop twice for two different files ("nodes", see the section on data catalogues below). On its third pass, PLOT will find the NODE

defaults exhausted and terminate. A semicolon is used as pass delimeter.
Window: /NOASK
The program must use its own default for this parameter.
Plot-format (halftone, contour): C
The parameter table defines this parameter as re-usable: Predefined defaults will be re-used as many times as required. (Note that this behaviour differs from that of the FILE parameter above). As a result, we will get contour plots of both files.
Limits:?
The extent of prompts varies according to the level at which the user has assessed himself ("beginner", "average" or "expert"). In case of confusion, users may transiently descend to a lower level by typing a question mark.
LIMITS specifies the white and black saturation limits in your half-tone plot.
Limits:
Having obtained this explanation, we realise that this parameter is irrelevant for contour plotting, so we skip it.
Contours: 10*PI, 20*PI, 30*PI; 30*PI.
We specify different sets of contours for the first and second passes, i.e. for files BIANCO and ROSSO. Expressions including DCL symbols may be used freely to define numeric values.
Plot-scale:
Leaving this parameter undefined, we will be prompted for a value on every pass-through the program.
Having run through the whole menu, for PLOT, SPECIFY terminates.
$ EXECUTE PLOT
This command starts PLOT's execution. Experienced users are likely to use abbreviated commands ("EXE PLOT") or define their own ones ("PLOT") through standard mechanisms in DCL.

"STREAMS": CREATING MULTIPLE VERSIONS OF A SINGLE PROGRAM

The concept of streams provides a very powerful means for a user to create multiple versions of one program, each version having its own functionality. As an example we shall define two versions of the PLOT program from the previous section, one for half-tone and one for contour plots:

```
$ SPECIFY/NOMENU PLOT$C
: PLOT_FORMAT=C
:
$ SPECIFY/NOMENU PLOT$H
: PLOT_FORMAT=H
:
$
```

Note that, by virtue of the stream names, these two SPECIFY operations do not in any way interfere with each other nor with that of the previous section. Thus if we now type
$ EXE PLOT$C
we will get a contour plot after having been prompted for all parameters except PLOT_FORMAT.
One may look at the act of "splitting up" a single program into several versions of it as the inverse of creating a command procedure or a "macro". We feel that this provides a very attractive alternative approach to varying the "aggregation level" or "granularity" of the program modules available to the user.

MESSAGES, LOGGING AND FILE HISTORY

Messages and logs are two fundamentally different things: Messages inform the usr at run-time on the performance of his program. The object of logging is to provide him an unambiguous retrospective survey of his actions. Both message output and the log should be as free as possible of spurious information. To achieve the latter, DWARF uses intermediate message buffering at two levels:

Firstly, when a subroutine detects an error, it only stores a message in a buffer and reports its findings back to its caller. The latter may decide either to report the error to the user and sollicit his help or to correct it by itself and simply discard the buffered messages, possibly outputting a different message of its own. In a similar way, log messages are not written directly to a log file, but to an intermediate buffer, which normally is appended to the user's log file only upon successful program completion.

A third method to suppress log file clutter is to assign "importance levels" to all messages. These allow the user to select only messages exceeding a certain importance threshold, both during program execution and when reading the log.

To provide for a correlation between data files and the log, each program run is given a unique "identifier", derived from the system time. The same identifiers recorded in the data files provide for the required cross-correlation.

THE DWARF DATA CATALOGUES

Each DWARF user has his own file catalogue, which is an administrative layer built on top of the VAX's file system. Rather than being a simple linear list, the catalogue has the form of a tree, to the branch points ("nodes") of which the data files are attached. All data access is through the names of these nodes. The user may control the lay-out of his tree in accordance with the relationships existing between his data files.

Presently, the facilities for catalogue manipulation are rather primitive; this observation confirms our reservations about maximally portable systems expressed earlier. In this particular case the decision not to rely on VMS's directory structure was motivated by expected run-time overhead as well as the desire to add some special features of our own design.

DATA I/O

Use of VMS's virtual mapping of entire files is extremely attractive from the programming point of view. There are however, serious limitations:
- very large array sizes (e.g. 64x512x512) cause practical problems;
- virtual I/O operations are strictly synchronous;
- one suspects that a price must be paid for programming simplicity in the form of considerable operating system overhead (although statements made in this conference seem to suggest the contrary).

DWARF offers an alternative programmed-I/O package that should be equally easy to use, yet -invisible to the programmer- provides the full power of asynchronous operation. We plan to make some benchmark tests to settle the efficiency issue of programmed vs virtual I/O.

COPING WITH DIFFERENT DATA FILE LAY-OUTS

Astronomical data very often have the same structure, -e.g. one-, two- or three-dimensional arrays-, yet the accompanying headers (containing array sizes, coordinates etc.) come in as many formats as there are data-producing systems. This makes the exchange of data between systems always problematic.

DWARF attempts to solve the problem through table-driven data access: File contents (both header and data) are addressed by their names rather than through their positions in the file. Thus, like FITS, DWARF relies on the definition of a set of standard names for the items to be retrieved. Tables provide the correspondences between these names and locations in the file for each type of file in the system. A call such as
   CALL GET_SCALAR ('RA',RA)
will cause the system to first refer to the appropriate table, then read right ascension from the proper place in the file.

Our limited experience so far indicates that this scheme is useful: It eases the transfer into the DWARF environment not only of data files, but also of application programs.

DYNAMIC CONTROL ARRAYS

Systems like DWARF require control arrays of various sorts for storing and communicating internal information. In the traditional approach, the applications programmer declares them; even though they are of no visible use to him, he must carefully work out how many of each type are needed, and in which modules they are to be placed. Later, when new system developments require extension of certain array types, he must revise and recompile all his programs. Although INCLUDE files and symbolic names for array sizes may help considerably, the whole matter of control arrays remains basically a nuisance.

DWARF does away with them completely by relying on the use of dynamic memory. All that the system needs are 4-byte variables, "identifiers", to hold addresses of dynamically allocated control blocks. One may liken their function to that of Fortran LUNs; they are even simpler to use, however, since the programmer does not have to keep track of their values. The application code no longer contains any references to the sizes of control arrays. Consequently, no revisions or recompilations are necessary to accomodate even major changes. (In DWARF we use a shared library for all system routines, so that even a re-link is superfluous.)

Obviously, we make use of VMS's dynamic core facilities. It is not at all difficult, however, to implement something similar in a traditional static Fortran environment: All it requires is a block of core from which control arrays can be allocated and a simple (de)allocating subroutine (the address-manipulating parts of which may have to be written in assembler). We think that the advantages spelled out above are very well worth the effort.

CONCLUSION

We have highlighted those aspects of DWARF which at this moment seem to us to be of greatest benefit to its users and programmers. What their actual value will be remains to be seen. At present, the main liability to DWARF's future is lack of manpower to support its further development and application. However, even if DWARF as a whole should turn out to be a transient phenomenon, we hope to see some of its ingredients perpetuated in other environments described in this conference.

ACKNOWLEDGEMENT

In the development of DWARF's basic concepts we profited from the experienced advice of W.N. Brouw and E. Raimond. The Netherlands Foundation for Radio Astronomy is financially supported by the Netherlands Organization for the Advancement of Pure Research (Z.W.O.).

SESSION

PARALLEL PROCESSING

CHAIRMAN:

J.H. FRIEDMAN

STEPS TOWARD PARALLEL PROCESSING

Stefano Levialdi

Dipartimento di Matematica
Università di Roma
00185 Roma, Italy

INTRODUCTION

The area of computation includes different, interrelated fields like computer architecture, language development, algorithm design and on a more formal level, computational models. Although the domain is certainly a large one, the inherent motivation (namely the achievement of efficient computation) helps in focussing all the related activity toward a converging end. Nevertheless, habits and tradition play a very important role in the use and understanding of new computational strategies for managing unconventional computers and different programming languages. For these reasons, it is generally hard to imagine the "best" approach in order to design a complete system requiring high speed and throughput: should we start from the architecture? from the algorithms? from the language that will be used to write the code?...or from a general model?

Perhaps we should also mention, as a starting point, the solution of a well defined class of problems assuming how the data will be collected, in which form the output is requested, etc. As it is easy to see, all these possible starting points are very intimately interconnected: this is one of the main reasons for continuing research in the understanding of the computational process, including interprocess communication, process synchronization, process execution and processor organization.

The attempt to improve computational efficiency and overall throughput by some form of parallelism is not a recent one, there are references to this even in Babbage's machine about one hundred years ago, but the particular emphasis on a new class of machines essentially based on the clever use of many processors, is relatively recent and a direct consequence of the technological breakthrough of VLSI circuitry.

The purpose of this talk is to show how different thinking attitudes and strategies may exploit parallelism in the design of classes of algorithms in general image processing applications (but not only) making use of new sophisticated multiprocessor machines.

GRAPH MATCHING STRATEGIES

Traditionally, the algorithms to be executed by a computer were designed on the basis of the sequential, uniprocessor structure of the machine. In most problems the solution is obtained by either repeating the same computation a number of times or by firstly computing a partial result and then proceeding to execute a different computation: in both cases the sequential nature of the von Neumann computer well matches the sequential process contained in the algorithm solution. We may therefore say that, in the past, algorithms were designed so as to properly use the resources of the conventional computer. The facts indicate that the technology is offering, approximately each couple of years, the possibility of doubling the number of devices that can be integrated into a single chip; a quoted figure for the end of the 80s is of 1,000,000 transistors per chip and, by extrapolation, we may reach the second millenium of our age with 10,000,000 transistors per chip! Although the problems connected to the use of this technology also grow exponentially (like reliability, fan in, interconnections, power dissipation, circuit layout and design, etc) it is easy to forecast a deep revolution in the structure of the new computing systems. The importance of the need for reformulating the existing algorithms or, better still, to model the new computer architectures to the algorithms so as to have an optimal resource utilization, cannot be overstressed.

If the solution to a problem (and therefore the algorithm used for such a solution) is represented by a graph (in a similar way to the flow graph of a conventional program) and if the processors and their interconnections are also represented by a graph, then the problem of evaluating the well matchedness of the solution on a computing system becomes a graph matching problem. (Refer to [1] for an interesting discussion on future multiprocessor configurations). In this way we can study whether all the processors are used during the execution of the algorithm if their communication introduces delays, if the overall throughput of the data is achieved at maximum speed. As in[1] ..."We can view our exploration of parallel algorithms and multicomputer architectures as one of choosing, modifying and conditioning information flow graphs and multicomputer graphs so that they can be mapped onto one another."

The area of image processing, pattern recognition and scene analysis appears particularly suitable for a good use of new computing structures where parallelism of some kind is introduced. Not only applications on structured data for fast processing is a driving force for the development of new algorithms and architectures based on the exploitation of parallelism but there are also other motivations like: a) the desire to have a better understanding of the computation process and b) by means of a full dissection of the algorithm, obtain an efficient serial solution for another related problem[2].

Wheras in the past much attention was payed to the actual operations performed on data, now growing interest is given to the different data structures, representation, flow and collection. The same is happening about the structure of the computer where memory, data paths, data security and reliability are becoming important items as much as the actual processing operations performed by the instructions.

The deeper understanding of the computation involved by an algorithm is directly related to the possibility of evaluating the communication and computation cost on a specific architecture the complexity measure in terms of time dependency on an input string of given length, the possible parallelization of the computation steps, etc.

CONTROL STRATEGIES

Since the physical implementation of parallelism plays

an important part in the development of algorithms and in their efficient execution, we may firstly consider, in a very general way, three main approaches to parallel computer architecture. These approaches are termed control flow, reduction (or demand driven) and data flow. The first kind, corresponding to the von Neumann architecture, uses centralized control which will schedule the work to be accomplished by the subsidiary processors. For instance in a typical application using only two processors, the controller may decide when and how the other processor may be devoted to the input output operations, if more processors were available to the controller, many input lines could be managed. The task of the central controller (the "boss") may become, in some complex situations, extremely difficult to solve efficiently.

In a second approach, the tasks are redistributed amongst other units (processing units) and then passed over so as to finally produce a result. These architectures are also termed reduction systems since they gradually reduce the expressions contained in the program until the final result is obtained. Another name is demand driven because they may be seen as a system where the units act only when requested. The problems that may arise here are those of communication between units (processors) especially due to the fact that the bottom units must send up, to the apex, the result of their computation. In this scheme the problem is broken down into subproblems and each subproblem is again broken down, so on recursively, until no further subdivision is possible. All processors may belong to the same kind, i.e. may have the same features, regardless of their level.

Finally the third, esoteric scheme, is called data flow and may be considered the most decentralized in nature. There are no hierarchical levels, every unit is at the same "level" and their operation is asynchronous, computing results as soon as data is available to the units. Whenever the computation unit has finished, it will send the obtained resluts to another processor. The problem here is of balanced partition of the work so that data are not left waiting and no processor is idle during a long time. Since this structure is not easy to grasp it may be worthwhile to expand on it. For a more detailed description and references see[4].

All data are tagged with information regarding the process that must be performed on them, these tags are identified by the processors so as to decide whether all the required data has arrived and then begin the computation; the compiler will distribute groups of instructions to the different processors before the execution of the program. When the task has been executed, the processor will update the tag so as to reroute the data to the new processor for the next computation on them.

There are two main ways for optimizing the communication path of the data and avoid idleness and bottlenecks: one is to have all equally powerful processors that will store tokens (which contain the tagged data) matching them and processing the data; the other possibility is to have two different kinds of processors, one for token matching and labelling and the other for data processing.

The M.I.T. machine, developed by Arvind, has only one processor type which contains a matching unit, an instruction fetch unit (to and from the memory) an arithmetic and logic unit, an output unit and a data storage (different from the instruction storage) as a final component of this unit. When the program is compiled, the tokens produced will go to the matching units of the processors that will store the tokens having identical labels (corresponding to the tags in their own associative memory). In some instances two tokens are required for the computations to take place, other times only one token is necessary; the ALU will perform the actual computation after the corresponding instruction is fetched from the program memory; finally by applying a set of specified rules, a new tag is attached to the output data so as to transmit the new token to the communication network for the next processing element to act on it.

The machine at MIT will have 64 units (or active processors) all of the same kind. A ring structure forms the base of another data flow project at Manchester University where a token firstly goes to a queue and then to a matching unit, when matched to the node store that contains the instructions for the processor, finally enters the processor and, when the computation has been accomplished, out to proceed to the next processor. Only one ring has been built, containing 12 processors, more rings may be connected and even if slow components have been used in the Manchester project (for economical reasons) the speed is of a few MIPS.

From some studies on the influence of the number of processors on the final speed, it seems that the obtained speed levels off at about 1 MIPS regardless of the number of processors; this is so because time is lost in the matching phase. If, on the other hand, a buffer of matched tokens is prepared so as to supply the tokens evenly, the expected results show that a 90% of the theoretical speed is reached (instead of 65%).

The other important problem is the connection between processors and the communication cost associated to it. If all the processors must be connected to all processors (as in Arvind's design) for n processors, n**2 links are required. Another possibility is to have the processors connected in an array mode (the direct communication is with the neighbors) so that the links will grow linearly with the number of processors and for many applications this communication pattern is adequate (image processing, simulation of physical problems in 2 and 3 dimensions electrical simulation). A radically different structure is the hierarchical one, tree like which has been chosen for the Data Driven Machine (DDM1/2), where each node has eight sons: the problem still remains of a correct allocation of tasks which may, in some instances, be performed in remote parts of the tree.

In conclusion, the important problem to be solved for a correct implementation of the data flow concept is the balanced distribution of the two main tasks: token matching and data computation. The Topstar machine[4] has separated these tasks introducing two different elements: the control modules and the processing modules. Each one of the first modules is connected to a number of modules of the second kind and viceversa: when a processing element requires an instruction to process, it asks for a matched pair of tokens form a control module, if the interrogated module has no pair, it asks another module and so on; when the token is found the instruction is executed and the results are sent back to one control module. The control modules store and match the tokens making them available when matched to the requests of the processing modules. There are, of course, a number of objections to this way of exploiting parallelism, the main criticsism regards: a) the problem of handling large arrays (suppose only one value in an array must be changed, all the array data would be circulating in the machine since there is no global memory), b) a high computational overhead (the matching and routing activities) and c) on some classes of serial problems this machine is not efficient. This last point is inherent to the structure of the algorithm and cannot be usd to evaluate a specific design; in general a parallel algorithm will be executed poorly on a sequential machine and viceversa: there is no way out of this, like the performance of a special purpose machine on its own job cannot be better on a general purpose computer...

CLASSIFICATION STRATEGIES

Taken as a special case where processors must be organized to make use of parallelism, the data flow scheme has shown that there are at least$_5$ three main dimensions by which any design can be characterized :
1) the processor type
2) the control of concurrent operations
3) the interconnection geometry

Regarding the processors we have noticed that if they must be alike (homogeneous) they are complex (for a general purpose machine) or very simple (for instance bit processors in cellular arrays). If they are different, they may specialize (like in Topstar) or in machines where the functions are devolved to special processors (comparators, accumulators, adders, etc).

According to^5 three main concurrency control schemes may be considered, generalizing the classification mentioned in the previous section, the distributed control may be synchronous or asynchronous and furthermore, partial automatic reconfiguration of the machine may split the control into the different groups of processors (each group may also execute the tasks in different modes: SIMD, multi SIMD, MIMD, etc.).

Finally the problem of interconnections has been superficially touched and deserves attention, specially if a match is desired between the data structure and the processor organization. A metric space is important in order to evaluate the maximum distances between two processors (for a given plane tesselation) so that the topology of the interconnection scheme is not sufficient. Many geometries have been suggested (some have actually been built) like the one and two dimensional arrays, the binary tree and its three dimensional extension, the pyramid, the cube and n cube, the shuffle and a number of interesting graphs like the Paterson one, which has the advantage of interconnecting all eight nodes with a maximum distance of two (only one node must be traversed in order to reach any node on the graph).

There are many ways in which the multiprocessor architectures may be described and classified. For instance in^6 four dimensions were chosen for the evaluation of the degree of parallelism on a multiprocessor architecture: (note that these dimensions are particularly suited for the class of machines that are oriented toward image processing and other pictorial applications)

1) the operator dimension (operator parallelism pipelining)
2) the image dimension (image parallelism)
3) the neighborhood dimension (neighborhood parallellism)
4) the pixel (picture element) dimension(conventional processing)

If all these dimensions would be used (certainly a theoretical case) we would have the maximum achievable parallelism, but this is not always the best solution since many algorithms are not fully parallel in nature and furthermore the number of processors required for the "image parallelism" would be prohibitive.

ALGORITHM STRATEGIES

A first distinction may be found on the algorithms running on multiprocessor machines, i.e. between those that do not need interprocessor communication and therefore each processor may work autonomously and the other algorithms requiring such communication. This first kind is referred to as having large module granularity^7 and is contrasted with the other kind where communication steps are often needed and the processors may work on their own for very short periods of time (small granularity algorithms). The first kind of algorithms are suited to run on a machine having many processors which execute different instructions with an asynchronous concurrency control. The specification and control of the interprocesses communication must be provided to the user, algorithms of this kind are, for instance, those for data base management. The second kind requires synchronous machines, direct hardware communication to reduce inherent delay and typical algorithms of this kind include matrix multiplication, solution of linear equations systems, image processing. Broadly speaking, for the serial algorithm where linear time is required, logarithmic time will be necessary on parallel machines with a linear number of processors; typical examples include the inner product, the evaluation of general expressions, the addition of N values, etc where the available processors are all used and the parallelism is therefore fully exploited by the nature of the computation and the use of all the resources.

Sorting has received wide attention since it is a very important task and the algorithm dependency becomes significant when the items to be sorted constitute a large number. Using the shuffle network (so called because it resembles the positions taken by the cards when reshuffled around the middle card in the deck) sorting with n processors may be accomplished in $O(\log^2 n)$ time. On an array machine, it has been shown that it would only take $O(\sqrt{n})$ for an $\sqrt{n}$ x $\sqrt{n}$ sized array while on a serial machine (one processor to sort n items) the time dependency would have been $O(n\log^2 n)$ using bitonic sort.

The sorting on the array processor may also be performed to obtain the sorted items in a snake like configuration so that the connected path that any item must travel on the array is at most n instead of 2n (distance from opposite vertices). A recent algorithm for sorting (due to Tanimoto[8]) is developed for a pyramid computer of non overlapped neighborhoods, where each processor has four sons recursively from a root (apex) until the bottommost layer. This algorithm is O(n) for n numbers at the base and uses in each cycle the even layers first and the odd layers next; the apex contains a highly negative value (that will gradually descend the pyramid to be substituted by the successive maxima obtained by the algorithm). In this way sorting on the pyramid will push the sorted values upwards and the negative dummy values downwards. The core of the algorithm is the compare and exchange process which takes place between the father processor and its children processors in a given sequence (NW, NE, SW and SE) in parallel for all fathers of the even levels (and then all odd levels repeat). The net result is to have the maximum value of the sons in the father's place, recursively until the maximum of the maxima exits from the root. An example (from[8]) is given below for the first three iterations:

```
            _99

         _99  _99
         _99  _99

         4   7  14  15
         2   5   1  17
         2   5   1   4
         6   9   1   4
```

iteration number 0

```
            _99

         7  17
         9   4


       _99  4 _99  14
         2  5   1  15
       _99  2 _99   1
         5  6   1   4
```

iteration number 1

```
         17

         5    15
         9     4


       _99 _99 _99   7
         2   4   1  14
       _99   2 _99   1
         5   6   1   4
```

iteration number 2: next element = 17

As may be seen from the example, the first maximum (17) is extracted from the root and, if the process would be continued the next value to come out would be 15.

In many image processing applications a common filter is the well known median filter. On an array, for instance, if we assume that each processing element is capable of computing the median, it will firstly fetch all the values from its neighbors, store them and then sort them to extract the median. For dxd neighbors (in a square window) the fetching will take $O(kd^3)$ since d neighbors are at an average distance of Ad (for A depending on the relative position of the neighbor) and k is the number of bits required to store the pixel value. The sorting takes $O(d^2)^2$ so that the final value for the order of this algorithm is $O(kd^4)$. A faster sorting algorithm cannot be used because all processing elements should be able to acess the same memory addresses simultaneously. Tanimoto has shown[8] that by using the pyramid sorting algorithm it may be possible to reduce the memory required (from $O(kd^2)$ bits to $O(k)$ bits). Danielsson's[9] median algorithm is a good example of the intricate relationship between algorithm and architecture. Suppose we have seven input arguments examined bit column wise, starting from the maximum significant bit. If the number of 0s is greater than 4 (rank of the median in a set of 7 values) then 0 will be the leading bit and all the arguments that have a leading 1 may be eliminated. Now, in the second step we require the second bit of the median value which is determined by looking for the number of arguments with the leading 01 combination. The process is iterated until no more digits are left. The algorithm may also be described by means of a binary tree which compares, at each node, the value of the bit combinations (msb downwards) from the root bottomward (i.e. their frequency) with a threshold that depends on the rank of the median. At the output of each node two possible answers (yes,no) will provide the two possible bit sequences (00,01) and so on until all the bit columns have been considered. In this way the median is found in $O(kd^3 logd)$. This algorithm may be performed efficiently only if the processor is able to perform bit column counting.

Another approach has been taken by Reeves[10] who has developed an approximate median algorithm, also called pseudo median, since it computes the median "horizontally" first (the median between the pixel and its horizontal neighbors) and "vertically" next (the median between the new pixel and its vertical neighbors): this value (for a three by three window) is within one ranked value of the true median and requires a shorter time since the sorting includes less values.

Although it is commonly believed that low level vision tasks are the ones more suited to be solved by using parallel computers like arrays and pyramids, a recent work on computational geometry performed by cellular computers[11] called mesh connected computers (mcc) shows that for a number of higher level tasks (like convexity detection of a component, counting and marking minimal internal paths in a component, computing the external diameter of a component as examples) the time order is only $O(n)$ (for an nxn mcc) instead of the worst case of $O(n^2)$ or the best case in a serial computer which is still $O(n^2)$.

A slight but important modification to the cellular computer is necessary: the introduction of the row major index which is used to identify each processor of the mcc (this index is given by n*i+j and is unique for each element of the mcc).

As we have seen the overall time dependency is essentially due to two separate mechanisms: computation and communication. It has been shown that no problem requiring data transfer between processors can be solved in less than $O(d)$ steps where d is the maximum distance between two processors. It would be important and useful to be able to classify algorithms in terms of these two requirements: computation effort and communication need. In this way, and adequate system could be developed to optimize the solution to both requirements. One measure which has been introduced to establish the match between an algorithm and its VLSI implementation is the areaxtime^2 which gives an idea of the complexity needed in terms of hardware resources.

INFORMATION TAPERING STRATEGIES

Contrary to the belief that cellular and pyramidal computers may also be conveniently used for pattern recognition and vision understanding are a group of researchers, Inagaki et al.[12] who have tackled a rather difficult practical problem from a very general starting point. They claim that the natural structure of patterns in real life is not only hierarchical (and this is well accepted) but it is also of the super exponential kind. There is a relation between the growth of technical innovation and that of pattern complexity, this relation may be described by the recurrence formula

$A(n) = A(n 1)^c$ (n=1,2,3,...; c is a constant)

The solution of this formula is $A(n)=A(1)c^{n-1}$ which is of

$O(2^{2^n})$; if the function $A(1)=32$ is plotted for increasing values n, using the x axis to represent the information technology level and the y axis to represent the amount of data, we may easily see that the character code requires 32 bits, that a character pattern 32x32 = 1024 bits, a document page $10^3 \times 10^3 = 10^6$ bits (pattern understanding) and, finally, a database containing documents will be of $10^6 \times 10^6 = 10^{12}$ bits (at the top of the curve, for n=4). This is obviously an empirical curve and uses few levels so that the corresponding "understanding system" (which must reflect this structure to be correctly matched) will also have few levels. For two dimensional images the number of pixels may be extracted simply by computing the square root of the formula: $A(n-1)=A(n)^{1/2}$.

The concept of event (a microprocess that can occur together with others, therefore paving the way to parallel execution of concurrent events) is a basic one in this project since the system has been thought as an event driven machine. These events will drive the processing by means of the interlayer interaction. The amount of data the system will process was estimated to be $10^6 \sim 10^7$ pixels. The maximum amount of time to perform the image analysis was of 1 ∼ 10 seconds and for a limiting speed of 100 MIPS (current value for a pixel by pixel processing), 10 to 100 instructions must suffice to accomplish the processing; 256 pixels in parallel can be processed by 16 processors which also have a 16 bit parallelism (the images to be treated are binary). The speed can be obtained, as discussed above, only if the algorithms are conscientiously analyzed since event driven computation may require the execution of different instructions by different processors (at different times) the system must operate in a MIMD mode, asynchronously.

Two top down approaches and two bottom up approaches are used: the first ones are the hierarchical treatment and problem decomposition which is a classical way to break down a large problem into smaller, easily solvable ones. The other is to use the properties of the pattern information source which means that, for instance, the data structure can justify coarse sampling and dimension reduction. The other two approaches are event driven processing which aims at avoiding useless processing only the information that requires processing will be treated, this saves time. Finally, parallel processing is strongly dependent on the structure of the hardware (available resources) and on the structure of the algorithm (if they may be conveniently parallelized or reformulated).

The bottom level was impossible to build at the required resolution so that an image memory of 10^7 bits exists and may interact with the next layer ($10^{3.5}$) which corresponds to one scanning line in parallel, yet only 256 bits in parallel may be read achieving the processing of the total $10^{3.5}$ bits by hardware and software together with a storage of the same amount. The next and last layer has a 16 bit microprocessor. This system is called MACSYM and has one master processor (top layer) and a maximum of 16 slave processors (in the second layer): programs are stored in the local memory, processors communicate via a ring (major ring) and also may communicate in a hierarchical order with local processors. Each processor may act as a personal computer for developing software or as a part of a system for contributing to the parallel execution of a task. With 4 processors, and image of 2.75 Mpixels can be processed to extract the borders in 1.57 seconds (6.15 seconds with only one processor and 0.4 seconds with 16 processors).

The processes are automatically assigned to the available processors via a message board; if the image would have been partioned mechanically the subdivision of tasks amongst the processors would have been straightforward, conversely, within an event driven mode, the processors waiting are found by interprocessor communication and processes are then assigned to these processors. The parallel process to be obtained must be independent from the order of processing, determinacy is not always a necessary condition and in many cases (fuzzy relations, relaxation methods) this constraint may be relaxed. An interesting example of an algorithm which uses knowledge about the problem and a recursive technique is the one used by EXPRESS (an application of MACSYM to newspaper page processing) for finding and extracting articles after their reconstruction (since they appear in blocks). Firstly, contiguous character lines in each column are merged into blocks column by column (headlines and photographs are also blocks). These columns are merged into articles by using the knowledge about the rules used for the layout of text on a newspaper page: articles are traced from the top to bottom and from right to left (remember we are reading Japanese!) block by block. The continuation of columns of a body is found by checking for the START and END characters that give the initium and terminus of a block. These characters are used to detect the beginning and the end of paragaphs, this program is recursively called at the beginning of the search of the next column where another article exists to the right of the "current" article. Finally the articles are reconstructed for future retrieval after normalizing the headlines, removing background noise and rearranging the headlines. An approximate figure is given for the time taken to extract all the articles in a page: 2.5 secs.

In conclusion we may note that the figures corresponding to the super exponential hierarchy are well matched to the amount of information to process on a page of a Japanese newspaper:

Level	Processing entity	Theor. Value	Experim.
1	Number of articles	6.8	6.1
2	Number of blocks	4.7×10^1	5.3×10^1
3	Symbolic map($\neq$ cells)	2.2×10^3	$8 \times 160 = 1.3 \times 10^3$
4	Image	5×10^6	5.2×10^6

Another significant point, concerning processing strategies for pattern recognition and image processing is the presence of recursion and iteration in patterns (as if one had a television viewer viewing in his set another television viewer...or in a fabric with a repetitive design). In fact there have been combinations of top down and bottom up algorithms (like the split and merge algorithm[13] which is used for segmentation) which try to fulfill the task at the level of best attainment (which generally depends on the size of the component which repeats itself). Another reason for the desire to use recursive algorithms[14] is because of the enormous variety of materializations of the same shape like for instance when trying to recognize a car out of a formal definition which contains features of a very abstract car. If the refinement of the definition could be driven by the recognition process (in a self calling mechanism) then different instances of a car could be checked against gradually increasing detail regarding the car, its components, subcomponents, etc. This approach is supported by the well known synctactical school of pattern recognition where the grammar rules are fixed and are completely used in each instance being therefore problem independent.

RULE STRATEGIES

An interesting example of the differences between sequential and parallel processing in order to achieve segementation of a grey level image may be found in [15] where the general strategy is given in terms of an ordered circular list of metarules which are called by their class:

1) PREVIOUS ACTION WAS START ⟶ ACTIVATE FOCUS RULES

2) PREVIOUS PROCESS WAS FOCUS
   PREVIOUS ACTION WAS NOT NULL ⟶ ACTIVATE REGION RULES

3) PREVIOUS PROCESS WAS REGIONS
   PREVIOUS ACTION WAS NOT NULL ⟶ ACTIVATE REGION RULES

4) PREVIOUS PROCESS WAS REGIONS
   PREVIOUS ACTION WAS NULL ⟶ ACTIVATE FOCUS RULES

5) PREVIOUS PROCESS WAS FOCUS
   PREVIOUS ACTION WAS NULL ⟶ STOP

The metarules are used to invoke the region analyzer and the focus of attention during processing: as soon as the system is invoked a START action is executed which will trigger rule 1). The focus of attention rules will be matched against the short term memory and an action will focus on the first region to be tested. If successful, rule 2) is triggered and the region rules will be matched to the current region, if the current region does not meet the requirements of any rule, the null action is executed. In this case rule 4) will be triggered and the focus of attention is brought in to obtain the next region, if it fails to provide a new region rule 5) will stop the system. A sequential strategy may be seen through rule 3) which will fire only if in the previous cycle a region analyzer rule fired and modified the current region, a new attempt to match the region rules to the current region will be tried (and so on): the strategy is to keep on matching the region rules until no more matches can occur. At this point the focus of attention will bring in the new region: as it can clearly be seen this is a sequential strategy. A parallel strategy may be obtained by changing rule 3) to ACTIVATE FOCUS RULES. In this second case the region is acted upon only once and then all other regions are tested. In the experimental work reported in this paper the parallel approach, although more matching effort was

required, produced better results since the analysis is more global.
The sequential strategy is more noise sensitive and this may be seen on the detected boundaries which, in some cases, are incomplete. Rules may be added or removed individually without having to modify the rest of the system, one or two rules may be incorporated into the merging process which may also be achieved by merging after splitting. The homogeneity predicates for merging are based on adjacency, area size, color feature difference, etc so as to approximate the segmentation process to the one performed by man; even if the results may differ the features are analogous.

CONCLUSIONS

We have seen a number of different attitudes regarding parallelism: a formal one considering computational models, their representation by graphs and the general idea that if computing structures are seen as graphs then a good computational match implies a graph match to the algorithmic solution. Another way of considering the problem of achieving parallelism is in the choice of the control structure which must underlie the multiprocessor system.

The classification strategies give some insight as to the signficant parameters which quantitatively describe the performance of the system while the strategies employed in designing the algorithms are intertwined with those used for the architectural design. Finally a hierarchical model of the information variation during analysis, detection and recognition of patterns is the foundation on which a number of tapering systems are based. The tapering may be exponential or superexponential (for an exponent greater than 2) and different consequences follow from this choice. At the end an example using the possibility of selecting some rules within a given set may, interactively, provide insight about the advantages of a parallel computation over a sequential one in the difficult general problem of image segmentation.

The ways to Heaven (parallelism) are paved by the Devil (bottlenecks, idle resources, contention, programming difficulties, reliability, etc.) but is parallelism really Heaven?

REFERENCES

1. L. Uhr, "Introduction: Toward Very Large Multicomputers", in Multicomputers and Image Processing, edits. L. Uhr, K. Preston, Jr., Academic Press, New York (1982).
2. N. Megiddo, "Applying parallel computation algorithms in design of serial algorithms", Proc. 22nd FOCS (1981).
3. E.J. Lerner, 1984,"Data flow architecture", IEEE Spectrum, 57:62.
4. N. Amamiya, "A Data Flow Processor Array", Int. Symp. in Applied Maths. and Inf. Science, Kyoto Univ.,(1983), 7.1:7.8.
5. G. Ausiello, P. Bertolazzi, "Parallel computer models: an introduction", IBM Parallel Scientific Computation Seminar, Rome (1982).
6. P.E. Danielsson, S. Levialdi, 1981,"Computer architectures for pictorial information systems", IEEE Computer, 53:67.
7. H.T. Kung, "The structure of parallel algorithms" in Advances in Computers, edit. M.C. Yovits, vol 19, Academic Press, New York (1980).
8. S. L. Tanimoto, "Algorithms for median filtering of images on a pyramid machine" in Computing Structures for Image Processing, edit. M.J.B. Duff, Academic Press, London (1983) 123:142.
9. P. E. Danielsson, 1981, "Getting the median faster", CGIP, 17, 71:78.
10. A.P. Reeves, 1982, "The local median and other window operations on SIMD computers", CGIP, 19, 165:172.
11. R. E. Miller, Q. F. Stout, 1984, "Geometric algorithms for digitized pictures on a mesh connected computer", internal report, State University of New York, Binghamton.
12. K. Inagaki, T. Kato, T. Hiroshima, T. Sakai, 1984, "MACSYM: a hierarchical parallel image processing system for event driven pattern understanding of documents",Pattern Recognition, 17, 85:108.
13. S. L. Horowitz, T. Pavlidis, 1974, "Picture segmentation by a directed split and merge procedure", Proc.2nd IJCPR.
14. M. Nagao, 1984, "Control strategies in pattern recognition", Pattern Recognition, 17, 45:56.
15. M. D. Levine, A. Nazif,"An experimental rule based system for testing low level segmentation strategies", in

Multicomputers and Image Processing, edits. K. Preston, Jr. and L. Uhr, Academic Press, New York, (1982),149:160

NEW ARCHITECTURES FOR IMAGE PROCESSING

Virginio Cantoni

Dipartimento di Informatica e Sistemistica
Università di Pavia
27100 Pavia, Italy

ABSTRACT

Image processing tasks usually require a power processing capacity higher than the one supplied by the Von Neumann sequential paradigm.

A number of "Non Von" multiprocessor structures have been proposed, designed, and sometimes built, since the early '60.

In order to point out the main features of these machines some characteristics will be introduced here in a dichotomic way: spatial and temporal parallelism; implementation by special (hardware oriented) function units and by general purpose (programmable) units; data driven or instruction driven solutions; with fixed interconnections or with dynamic permutation capabilities; whether they follow or not the linear scaling assumption; whether automatic or interactive.

INTRODUCTION

It is well known that "fifth generation" computer systems must have image and speech processing capabilities. These tasks usually require computation power of over 100 MOPS (Reddy and Newel, 1978), and the only way to obtain such performances is to exploit extensively parallel computations.

Since 1958 (Unger, 1958) new hardware solutions have been proposed, but it is only in recent years that a few commercial systems approach these performances.

A large class of image processing tasks can be tackled with the most recent machines: usually only when local, and data independent computation is required. In these cases it is possible to exploit parallel executions at three levels of processing (Handler, 1977). Namely, going from the highest to the lowest level: the tasks or subtasks level (more than one Program Control Unit (PCU), included in the system, interprets a program, instruction by instruction); at the instructions level (more than one Arithmetic and Logical Unit (ALU) executes an instruction according to the interpretation process performed by the PCU); at the suboperations level, each ALU may contain more than one Elementary Logic Circuit (ELC), with each one dedicated to one bit position.

For each level, both "spatial" and "temporal" parallelisms can be introduced (Hwang and FU, 1983). In the first case PCU, ALU, ELC can be replicated "horizontally", executing respectively tasks, instructions and suboperations on different data (obtaining respectively a multi-processor machine, an array computer, and a wordlength greater than one). In the second case a pipeline of PCU, ALU, ELC, in order to overlap tasks, (macro-pipeline), instructions (instruction pipelining look-a-head), and suboperations (arithmetic pipeline).

The quoted solutions are more effective when the task can be resolved by being "tight orchestrated" in advance. In the image "understanding" or "interpretation" field it is often impossible to plan in advance which procedure will need to be run, so the architecture must be able to allocate and re-allocate computational resources as required by ongoing empirical computations. Some machines that have this capability have been proposed and in a few research groups the first prototypes have been built.

In what follows the main features of the computer architectures for image processing that have been till now proposed will be discussed pointing out five different clear-cut choices: whether dedicated or general purpose; whether spatial parallelism or temporal parallelism; whether data driven or instruction driven; whether fixed or reconfigurable interconnections; whether interactive or automatic.

At the end of this paper a new machine proposed and designed by a group of italian universities will be discussed. It is a general purpose, fixed interconnection system that has some peculiarities with respect to the previous "Non Von" architecture projects: the structure has been designed to exploit a "planning" strategy (Kelly, 1971). This allows a computation speed higher than the one obtainable by a "linear scaling" (in which the computation speed is increased in direct proportion to the number of computing elements).

SPATIAL AND TEMPORAL PARALLELISMS

The first classes of image processing machines were oriented to "low level" vision problems (Cantoni, 1983); the algorithms of these tasks are characterized by the repetition of the same computation sequence on every pixel of the matrix representing the image. If N is the number of instructions of the sequence required by a given algorithm, and M the number of pixels of the image, N*M instructions are necessary in the Von Neumann sequential machine.

Two well known architectures match well with the computational characteristics quoted above: the array of processor elements (PEs), in which, by having at the most one PE assigned to each pixel of the image, requires N number of instruction cycles, (the computation time depends linearly on the computational depth); the pipeline of PEs, in which, by having at the most one PE assigned to each elementary instruction, requires M number of instruction cycles (the computation time depends linearly on the amount of data).

The former architecture is composed by a PCU and a number of ALUs (the PEs) which execute the same function according to the interpretation process performed by the PCU, on different pixels.

Two implementation approaches have been followed; the most popular is to realize the larger array of PEs, compatible with technological and economic constraints, by simplifying (reducing the processing power) the elements. In these cases the PEs have roughly one hundred bits of local memory; the arithmetic is bit serial (the ALU is composed by a boolean processor (CLIP 4, BASE 8) or a full adder (MPP, PAPIA) and sometimes includes shift register(s) (MPP,LIPP,PAPIA, etc.). The interconnections between PEs is limited to the near neighbors (4 cardinal connectivity in MPP, 8 and sometimes 6 connectivity in all the others).

As has been pointed out (Wood, 1983), following the Flynn taxonomy (Flynn, 1972), these systems can be considered SISD machines (Single Instruction stream - Single Data stream) when the image bit plane is considered the elementary data of the system, or SIMD machines (Single Instruction stream - Multiple data streams) when the pixel is the input data.

The largest system of this class is the MPP (Massively Parallel Processor) (Batcher, 1980) composed of 128 x 128 PEs. When the image size is larger than the size of the array of PEs (and this happens very often) only parts of the image can be processed simultaneously. In these conditions the execution of common algorithms, as for example in the case of iterative local computations, becomes cumbersome and time consuming.

A second approach to overcome this bottleneck has been proposed by some authors (Preston, 1983) and some industrial realizations have appeared. In this approach the PEs are distributed over the image (see fig. 1) and each one deals with a segment of image. The processors are more powerful and with greater local memory, but also here the execution has to be orchestrated in advance, as in all the machines of the SIMD class.

The latter architecture is shown in fig. 2; data streams from the input TV-camera in raster scan format into a cascade of processing stages. The computation is decomposed a priori in a sequence of subtasks that are executed in the linear chain of processors, each one having its own subprogram. After an initial phase, as one raw datum streams in, one processed datum streams out. Usually each stage is composed by a simple PE that executes a local operation over a limited area (generally a 3x3 subarray obtained by a shift register of length 2N+3, where N is the linear size of the image). The largest system of this class has been the Cytocomputer (Lougheed et al., 1980) that is composed of hundreds of stages.

When the number of stages is not matched to the number of elementary operations required (it is not a submultiple), the hardware of the machine is not fully exploited: if it is lower a number of idle stages are introduced; if the PEs are not enough, data has to be recycled after having loaded the successive set of instructions.

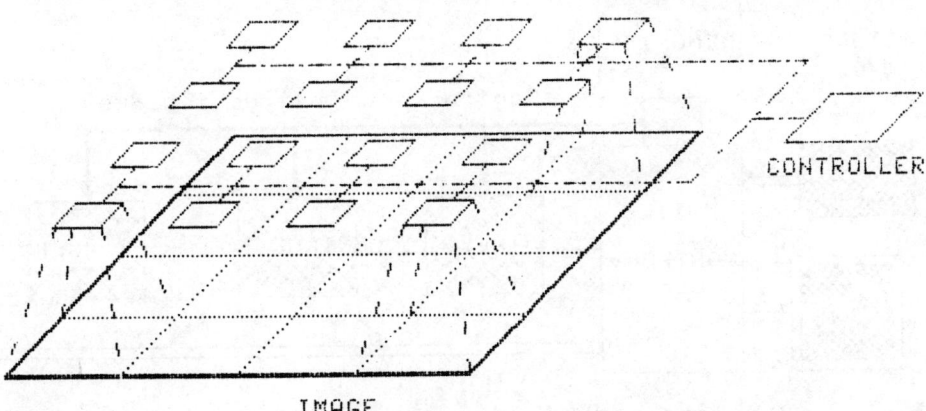

Fig. 1. SIMD machine for image processing. Array of processors following the PEs distributed over the image approach. The dasced and dotted line represents the bus for instruction broadcasting to PEs

Some solutions have been considered and built (PICAP II, GOP, etc.) in which both the temporal and spatial parallelisms are utilized. In this mixed approach the number of processors is usually limited, but flexible structure can be developed.

DEDICATED OR GENERAL PURPOSE

A great proportion of the proposed systems has been developed in research groups in order to obtain efficient machines for algorithm development and for performance analysis and comparison of discussed methods. A number of machines have been devoted instead to particular applications and routine tasks. Generally, in the former case a set of homogeneous, programmable units form the machine, in the latter case, hardware oriented solutions of common tasks have been used, in order to gain in throughput rate, in spite of a lower flexibility to different tasks.

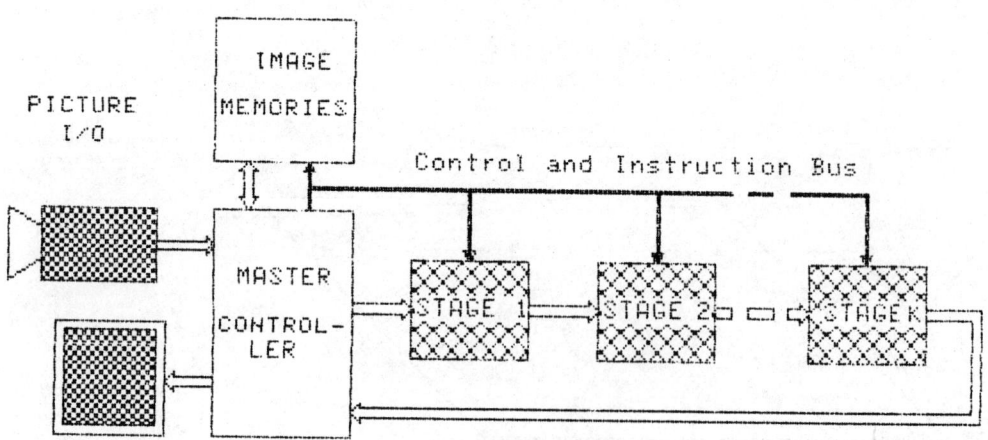

Fig. 2. Pipeline machine for "low level" image processing. Data streams from/to the I/O unit in serial mode. After an initial phase as one datum streams in, one processed datum streams out.

An important case that belongs to the second class is the Dynamic Spatial Reconstructor (Gilbert et al., 1979), developed at the Mayo Clinic for application in computerized tomography. In order to reconstruct human organs from a set of their projections a higher number of inner products are needed; to suit this need a family of hardware units (Swartzlander et al., 1978) have been developed to perform this operation. It is worthwhile to note that convolutions and digital transforms can be easily implemented by means of inner products, so that systems realized around IP computers can be effective in common operations both in time and in frequency domains.

In the DSR there is just one operation that for its particular application is executed a great number of times. In general a subset of image processing tasks can be found, and efficient (usually hardware oriented) solutions can be implemented for these tasks and then jointed together (usually by a bus structure) to realize a general purpose system. This approach which utilizes so called "Special Function Units (SFU)" is the one that has the highest number of industrial implementations.

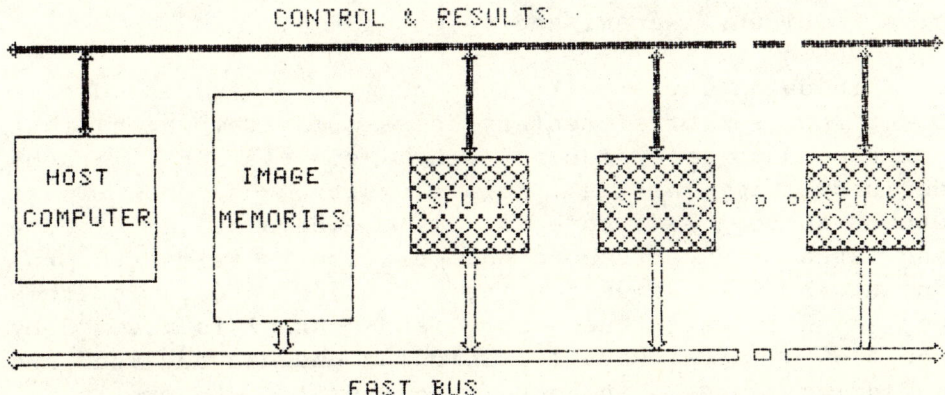

Fig. 3. Special Function Units (SFUs) image processing system. Each unit (connected to the system via data (empty) and control buses) is devoted to a particular function or to a set of related functions.

In this case the system is normally composed by an host computer (see fig. 3), high speed storage unit for the images, and a high speed bus to interconnect the SPUs, each one implementing a particular function or a set of related functions. The image processing task is then decomposed in the sequence of subtasks that better matches the SFUs included in the machine. The host computer controls the synchronization and the execution of the sequence programmed. The SFU set usually includes: video and display processors, two-dimensional convolvers, filter processors, segmentation processors, region labeling modules, look-up tables, and local processors. An example of these machines is TOSPICS by Toshiba (Mori et al. 1978). The systolic design for VLSI is particularly suited to the implementation of SFUs (Kung, 1984).

DATA OR INSTRUCTION DRIVEN

An architectural solution that differs consistently from the Von Neumann machine and from the paradigms considered by Flynn is the one which follows data driven computation concepts. Some systems for image processing applications following this approach have been developed. The most known among these is the Template-controlled Image Processor (TIP) by the Nippon Electronic Company (Hanaki and Temma, 1982).

In data flow architecture a given task is decomposed in a set of elementary operations (templates) that correspond to the node of the "computational graph" defined from the task. Data flows from one node to the other, as soon as a node has all the data of the incoming arcs; the corresponding operation is triggered producing the data value for the outcoming arcs. In this way a PCU is not necessary, and control is assumed by the flow of data. Parallel computations can be realized including more than one function or ALU units in the system.

In TIP a set of units including adders, multipliers, table memories, etc. permit the parallel execution of several templates; data flows in a circular pipeline and when it reaches the "right" processing unit it disappears; a new datum (the result of the operational unit) is inserted into the pipe when of input data set is completed. Other two rings complete the system (see fig. 4): an addressor ring that has I/O functions and provides data and destination flags; and the "main ring" which interconnects the other two.

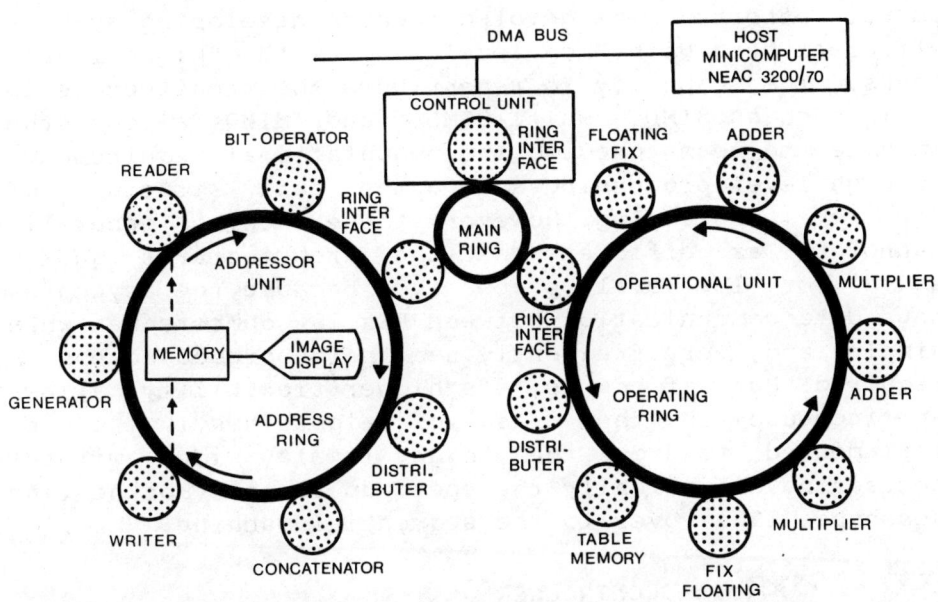

Fig. 4. Simplified schema of an image processing dataflow machine: the Template-controlled Image Processor. The addressor ring supplies, via the Main ring, data to the Operational ring for execution.

FIXED OR RECONFIGURABLE INTERCONNECTIONS

The systems described so far are characterized by fixed, well defined interconnection structures. The image processing tasks suiteble for implementation in these machines are those which have well defined algorithms, for which properties such as efficiency can be proved mathematically and the computation can be "orchestrated" among the processing units in advance, before the execution. This class corresponds to "low level" processing in a well Known dichotomy.

487

The other class, the "high level", in general requires empirical procedures, and often, depending on ongoing computations, needs to allocate dynamically and to re-allocate computational resources.

In recent years some solutions have been investigated by researchers in order to permit a dynamic reconfiguration of the interconnection network. In the first phase, efforts were devoted towards developing systems efficient both with "low level" and with "high level" tasks. The capacity to reconfigure the architecture in forms such as SIMD, Multi-SIMD, and MIMD allows the machine to be matched to the computational requirements of high level processing and to the data structure of low level processing. However, these machines generally cannot be, as efficient as the architectures quoted above for "low level" processing, because the asynchronous data communication between PEs is cumbersome when simple and very frequently near neighborhood access is required. But, of course, its higher flexibility allows an increase in the class of algorithms that can be implemented towards the "high level" side without decreasing dramatically the speed-up factor (Hockney and Jesshope, 1981) over to the sequential machine.

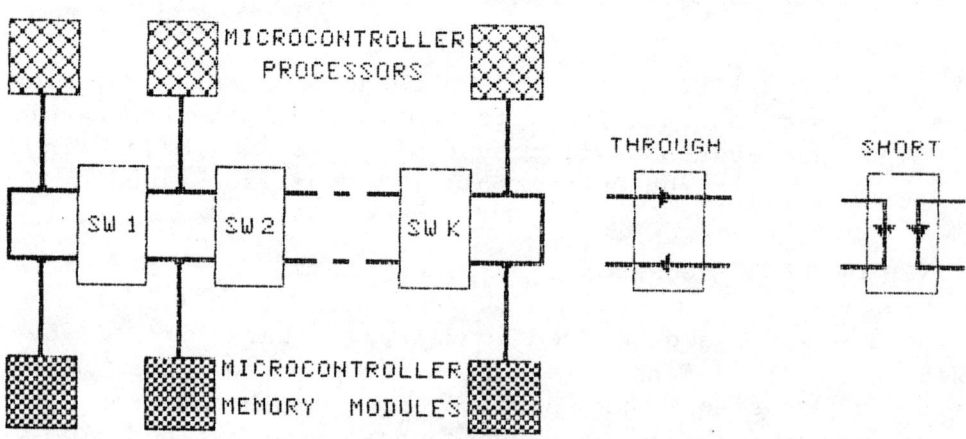

Fig. 5. Simplified schema of a reconfigurable system: the Partitionable SIMD-MIMD machine. It is a Multi-SIMD architecture based on a network including two status switching elements: through and short. The MCPs can access every memory modules of the system.

An example of these systems is the PASM (Siegel, 1981). In this case 1024 eight-bit processors are connected under the control of 16 MicroController Processors (MCPs) in up to 16 groups of 64 processors each working independently in SIMD mode (alternatively all the 1024 can be configured in a single SIMD machine). The MCPs are also interconnected to 16 microcontroller memory modules (see fig. 5) by a particular reconfigurable bus structure.

A different approach has been followed by Briggs et al. (1981). In this case the system contains parts devoted to "low level" processing (for example hardware oriented as in systolic units) and some Task Processing Units (TPUs). These TPUs can be connected by a reconfigurable network to the peripheral "low level" part (by a crossbar), and to a shared memory (by a delta network), and via TPU bus can communicate with each other.

Several interconnection networks have been proposed; from the crossbar that has NxN switching elements and one level of switching, to the Banyan class (including delta and omega networks) which has $0.5 \lg N$ switching elements and $\lg N$ levels (for a review see Gottlieb and Schwartz, 1982 and Haynes et al., 1982)

INTERACTIVE OR AUTOMATIC

The machines described till now follow the "closed world" assumption. That is, the information necessary for the development and the completion of a task has to be completely derivable from the local information possessed by the system.

As described in the previous paragraph, "high level" processing of images cannot be completely planned in advance; this make the automatic approach for the solution of these tasks difficult. It is worthwhile pointing out the effort that several researchers have made in order to implement "high level" capabilities using algorithms that possess the characteristics of "low level" computations (fixed data structures, data independent computations, operations involving only local neighbors). A significative example is the Hough methodology for shape recognition (Ballard, 1980), in which the analysis of a scene to detect the presence of a given object is performed by preliminary local feature extraction and fixed mapping in a suitable parameter space.

An alternative is to introduce in the system real time interaction capabilities with an expert human operator. In this connection specialized hardware options have been developed (as also in the industrial environment) in order to facilitate the intelligent man-machine interaction, (Kidode, 1982), namely: display oriented image memory units furnished with real time video processing modules for zoom, roam, pan, lookup tables for grey scale modification, pipeline arithmetic processors, etc.

An example of this machine is given by Hanaki (1981). In this case the image memory unit has two memory banks A and B of 256 Kbytes and 64 Kbytes respectively. When data is read out for display they can be hardware modified for shifting, zooming, and segmenting performed by range values on the R G B components, etc.

LINEAR SCALING AND PLANNING

The development of image processing architectures has been characterized by efforts to reach, for typical algorithms of the low level area, a speed up factor equal to the number of processor elements (linear scaling assumption according to Hewitt and Liebeman, 1983). The goal was to reduce as much as possible the overhead time that is spent performing operations to "manage" the parallelism. The maximum efficiency (Siegel et al., 1982) has a limit value equal to one when there is the ideal speedup (that is equal to the number of processors), it can almost be achieved only when there is a perfect match between algorithm and architecture.

Some architectures that support capabilities to overcome linear scaling have been proposed with reference to artificial intelligent and "high level" vision tasks. Among these the one most commonly considered by the pattern recognition people is the one based on the pyramid structure. Pyramidal architectures supply the same image at different resolution levels ensuring the use of the most appropriate resolution for the operation, the task and the image at hand. Furthermore, the interplane communication allows the implementation of a general "planning" strategy (Kelly, 1971 and Tanimoto, 1980), in which the basic idea is to proceed to the upper (low resolution) levels when a homogeneous region is detected or coarse computations are required; and descent to the higher resolution planes when finer details are needed. When going from the base to the apex, not only is there a reduction in the number of data recursively from one plane to the other, but also the object description is usually smaller, and in general the computations required can be reduced consistently (Cantoni et al., 1984). In this way speed-up factors greater than one can also be obtained for low level vision tasks at the expense of a moderate increase in hardware both in interconnections and in the number of processors required (the total sum of extra processors amounts to 1/3 of those present in the base having each layer half the size of the previous one).

Different pyramidal machines have been proposed, with reference to the ALU of the processing elements, the amounts of local memory, the connectivity and the interconnections modality (Dyer, 1982).

An example of these machines is PAPIA (Cantoni et al., 1984) designed by a group of italian universities. In fig. 6 the interconnections are shown at the chip level, 5 PEs are included: one in the upper layer (PEf) and four in the lower plane (PEs1-4). As can be seen each PE is horizontally 4-connected to the PE of the same plane (straight lines and dasced and dotted lines for the upper and the lower planes respectively), and vertically to a "father" in the upper plane, and to 4 "sons" in the lower layer (dotted lines). In order to fully exploit the advantages quoted above two operative horizontal and vertical modes have been considered: in the former mode each layer is an independent SIMD

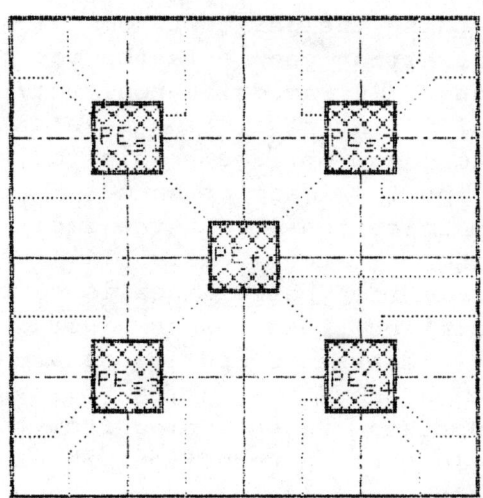

Fig. 6. Multiprocessor chip schema of a pyramidal structure: the Pyramidal Architecture for Parallel Image Analysis. Straight and dasced-dotted lines represent horizontal connections (upper and lower planes respectively). Dotted lines represent interplane connections.

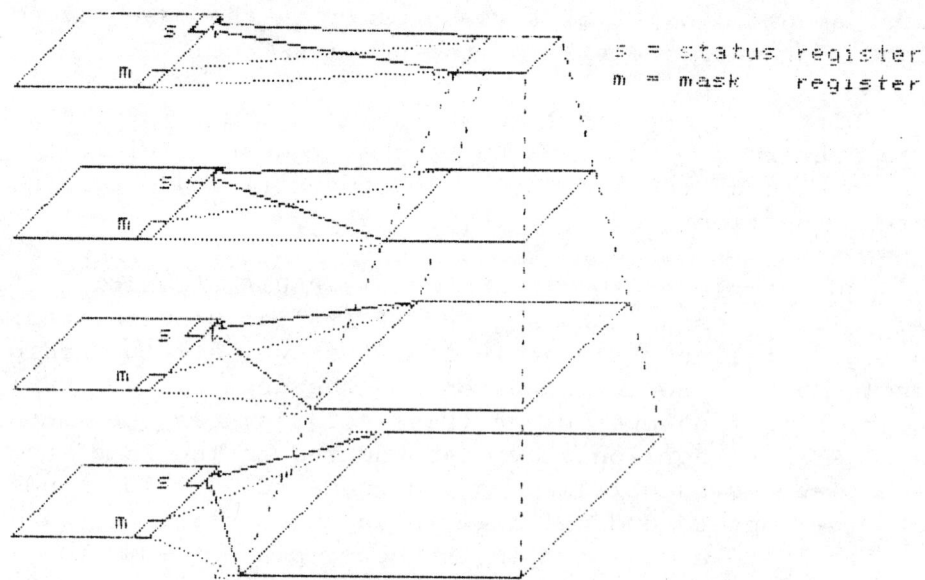

Fig. 7. Simplified schema of the PAPIA architecture. The Multi-SIMD mode is highlighted by the global features of the system: the mask register that enables the plane and the status plane register.

machine (the system becomes Multi-SIMD, as shown in fig 7, where the masking and the global status plane controller capabilities) are highlighted; in the latter mode the pyramid can be considered as a pipeline of array of processors. In this machine (Cantoni et al., 1984) other than adopting the planning strategy, the advantages of the pyramid topology, and the pipeline of array capabilities can be effectively exploited.

REFERENCES

Ballard, D. H., and Sloan, K. R., 1980, Experience with the Generalized Hough transform, Proc. 5th Patt. Recog. Int. Conf., Miami, 174:177.
Batcher, K. E., 1980, Design of a Massively Parallel Processor, IEEE Trans. Comput., C-29, 836-840.
Briggs, F. A., Hwang, K and Fu, K. s., Dubois, M., 1981, PUMPS architecture for pattern analysis and database management, Proc. Patt. Recog. Im. Proc., Dallas, 178:187.
Cantoni, V., 1983, Organization of multiprocessor systems for image processing, in: "Parallel processing: logic, organization, and technology," J. Becker and I. Eisele, eds., Springer-Verlag, Berlin, 145:157.
Cantoni, V. and Levialdi, S., 1983, Matching the task to an image processing architecture, Comp. Vis. Graph. Im. Proc., vol. 22, N. 2, 301-309.
Cantoni, V., et al., 1984, Image processing in hierarchical computer structures, Proc. Digital Signal Processing Conf., Florence, in press.
Cantoni, V., Ferretti, M., Levialdi, S., and Maloberti, F., 1984, A pyramid project using integrated technology, in "Parallel integrated technology," S. Levialdi, ed., Academic Press, New York, in press.
Danielsson, P. E., 1984, Vices and virtues of image parallel machines, in: "Digital image analysis," S. Levialdi, ed., Pitman, 47:59.
Dyer, C. R., 1982, Pyramid algorithms and machines, in: "Multicomputers and Image Processing," K. Preston and L. Uhr, eds., Academic Press, New York, 409:420.
Flynn, M. J., 1972, Some computer organizations and their effectiveness, IEEE Trans. Comput., C-21, 948:960.
Gilbert, B. K., Chu, A., Atkins, D. E., Swartzlander, E.

E. jr, Ritman, E. L., 1979, Ultra high-speed transaxial image reconstruction of the heart, lungs, and circulation via numerical approximation methods and optimized processor architecture, Comp. and Biom. Res., 12, 17:38
Gottlieb, A. and Schwartz, J. T., 1982, Networks and algorithms for Very-Large-Scale parallel computation, Computer, vol. 15, N. 1, 27:36.
Hanaki, S., 1981, An interactive image processing and analysis system, in Real time/Parallel computing, M. Onoe, K. Preston, jr., and A. Rosenfeld, eds., Plenum Press, New York, 219:226.
Hanaki, S. and Temma, T., 1982, Template-controlled Image Processor (TIP project), in: "Multicomputers and Image Processing," K. Preston and L. Uhr, eds., Academic Press, New York.
Handler, W., 1977, The impact of classification schemes on computer architecture, Proc. IEEE Int. Conf. Parallel Processing.
Haynes, L. S., Lau, R. L., Siewiorek, D. P. and Mizell, D. W., 1982, A survey of highly parallel computing, Computer, vol. 15, N. 1, 9:24.
Hewitt, C. and Lieberman, H., 1983, Design issues in parallel architectures for artificial intelligence, A.I. Memo No. 750, 1:14.
Hockney, R. W. and Jesshope, C. R., 1981, "Parallel computers", Adam Hilger, Bristol.
Hwang, K., and Fu, k, 1983, Integrated computer architecture for image processing and database management, Computer, vol. 16, N. 1, 51:61.
Kelly, M. D., 1971, Edge detection in pictures by computer using planning, Machine intelligence 6, Edinburgh University Press, 397:409.
Kidode, M, 1983, Image processing machines in Japan, Computer, vol. 16, N. 1, 68:80.
Kung, H. T., 1984, Systolic algorithms for the CMU warp processor, Proc. 7th Int. Conf. Pattern Recognition, Montreal, 570:577.
Lougheed, R. M., Mc Cubbrey, D. L., and Sternberg, S. R., 1980, Cytocomputers: architectures for parallel image processing, Proc. Workshop on Picture data description and management, Pacific grove, California, 281-286.
Mori, K. I., Kidode, M., Shinoda, H., and Asada, H.,

1978, Design of local parallel processor for IP, Proc. AFIPS Conf., vol. 47, 1025:1032.
Preston, K. jr., 1983, Cellular logic computers for pattern recognition, Computer, vol. 16, N. 1, 36:50.
Reddy, D., and Newel, A, 1978, A multiplicative speed-up of systems, in :"Perspective on computer science, A. K. Jones, Academic Press, New York.
Siegel, L. J., Siegel, H. J. and Swain, P. H., 1982, Parallel algorithm performance measures, in: "Multicomputers and Image Processing," K. Preston and L. Uhr, eds., Academic Press, New York, 241:252.
Siegel, H. J., 1981, PASM: a reconfigurable multimicrocomputer system for image processing, in "Languages and architectures for image processing," M. J. Duff and S. Levialdi, eds., Academic Press, New York, 257:266.
Swartzlander,E. S. jr., Gilbert, B. K., and Reed, I. S., 1978, Inner product computers, IEEE Trans. Comp., vol. C-27, 21:31.
Tanimoto, S. L. and Klinger A., 1980, Structured computer vision: machine perception through hierarchical computation structures, Academic Press, New York.
Unger, S. H., 1958, A computer oriented towards spatial parallelism, Proc. of IRE, 1744:1750.
Wood, A. M., 1983, The organization of parallel processing machines, in: "Parallel processing: logic, organization, and technology," J. Becker and I. Eisele, eds., Springer-Verlag, Berlin, 132:144.

DATA STRUCTURES AND LANGUAGES IN SUPPORT OF PARALLEL

IMAGE PROCESSING FOR ASTRONOMY

Steven L. Tanimoto

Department of Computer Science, FR-35
University of Washington
Seattle, WA 98195 U.S.A.

ABSTRACT

Data structures, and aspects of programming languages and programming systems that are relevant to image processing of astronomy data are discussed. Since parallel processing promises to greatly increase the speed with which images can be analyzed, techniques that will help achieve this speed and new methods that may become possible because of it are of interest.

INTRODUCTION

Scientists in many fields would like to compute faster so that they can test more hypotheses, better validate their hypotheses and more quickly gain new insights. Astronomy is such a field. It is fitting that astronomers begin to consider using parallel processing because it is becoming more economical to do so, and this will probably lead to dramatic speedups. Microprocessors are now available at very low prices and custom VLSI circuitry is feasible and becoming common.

In the design of systems that use parallel processing, however, there are some technical problems. Many of these problems are pointed out in [Levialdi 85]. The kind of problems can be divided into two classes: hardware and software. Although this paper is mostly concerned with software, it must be clear that hardware architecture is intimately related to data structure, operations, algorithms and language of programming. Some ways in which hardware and software impact algorithms are described in [Wood 81]. To organize these software issues and the hardware architecture one, we suggest a tetrahedral framework. It can be used to relate, at an abstract level, the critical design issues for new parallel computing systems.

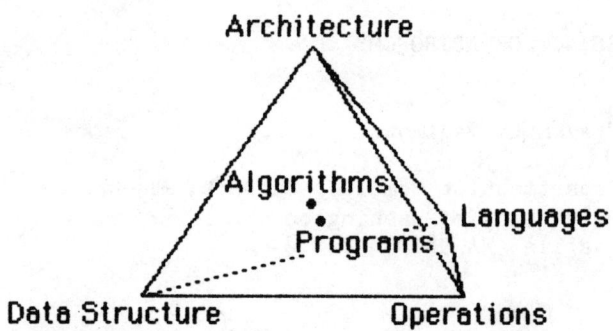

Fig 1: A tetrahedral framework for computing technology.

All these issues are interdependent. Data structure focusses on the forms of the objects on which operations will be performed. Languages provide means of representation for the operations and transforms on the data objects. In a sense, data structures suggest most of the nouns (subjects) for the languages. The desired changes to data suggest the verbs (operations) of the languages. Conversely, the data-definition capabilities of languages allow new data structures to be created by the programmer, and new operations can be built up from those given.

Some forms of parallel processing differ from uniprocessing by performing operations on entire data objects (possibly having many components) at once [Hwang&Briggs 84], [Preston et al 79]. Data structures may therefore serve as the organizational basis for parallelism in a system. The operations of a language are, ideally, close to the instructions for a parallel machine, so that the machine executes the operations efficiently.

One cannot begin a discussion such as this without mentioning the role of algorithms. It seems reasonable that algorithms be put right in the center of the Data Structure - Operations - Architecture triangle. Algorithms are tied to all three of these corners but are not dependent upon a particular language.

An algorithm is a specification for a process, manipulating data in certain structures, assuming a particular computer architecture. In addition to these three aspects, an algorithm also is related to a problem--the algorithm solves some problem. One may then ask what the place of a program is. As an implementation of an algorithm a program is equally tied to data structure, operations, architecture and language, and therefore the issue of programming resides in the center of the tetrahedron.

The purpose of this paper is to survey some of the ideas about data structures, languages and programming in general from current computer science and technology and to suggest how they might be of use in astronomical data processing. Emphasis is on image processing computations, because this kind of data processing is obviously a ripe one for parallelism and it is important in astronomy. However, some discussion of more general possibilities is given in an attempt to stimulate some thinking about computing environments for astronomy.

DATA STRUCTURES

Ideally, a data structure makes the salient aspects of some data explicit. A two-dimensional array may make explicit some image intensity information as a function of position. In addition, data structures provide units of description for data in algorithms, making it convenient to write programs and other descriptions of processes. A third aspect of data structures is that they may help achieve efficiency in algorithms.

Primary forms of image data are various kinds of two dimensional arrays [Tanimoto&Klinger 80]. Most often, an image array represents a sampling of a continuous image according to a rectangular mesh. However, hexagonal, triangular and irregular sampling patterns have also been used. While the form of an image is a 2D array, the contents of such an array might represent any of many kinds of data: visible intensity, multispectral intensities, edge data, texture data, velocity vector field, etc.

From 2D arrays containing image data, several derived structures may easily be obtained. One class of these are the pyramids, which are collections of images at different degrees of spatial resolution. Quadtrees are another representation for image data, where some compression may be gained if there are large blocks of pixels with equal values in the image. It should be mentioned that some of these data-structuring methods work well in three dimensions also: the octree is finding applications in computer aided design of 3-D objects and in medical 3-D image processing. Octrees have not yet, but might be applied in astonomy, as well.

Another kind of derived structure, close to the original image, is the "patched database" representation of an image [Tanimoto&Fowler 80]. Such a scheme depends on the assumption that the original image contains significant areas that are of little or no interest. The portions which are of interest are covered by a series of rectangular patches, and these patches are catalogued with a quadtree index, and superimposed upon a low-resolution version of the original image, so that the result is a space-efficient, multi-resolution approximation to the original. With software that swaps patches in and out of active memory, as the pages of a virtual memory system, one gains not only the savings in storage space in the archive but also the possibility of processing the image without having to store the entire image in main memory. With algorithms such as boundary following or local property extraction, the page-fault rate is low because of the geometric locality of the data organized as patches. Astronomy images seem to meet the assumption that certain areas are of interest (e.g. areas containing certain kinds of stars) while others are not (e.g. interstellar space), and thus, it may sometimes be useful to have a patched image representation in image processing workstation systems for astronomy.

Data structures for pictorial information that are further from the original array representations are the boundary-based structures. For these, it is assumed that the major pictorial objects in the image may be described in terms of the shapes of their boundaries. The Freeman chain code gives a string of elementary vectors corresponding to a clockwise walk around each object in the image. Polygonal approximations provide more compact representations although some error generally is introduced. Greater flexibility in trading off error and data reduction is afforded if one permits curves, such as splines or arcs of ellipses, to be used in describing the contours. Boundary representations lead more readily to shape features than do 2D array representations.

Yet further away from the original image arrays are data structures that are symbolic or highly abstracted parametric descriptions. A feature vector such as a list of the measurements for average intensity, centroid x and y, and "busyness" value, is a good example of a parametric description. Symbolic representations include lists of the names of objects present in an image together with their spatial relationships ("star number five northwest of galaxy number two"). Relational databases offer one kind of representation for such symbolic information. If the relations concerned are never of order higher than binary, then graphs with nodes and arcs can represent them effectively. More complicated symbolic data structures are semantic nets and schemata. With these, procedures may be attached to graph nodes or to arcs and the procedures may be included as part of the data structure. Such schemes have been used in computer vision [Ballard&Brown 82].

An aspect of computing with images that will be increasingly important for astronomy is the structure of collections of images. A system that uses a substantial collection of images is called a pictorial database system (PDBS). Researchers have studied several aspects of PDBS other than the image data structures used. Some of these issues are the following: data integrity, data compaction, image catalogues, indexing by content, by pictorial features and by image source, management of image sequences, graphical annotation to images, pictorial query systems, integration of optical disk technology, distributed databases, progressive transmission, and special processors to support pictorial database operations.

Sometimes a collection of images should be treated as a stream of images. An ordered sequence of telescope images of one portion of the sky, taken at regular intervals, constitutes such a stream. Temporal analysis operations are applicable to such a stream, and it may well be helpful to consider the stream as a single data structure to which the appropriate operations may be applied. In the last decade, researchers in image processing have made substantial progress in dynamic scene analysis, change detection, and motion description. Some of this work makes it clear that meaningful operations on image steams do exist indeed.

Data structures provide logical units for data description and manipulation. How the data structure and operations are specified to a computer is in the realm of programming languages and systems. The use of data structures such as images, lists, graphs and collections of these generally provides opportunities for parallelizing computations. Any operation that can be applied to a data structure in a highly parallel way will yield such parallelism every time it is used. If the data structure were not used, or used only in an implicit way, the potential parallelism would be much harder to find.

LANGUAGES

While a data structure provides the organization for a collection of related constituents of an image, it also provides a unit of description for describing operations and algorithms. In order to accurately specify the transformations and algorithms, a particular means of representation is required. The purpose of a programming language is to provide such a means of representation.

There are many expectations or requirements of programming languages. A language must provide adequate facilities so that an algorithm can be implemented; it needs enough representational power and the ability to be translated into machine language or to be interpreted by a computer program. A language should encourage correctness of programs through cleanliness of constructs, simplicity of structure, and possibly through the provision of assertion mechanisms. Also, the constructs should fit the problem so that great effort is not needed to program a solution. A programming language often is a means of communication between humans: a clearly written program helps smooth subsequent maintenance; a readable language helps make readable programs. Languages can provide a measure of portability for programs. If a language's compilers are targeted on many different machines, a program will enjoy a longer and richer life.

Perhaps most important, a language should provide the conceptual building blocks that lead to insights for solving problems. The language should support modular software development so that new problems can be solved, standing on the shoulders of previous investigators.

A language for image processing should provide, as primitive operations, transforms on entire images. Traditionally, programming languages are procedural, textual schemes whereby a program comprises a list of statements, each of which describes either the form of some data or some steps of a computation. Examples of traditional, general-purpose languages are FORTRAN, PASCAL, C and LISP. A machine-specific programming language for image processing is the CAP4 language for the CLIP4 (cellular logic image processor [Duff 76]) at University College London. A general purpose language and a special-purpose, machine-dependent language may be combined to provide a powerful facility for describing image processing algorithms. An example of this is the IPC language, also developed for the CLIP4.

A number of language extentions for parallel processing of images have been proposed. A notable example is PIXAL [Levialdi et al 80], which provides a means for expressing parallel image processing algorithms in terms of the computation at a typical cell. Another example is PascalPL, a version of Pascal with special constructs for parallel operations [Uhr 81]. Other examples are cited in [Tanimoto 80]. Languages for describing cooperating concurrent processes include Ada and Concurrent Pascal. Because of the heavy overhead in interprocess communication with systems like Ada, they are suitable for "large-grain" parallelism. They are not very suitable for pixel-level parallelism.

The cost of developing software can be reduced when new applications can make use of modules previously produced. Certain programming languages such as Smalltalk encourage this by making modules called "objects" the focus of the language. Object-oriented languages help to package the data and the operations that go with it together. It is easy to create new objects that are modifications of old ones, and not have to rewrite the procedures for the operations. Objects are maintained in a class hierarchy that provides "inheritance" of properties from any object to any other object in a subclass of it. The language Simula provided the class feature, but in a more cumbersome way than Smalltalk. Some LISP dialects such as MIT Lisp Machine LISP support object-oriented programming through enhancements, such as the "Flavors" package. In the future, object-oriented programming is likely to become more widespread.

While programming languages have been getting ever more sophisticated and useful, in some sense the need for them has been gradually evaporating. More and more canned programs are becoming available. More importantly, there are new kinds of problem-solving software tools that provide much of the flexibility of a programming language without the complexity of one. A good example of such a tool is the popular program "VisiCalc" developed to provide an instantly-recalculating electronic spreadsheet. Before VisiCalc, even more sophisticated software systems based on the same principle (constraints between data objects) were developed experimentally. The ThingLab system [Borning 81] is such a system. In these systems, the user creates a model of his problem on the screen, using interactive graphical editing. The system provides computation to enforce relationships among the parts of the model. The process of specifying the constraints is a kind of programming, but it is a much more natural one for most users than writing in BASIC, Pascal or FORTRAN.

Experimental programming environments are an important discussion topic in any survey of the prospects for software on parallel computers. Methods that simplify programming or eliminate the need for it may help to solve the problem of programming parallel computers.

Demonstrational programming is a way for the user to specify an algorithm to a machine by going through a sample computation. In order to do this, some sample data must be arranged on the screen, so that the user can manipulate it interactively. The computer records the sequence of interactions and performs some simple generalization to obtain a program. For example, the computer can create variables to represent data that was constant during the demonstration run. Two systems which permitted such programming are Pygmalion [Smith 75] and PAD (Programming by Abstract Demonstration) [Curry 78].

Another approach to improving the programming process involves using graphical representations of programs rather than textual ones. Iconic programming typically allows the user to draw a control flow diagram, in which operations represented by icons are linked together into a graph. The machine supports the editing of these diagrams, and it can execute the diagrams directly without translation (i.e. an interpreter processes the diagrams). A system that supports this approach is PICT [Glinert&Tanimoto 84].

A slightly different approach to modular software is the development of "tool kits". In addition to editors and compilers, it is helpful to have parser generators, frameworks of user interfaces, systems packages with user "hooks" and other software generators. The Unix environment has embodied this philosophy to a large extent. Development tools such as visual debugging packages and error-checking interpreters also help. One can imagine a toolkit for parallel programming including graphics routines that can help monitor the activities at the many processors in the system, and display the most commonly used data structures in meaningful ways.

CONCLUDING REMARKS

Data structure, languages, and parallel processing are interdependent and tied to architectures, algorithms and programs. If significant progress is to be made, innovations in all three areas are needed. Improvements in one without the other two will only make small gains.

Computing environments for astronomers would, ideally, be jointly developed by astronomers and computer scientists. This happens to a certain extent already, since astronomers do not start from scratch in building their systems, but start with purchased systems that were designed by computer scientists and engineers. However, astronomers are not typical business users of computers. As a particular kind of scientist, an astronomer has needs that may not be addressed in the marketplace. There are two obvious undesirable results that we may get if only one of these two groups designs the system: (a) systems that solve the wrong problems, or (b) systems that are obsolete before they are finished and tested. Quality is more expensive in the beginning, cheaper in the long run. Computer technology is evolving rapidly. When moving up to a new system, one should usually make as big a leap as possible, so as to not have to do it again for a long time.

Future developments in data structures, languages and parallel processing that will impact image processing are likely. Some current directions of research in these topics are worthy of note.

In data structures, we can expect significant improvements to result from multiresolution methods of image representation [Rosenfeld 83], iconic/symbolic approaches, semantic networks, and organizational methods for pictorial database systems.

In languages, object-oriented languages will proliferate, as will tool kits, screen-based, iconic languages, constraint-based languages, and languages for building expert systems.

In parallel processing, image-based architectures including pyramid machines will soon be available. Local area networks and global information networks will bring resources that are physically remote to bear intimately on solving an astronomer's problem. Symbolic processing is making artificial intelligence approaches computationally more attractive for complex image interpretation problems. Integrated architectures that couple parallel image processors with efficient symbol processors will allow relatively high-level visual analysis tasks to be automated. A class of architectures called reconfigurable, which includes systems like the Pringle [Snyder 84] and PASM [Siegel 81], promises to make parallel processing both efficient and flexible. The software problems of reconfigurable systems are not yet well understood.

A rapidly developing area of software technology that should not be overlooked by any scientist is artificial intelligence. Programming methods are being improved that allow the codification of many types of knowledge, and that allow automatic inference from that knowledge [Rich 83]. Artificial intelligence will probably have applications to scientific theory formation, to resource optimization (e.g. telescope scheduling), and the formalization of astronomical knowledge. In the future, it may be possible to have integrated systems comprising controlled sensors, facilities for interactive experiment design, execution, analysis, and reporting. With astronomy laboratory computers networked together and tied into space-based telescopes, it may be just a matter of designing the right software to create a solar-system-scale distributed research machine that, with a little human guidance, would plan and carry out the experiments to develop a mature cosmology.

REFERENCES

Ballard, D.H. and Brown, C.M. Computer Vision, Englewood Cliffs, NJ: Prentice-Hall, 1982.

Borning, A. The programming language aspects of ThingLab, a constraint-oriented simulation laboratory, ACM TOPLAS, Vol. 3, pp353-387, 1981.

Curry, G.A. Programming by abstract demonstration, Ph.D. Thesis, Dept. of Computer Science, Univ. of Washington, Seattle, 1978.

Duff, M.J.B. CLIP4: a large scale integrated circuit array parallel processor. Proc. Third Int. Joint Conf. on Pattern Recognition, pp728-733.

Duff, M.J.B. and Levialdi, S. (eds.). Languages and Architectures for Image Processing, London: Academic Press, 1981.

Glinert, E.P. and Tanimoto, S.L. PICT: Experiments in the design of interactive graphical programming environments. (To appear in IEEE Computer, 1984).

Hwang, K. and Briggs, F.A. Computer Architecture and Parallel Processing, New York: McGraw-Hill, 1984.

Levialdi, S. Steps toward parallel processing. In Data Analysis in Astronomy, New York: Plenum Press, 1985.

Levialdi, S., Isoldi, M., and Uccella, G. Programming in PIXAL. Proc. Workshop on Picture Data Description and Management, Pacific Grove, CA, 1980, pp74-79.

Preston, K., Jr., Duff, M.J.B., Levialdi, S., Norgren, P.E., and Toriwaki, J.-I. Basics of cellular logic with some applications in medical image processing. Proc. of the IEEE, Vol. 67, No. 5, pp826-855, 1979.

Rich. E. *Artificial Intelligence*. New York: McGraw-Hill, 1983.

Rosenfeld, A. Pyramids: Multiresolution image analysis, *Proc. of The Third Scandinavian Conf. on Image Analysis*, Copenhagen, Denmark, July 12-14, 1983, pp23-28.

Siegel, H.J. PASM: A reconfigurable multi-microcomputer for image processing. In [Duff&Levialdi 81], pp257-265.

Smith, D.C. Pygmalion: A creative programming environment, Ph.D. Thesis, Dept. of Computer Science, Stanford Univ. 1975.

Snyder, L. Parallel programming and the Poker programming environment. *IEEE Computer*, Vol. 17, No. 7, July, 1984, pp27-36.

Tanimoto, S.L. Advances in software engineering and their relations to pattern recognition and image processing, *Proc. Fifth International Conference on Pattern Recognition*, Miami Beach, Florida, Dec. 1-4, 1980, pp. 734-741.

Tanimoto, S.L. and Fowler, R.J. Covering an image subset with patches, *Proc. Fifth International Conference on Pattern Recognition*, Miami Beach FL, Dec. 1-4, 1980, pp 835-839.

Tanimoto, S.L. and Klinger, A. (eds.). *Structured Computer Vision: Machine Perception Through Hierarchical Computation Structures*. NY: Academic Press, 1980.

Uhr, L. A language for parallel processing of arrays, embedded in PASCAL. In [Duff&Levialdi 81], pp54-67.

Wood, A. The interaction between hardware, software and algorithms. In [Duff&Levialdi 81], pp1-11.

MORPHOLOGY AND PROBABILITY IN IMAGE PROCESSING

Andrea G. Fabbri

Istituto di Geologia Marina
Consiglio Nazionale delle Ricerche
Via Zamboni 65, 40127, Bologna, Italia

ABSTRACT

This paper analyzes some concepts that relate morphological attributes of digital objects to statistically meaningful measures. Few elementary transformations of binary images are described first. Then, examples of applications are drawn from the geological and image analysis domains. Some morphological models are applicable in astronomy. The development of new spatially oriented computers leads to more extensive applications of image processing in the geosciences.

INTRODUCTION

The term "morphology" was coined by Geoethe in 1800 to comprehend the study of formation and transformation, whether of rocks, clouds, colors, plants, animals or cultural phenomena of human society, as those presented themselves to sentient experiences. Since that time, a great deal of research work in the geosciences is based on quantitative morphological studies, from the modal analysis of rocks (to estimate the volume occupied by different crystal phases) to the description of ore bodies or the comparison of continental plates.

Presently, shape analysis is a field of great concern among geologists who use quantitative and statistical methods. Fabbri (1984a) employed digital image processing in studies of spatially distributed data obtained by different techniques of digitization from geological and ancillary geophysical maps or from microscopic images of rocks. Most computations to be performed on digital images are of local nature, e.g., edge detection, thresholding, line thinning, skeletonizing, and object counting. This means that a function is evaluated which, for each picture element or pixel (in a regular two-dimensional arrangement or raster) takes into account the values of a subset of neighboring pixels (not excluding the pixel itself).

Digital morphology can be more easily exemplified by considering binary images, where there are only two gray levels, black (the objects) and white (the background). The description and quantitative charaterization of binary images involve the geometrical properties of objects and relationships among objects. The set of these properties, however not necessarily limited to binary images, is termed texture. Some of the first attempts to study digital pictures as sets of binary images by "binary transformations" were made by Kirsch et al. (1957), Uhr and Vossler (1961) and Moore (1968). Those works led to the programming of special image processing software, to the development of image analyzers, and to the design of parallel computers.

Recently, a renewed interest in the analysis of binary image information was generated partly because of the work for a theoretical background to geometrical probability concepts and applications done by the French School of Mathematical Morphology and also the work of stereologists in both theoretical and practical aspects of quantitative microscopy (Matheron, 1972, 1975; Serra, 1982).

An illustration of simple binary transformations is provided in this contribution. It is followed by examples of probabilistic representations that can be related to problems in astronomy. Comments are also made on recent developments in the design of spatial computer architectures that are cmputationally efficient for processing large gray level images.

BINARY TRANSFORMATIONS

Let us consider the two-dimensional binary image space set I containing 9x9=81 pixels shown in Figure 1. The pixels in I are either black or white, and represent the set A of black pixels, and the set A^c of white pixels. We shall consider the set A as the object contained in our image set I, and the set A^c as the background. We can use a second image set B, that consists of 3x3=9 black pixels, to scan the set I in search of the set A. Because B is isotropic in our example in Figure 1, its reflection $\check{B}$ is identical to B. Of all possible translates of B across I, B will be fully contained, only partly overlapping, or just outside I, A, or A^c, a finite number of times. In addition to counting these "hits or misses" (for example we might consider the procedure as a crosscorrelation between **A** and **B**) new images can be computed that represent spatially those counts.

If **B** is a set of only one black pixel, it can be used to count the number of black pixels in **A**, or the number of white pixels in A^C, the complement of **A** in **I**. If **B** is a set consisting of a couplet of pixels, one black and one white, adjacent in the horizontal direction, it can be used to count all transitions from **A** to A^C, in that direction and sense of direction (and to compute the corresponding image of transitions) or to compute a translate of **A** one pixel to the right.

Let us also consider a set **T** of size 9x9 pixels (same as the set **I**) of all black pixels. **T** can be obtained by computing the logical union of the set **A** and its complement A^C. Any transformation of **A** or of its complement must be intersected with the set of black pixels **T** or with a corresponding transformation of **T**, for proper statistical weighting.

The transformation obtained by changing to black all white pixels in **I** that correspond to the center pixel in **B̆** when **B̆** overlaps **A** only in part and leaving unchanged all black pixels in **A**, is termed a "dilatation" of the set **A** by the set **B̆**, and is indicated by the expression **A ⊕ B**, where the symbol ⊕ is the operator of an "expansion" type of transformation (in which the number of black pixels is greater or equal to the one in the original image before the transformation).

The transformation obtained by changing to white all black pixels in the set **I** corresponding to the center of **B̆**, when **B̆** overlaps only in part with **A**, is termed an "erosion" of the set **A** by **B̆**, and is indicated by **A ⊖ B**, where ⊖ is the operator for a "shrinking" type of transformation (in which the number of black pixels is smaller then or equal to the one in the original image). The image set **B** used for the transformations is termed "structuring element" in mathematical morphology.

We can consider the corresponding transformations of the set **T**, by expanding or shrinking (i.e., dilatating or eroding) **T** by **B̆**, accordingly. **T** and its transformations represent "masks" that identify the image area to which the results of the transformation must be related. For example, the set **I** can be a sample of a larger image space from which information is required to avoid biasing the transformations of **A** or of A^C at or past the edges of the image set **I** ("frame bias").

Fig. 1. Simple example of a binary image set **I** (top left, with image coordinates) containing set **A** and set **A**^{C}, of a mask set **T**, and of erosion and dilatation transformations by a structuring element set **B** (pixel at origin or center pixel is underlined). Asterisks indicate black pixels, dots are white pixels. For erosions, the lighter asterisks indicate the black pixels that became white; the darker asterisks indicate the white pixels that became black during the dilatations. Similarly, lighter and darker dots indicate the corresponding shape and size changes of the mask set **T**. On the right side, the black pixel count (mes) is added for each image or transformation. The expression of the probability of hitting the set **A** by the set **nB** is at the lower right of the illustration, followed by the computations for: one dilatation, no transformation, and one erosion (in parentheses the values not corrected forthe transformations of the set **T**).

```
    123456789
   +---------+
 1 !.........!      ***
 2 !.........!      ***      B = B̌
 3 !.........!      ***
 4 !....**...!
 5 !...***...!   set A  (*'s)    mes (A) = 13
 6 !..*****..!        c                 c
 7 !...***...!   set A  (.'s)    mes (A ) = 81 - 13 = 78
 8 !.........!
 9 !.........!
   +---------+
     set I                        mes (T) = 81
                 ***********
 *********       ***********      mes (T ⊖ B̌) = 49
 *********       ***********
 *********       ***********      mes (T ⊕ B̌) = 121
 *********       ***********
 *********       ***********
 *********       ***********
 *********       ***********      mes (A ⊖ B̌) = 1
 *********       ***********
 *********       ***********      mes ((A ∩ (A ⊖ B̌)^c ) = 12
                 ***********
                                  mes (A ⊕ B̌) = 35
  set T ⊖ B̌      set T ⊕ B̌
                                  mes ((A ⊕ B̌) ∩ A^c ) =
 .........                                = 35 - 13 = 22
 .........
 .........
 ....**...
 ...***...       set A ⊖ B̌
 ..*****..
 ...***...                                    mes (A ⊕ nB̌)
 .........                        P(nB) =  ---------------
 .........
                                              mes (T ⊕ nB̌)
 ...........
 ...........
 ...........
 ....****...
 ...*****...                      35/121 = 0.2893   (0.4321)
 ..*******..     set A ⊕ B̌
 ..*******..                      13/ 81 = 0.1605
 ..*******..
 ...*****...                       1/ 49 = 0.0204   (0.0123)
 ...........
 ...........
```

513

Fig. 2. Upper half: illustration of Minkowski additions and subtractions of binary image sets, and expressions for the corresponding transformations. To the left two identical copies of a simple binary image are shown after a diagonal shift of one pixel. Asterisks indicate black pixels (unshifted image), crosses indicate black pixels in the shifted image. Darker asterisks indicate the black pixel overlap; darker dots indicate the expansion of the image space due to the shift. In the expressions below, b and a, and (a+b) indicate two points (pixels) belonging to the sets **B** and **A**, and a segment between them in Euclidean space, respectively. Lower half: comparison between the Minkowski operations and the transformations used in mathematical morphology. In the structuring elements sets **B** and **B1**, the pixels at the origin are underlined. In the transformations, the darker asterisks indicate white pixels that turned black; the dots indicate black pixels that turned white during erosions or Minkowski subtractions. The closing exemplified at the lower right shows the merging of two close objects.

```
  . . . . . . . . .         . . . . . . . . .         . . . . . . . .
  . . . . . . . . .         . . . . . . . . .         . . . . . . . .
  . . . + + . . . .         . . . . * * . . .         . . . . . . . .
  . . . + + * * . .         . . . * * * * . .         . . . . * . . .
  . . + + * * * . .         . . * * * * * . .         . . . * * . . .
  . . . * * * * . .         . . * * * * * . .         . . * * * * . .
  . . . . * * * . .         . . . * * * * . .         . . . . * * . .
  . . . . * * . . .         . . . . * * . . .         . . . . . . . .
  . . . . . . . . .         . . . . . . . . .         . . . . . . . .
  . . . . . . . . .         . . . . . . . . .         . . . . . . . .
                                    !                         !
                                    v                         v
```

Minkowski (1911):   addition   subtraction

$A \oplus B = \bigcup_{b \in B} \left(\bigcup_{a \in A} (a+b) \right) = \bigcup_{\substack{b \in B \\ a \in A}} (a+b)$ addition

$A \ominus B = \bigcap_{b \in B} \left(\bigcup_{a \in A} (a+b) \right)$ subtraction

---

	B ≠ B̌		
Dilatation	``*** ** **``	``.*``	Erosion
	``****``	``.**``	
A ⊕ B̌	``*****``	``.***``	A ⊖ B̌
	``*****``	``.***``	
	``***``	``.*``	
Minkowski	``***``	``*.``	Minkowski
addition	``****``	``**.``	subtraction
	``*****``	``***.``	
A ⊕ B	``*****``	``***.``	A ⊖ B
	``***``	``*.``	

	B1 ≠ B̌1		
Closing	``**.. *** ***``	``..``	Opening
	``***..``	``***``	
(A ⊕ B̌1) ⊖ B1	``****..``	``****``	(A ⊖ B̌1) ⊕ B1
	``****..``	``****``	
	``**..``	``..``	
Minkowski	``**..``	``**.. **..``	Closing
addition +	``***..``	``***.. ***..``	
subtraction	``****..``	``****.. ****..``	
	``****..``	``**********..``	
(A ⊕ B1) ⊖ B1	``**..``	``**.. **..``	

Fig. 3. Illustration of the geometrical covariance function computed for a simple binary image set **A**. Structuring elements are exemplified that consist of two black pixels placed at distance l in the horizontal direction **alpha** (the pixels at the origin are underlined). To the right of the illustrations the value of l is indicated, and below it the black pixel count is added. The asterisks indicate the black pixels coinciding after the shifting of two identical copies of the image. Two expressions are added below for $K(l)$ and $K(l)^*$ (uncorrected and corrected covariances, respectively). The illustration to the lower left shows the two binary images and their overlap for shift l = 10, by darker dots. The crosses indicate the black pixels in the shifted a image. In the small table to the lower right, the following values are listed in columns (1) to (5): (1) l, (2) mes **A ⊖ B̃**, (3) $K(l)$, (4) mes **T ⊖ B̃**, and (5) $K(l)^*$.

```
                    B ≠ B̌
  .*      .*       **    **         ..     .*        *       *
 .**     .**       ...   ***        ..     .*        -       -
 .***    .***      l = 1            ....   ***.      l = 6
 .***    .***                       ....   .***.
  .*      .*        20              ..     *.         1 1

  ..      ..       *  *             ..     ..         *       *
  ..*     ..*         -             ...    ..**       -       -
  ..**    ..**     l = 2            ....   ****       l = 7
  ..**    ..**                      ....   ..**
  ..      ..        1 1             ..     ..         8

  ..      ..       *  *             ..     ..         *       *
  ...     ...         -             ...    ..*        -       -
  ...*    ...*     l = 3            ....   .***       l = 8
  ...*    **.*                      ....   ...*
  ..      ..        6               ..     ..         5

  ..      *.       *  *             ..     ..         *       *
  ...     *..         -             ...    ...        -       -
  ....    *...     l = 4            ....   ..**       l = 9
  ....    ***.                      ....   ....
  ..      *.        7               ..     ..         2

  ..      **       *  *             ..     ..         *       *
  ...     **.         -             ...    ...        -       -
  ....    **..     l = 5            ....   ...*       l = 10
  ....    ****                      ....   ....
  ..      **        12              ..     ..         1
```

$$\text{---------} \rightarrow \text{alpha}$$

$$K_{\text{alpha}}(l) = \text{mes}(A \ominus l\check{B}) / \text{mes } A$$

$$K_{\text{alpha}}^{*}(l) = \langle \text{mes}(A \ominus l\check{B}) / \text{mes } A \rangle \cdot \langle \text{mes } T / \text{mes}(T \ominus l\check{B}) \rangle$$

```
first image
 !
 !       second image (shifted)
 !        !
 v        v

.........................
.........................
...**...**...++...++...
..***...***..+++..+++..
.****...****#+++...++++.
..****.****.++++.++++..
...**...**...++...++...
.........................
.........................
```

mes A = 30, mes T = 117

(1)	(2)	(3)	(4)	(5)
0	30	1.00	117	1.00
1	20	.67	108	.71
2	11	.37	99	.43
3	6	.17	90	.21
4	7	.23	81	.33
5	12	.40	72	.65
6	11	.37	63	.67
7	8	.27	54	.56
8	5	.17	45	.41
9	2	.07	36	.19
10	1	.03	27	.13
11	0	.00	18	.00

Figure 1 lists various measures (mes) that are counts of black pixels (or areas) for each transformation. Such counts, and the area proportions that are computed, can be used to estimate the probabilities associated to many geometrical attributes of the objects contained in digital binary images. Examples are: elongatedness, roundness, orientation, vicinity, curvature, and periodicity. The underlying model is that **B** is translated at random throughout **I** to estimate the characteritics of **A** in terms of the observed hits in relation to all possible hits. In the model, the set **I** must be a representative sample of a larger image set space.

In Figure 2, a distinction is made between the set transformations proposed by Minkowski (1911) for the approximation of volumes in solids (addition and subtraction of sets), and the ones proposed in Mathematical Morphology (Serra, 1982). Because, in general, the set **B** can be anisotropic, it is necessary to define dilatations and erosions, i.e., transformations that require **B̆**, the reflection of **B**.

Also, because many transformations are sequential to estimate density functions that can be in one or more directions, compound transformations are needed. A "closing" transformation consists of a dilatation followed by a Minkowski subtraction, while an "opening" transformation, consists of an erosion followed by a Minkowski addition.

As seen in Figure 2, these set operations can be formalized by Boolean operations between sets with shifts. Because of the definition of erosion (by the reflection of B) it follows that any transformation which is "antiextensive", "increasing" and "idempotent", must be made up by erosions (Matheron, 1975).

An erosion function can be seen as the decreasing proportion of black pixels over white pixels. It is generated by successively eroding previously eroded images, until the ratio of black over white pixels reaches zero (or until all black pixels in the original image have become white). Similarly, a dilatation function can be seen as an increasing proportion of black over white pixels until a given threshold value is reached for that ratio (or until all white pixels in the original image have become black).

A different function is the "geometrical covariance" that leads to estimating the autocorrelation. This function is illustrated in Figure 3, where it is generated by linearly eroding a binary image by a structuring element set consisting of two black pixels spaced 1 pixels apart in a given direction **aplha** (horizontal in the illustration).

Alternatively, this transformation can be computed by using two identical copies of a binary image set, translating one copy 1 pixels in direction **alpha** relatively to the other copy, and then computing the intersection between the two images. A value for the function is obtained by counting the number of black pixels in the intersection image set.

In each step, it is the original image that is eroded by a longer set B, or alternatively, it is the two identical copies that are intersected after a larger shift. Clearly, the transformation of the geometrical covariance estimates the probability of occurrence of a black pixel at successively larger distances in a given direction **alpha**. This function can easily be computed in the two directions of the raster (two-dimensional geometrical covariance function).

EXAMPLES OF PROBABILISTIC REPRESENTATIONS

Example 1: Crosscorrelation of map units from geological maps

Agterberg and Fabbri (1978) used transformations of binary images of stratigraphic units digitized from geological maps and ancillary mineral occurrence distribution maps (areal data and point data, respectively). They derived the frequency distributions of the areal amount of one lithologic unit per random cells of variable size. The areal amount of that unit in cells that contain one or more mineral deposits was related to the moments of this frequency distributions. The purpose of the technique applied, was to estimate in a quantitative manner the correlation existing between mineral deposit locations and the units represented on geological maps. The method was used for modeling quantitative representations of the mineral potential in mineralized areas in Canada by the geometrical attributes of map features.

Example 2: Two dimensional geometrical covariance functions of crystal phases in a thin section of a metamorphic rock

Fabbri (1984a) used a two-dimensional geometrical covariance function computed from the binary images of two crystalline phases digitized from a thin section of a metamorphic rock (granulite) to bring out the tendency of grain profile clusters to be distributed in some preferential direction. Two such directions, not detectable by eye, were clearly identified. They were interpreted as due to the combined effect of grain boundary orientation and of preferential distribution of grains in elongated clusters. Quantitatively, the distribution patterns for the different crystalline phases are dissimilar and can be used to classify textures in terms of spatial characterization of shape and distribution anisotropy.

Example 3: Models for characterizing point distributions

Serra (1982) studied point and line pattern distribution functions obtained by sequences of closing and dilatation transformations by circular structuring elements or disks. The following five models ware considered as realizations of random sets: (1) Poisson point models, (2) Neyman-Scott clustering models, (3) hard-core model, (4) Poisson alignment model, and (5) Brownian motion model. Different tessellations of space can be modeled and studied by simulating point distributions and by studying the morphological attributes of tessellations that correspond to particular growth patterns.

CONCLUDING REMARKS

The concepts reviewed in this paper, exemplify how to obtain morphological measures that lead to probabilistic estimates. Methods of mathematical morphology in processing astronomic images can be used for: (1) object identification (i.e., object characterization, elimination of unwanted objects, extraction of relationships among objects, etc.), and for (2) the study of object distribution in space (i.e., detection of distribution patterns, characterization of distributions, comparison and prediction of distributions, etc.). For example, given a pattern of distribution of objects in space, that is similar to a known one, can we predict the existence of particular kinds of objects at particular locations ? Or, can we produce probability contour maps of parts of the sky ?

As is in various fields of the earth sciences, much information is presently made available in digital form. In general, several gray level images are used to represent an area, e.g., the images corresponding to different bands used in remotely sensed imagery. The extension of mathematical morphology to gray tone functions has been developed (Goetcherian, 1979; Sternberg, 1982; Serra, 1982). However, it requires special processors to be computationally efficient in many realistic applications that involve images with over a quarter million pixels.

Some desirable computer architectures for spatial analysis of geoscience data by image processing have been considered (pipeline, parallel or pyramidal computers) by Kasvand and Fabbri (1983) and by Fabbri and Levialdi (1984). Two of these computers already exist as commercial products, for which both a programming language and image processing software have been developed. A review by Fabbri (1984b) described the General Operator Processor, GOP, designed at the University of Linköping, Sweden, and the GENESIS 2000 built by Machine Vision International in Ann Arbor, Michigan. Such machines represent a considerable forward leap in a more extensive application of image processing in the geosciences.

REFERENCES

Agterberg, F.P., and Fabbri, A.G., 1978, Spatial correlation of stratigraphic units quantified from geological maps. Computers & Geosciences, v. 4, p. 285-294.

Fabbri, A.G., 1984a, "Image Processing of Geological Data." Van Nostrand-Reihnold, Stroudsburg, (book in press).

Fabbri, A.G., 1984b, Promising aspects of geological image analysis. Proc. of IGCP-148 Meeting held during 27th Int. Geol. Congress, Moscow, August 4-14, 1984, (unpublished manuscript).

Fabbri, A.G., and Levialdi, S., 1984, New computer architectures suitable for spatial analysis in the earth sciences. Proc. Int. Coll. on: "Computers in Earth Sciences for Natural Resource Characterization." Nancy, France, April 9-13, 1984, Sciences de la Terre (in press).

Goetcherian, V., 1980, From binary to grey tone image processing using fuzzy logic concepts. Pattern Recognition, v. 12, p. 7-15.

Kasvand, T., and Fabbri, A.G., 1983, Considerations on pyramidal pipelines for spatial analysis of geoscience data. In, H. Freeman and G. Pieroni, Eds., 1984, "Computer Architectures for Spatially Distributed Data." Proc. NATO ASI, Cetraro, Italy, June 6-17, 1983. Springer-Verlag, New York (book in press).

Kirsch, R.A., Cahn, L., Ray, C., and Urban, G.H., 1957, Experiments in processing information with a digital computer. Proc. Eastern Joint Computer Conf., p. 221-229.

Matheron, G., 1972, "Elements pour une theorie des milieux poreux." Masson & Cie., Paris, 166 p.

Matheron, G., 1975, "Random Sets and Integral Geometry." John Wiley and Sons, New York, 261 p.

Minkowski, H., 1911, Theorie der Konvexen Korper insbesondere Begrundung ihres Oberflachen Begriffs, Gesammelte Abh. 2, p. 131-229.

Moore, G., 1968, Automatic scanning and computer processes for the quantitative analysis of micrographs and equivalent subjects. In, G.C. Cheng, R.S. Ledley, D.K. Pollock, and A. Rosenfeld, Eds., "Pictorial Pattern Recognition." Thompson Book Co., Washington, D.C., p. 275-326.

Serra, J., 1982, "Image Analysis and Mathematical Morphology." Academic Press, New York, 610 p.

Sternberg, S.R., 1982, Pipeline architectures for image processing. In, K. Preston jr., and L. Uhr, Eds., "Multicomputers for Image Processing." Academic Press, London, p. 291-306.

Uhr, L., and Vossler, C., 1961, A pattern recognition program that generates, evaluates, and adjusts its operators. Proc. Western Joint Comput. Conf., p. 555-569.

TREND IN PARALLEL PROCESSING APPLICATIONS

Stefano Levialdi

Dipartimento di Matematica
Università di Roma
00185 Roma, Italy

PANEL POSITION

After listening to the many contributors of this Workshop and noting how hard it has been for many authors to prove their system works, and works correctly, it is natural to expect that those systems will continue in operation for some time. A great deal of effort cannot be thrown away just because these systems are not the "best" in today's terms.

At the same time, the experience gained in designing and operating the existing systems for data analysis and processing can be usefully employed in the evaluation and design of new, different, multiprocessor computers. The big problem that computer scientists must face, and try to give an exhaustive answer, is whether the tasks of Astronomy have some generality and can be solved by improving the computational efficiency of general purpose machines or, conversely, if Astronomical data analysis and processing requires specific features directly related to the way data are collected, processed and visualized. Broadly speaking it seems hard, today, to suggest very high speed computing (hundreds of MIPS) independently from the tasks (nature of the data, algorithm structure, system reliability and friendliness, etc).

Perhaps the already long experience of cooperation between biologists and computer scientists may teach us something: at the beginning simple tasks were to be automated, then more elaborate tests were chosen to be fully computerized, eventually interactive schemes were selected so that the power of the computer was fully employed (exact repeatability, large memory for past events, broadcasting of results from different sources) and the final decision was left to the expert in the dubious cases pointed out by the machine. the best results were obtained by common work of problem analysis so as to pose reasonable goals to the future machines using the best technology as well as the latest know how from the biological world.

The emphasis of our workshop has been to try to bridge the gap between the experimental astronomers looking for new and better tools and computer scientists offering powerful systems, yet requiring the reformulation of algorithms, requesting therefore more work rather than offering an available solution.

Yet in some instances only a very fast computation may yield useful results (like for instance in the case of real time air traffic control) so that new hypothesis may be tested or simulated on very large amounts of data if the new computing systems outperform the existing ones by one or two orders of magnitude. it is essentially for this reason that an eye must be kept on those pioneering projects that aim towards the speeding up of present computation, either by using novel architectures based on a great number of processors (each one actively engaged in accomplishing part of the algorithm) or by providing the user with a fast communication means that enables him to visualize partial results in picture like form (both in the case of image processing and in the case of data represented iconically).

We are still at the beginning of the information revolution era, much will come in the next few years and we should be, as in the old electronic (tube) times, in a stand by position.

CONCLUDING DISCUSSION

CHAIRMAN

L. SCARSI

PANEL DISCUSSION ON " HOW CAN COMPUTER SCIENCE CONTRIBUTE TO THE SOLUTION OF PROBLEMS POSED BY ASTRONOMERS ?"

Chairman: L.Scarsi, Istituto di Fisica Cosmica e di Applicazioni dell'Informatica-C.N.R.-Palermo,Italy

Participants: M.Disney, Univ.College-Cardif,U.K.
J.Friedman, S.L.A.C. Stanford Univ.-Stanford U.S.A.
E.Groth,Princeton Univ.-Princeton U.S.A.
S.Levialdi, Dipart. Matematica,Univ.Roma-Italy
S.Tanimoto, Univ.of Washington-Seattle U.S.A.
D. Wells, N.R.A.O.-Charlottesville U.S.A.

A Panel was hold on June 3rd summarizing, in a way, the guidelines and the aims of the Workshop. General questions were addressed to M.Disney, E.Groth and D.Wells, who have expressed in the Workshop the point of view from Astronomy in the Sections "Data Analysis methodologies", "Image processing" and "Systems for Data Analysis" respectively:
"What they expect from Computer Science ?" "What is their field of interest in this respect?" " What they would like to be done in this field?"
J.Friedman, S.Levialdi and S.Tanimoto have given their point of view from the side of Computer Science. S.Levialdi and S.Tanimoto have contributed to the Session on "Parallel Processing", while J.Friedman has partecipated to the Session on "Data Analysis methodologies" from a general point of view, independently from the special applications to problems related to Astronomy.

M.Disney: There are one to two things which seem of interest to me. The first is that I would like to see more available to us sophisticated tools for examining large noisy data sets. There is no doubt that Astronomy, more than most other subjects, is statistical in nature and I think is very interesting to make up a list of what you consider to be the twenty top discoveries that have been made in Astronomy during the last 50 years or so, and decide for each of them to put down the number of observations which have been required before the final discovery to be made. For example let's just take one case early on. We could not have understood stellar structure and evolution without the H.R. diagram, there is no doubt. Well, it required something like a hundred parallax measurements to be made, each of them very difficult and complex, before suddendly "the light was on". On a larger scale, one think of the luminosity function for stars, in which one deals with a data set of the order of a million stars. Going on to talk about Galaxies, which seem to be a complex phenomenon subject to different forces and possibly born at different times, then clearly when we are looking for correlations which are actually crucial to theoretical understanding, we may be dealing again with data sets of hundreds of thousands of objects; to do that objectively we are going to need a sort of tools like "Cluster analysis" and so on, which we all know about, but we want to be able to use them easily and quickly in the sort of way that J.Friedman showed us.

A second thing I would like to emphasize is the capacity of data analysis systems to make objective selections of objects or phenomena. Anyone who is interested in the history of Astronomy knows that all subject has always been bedeviled by observation selections and it is always bedeviled by the same today and I believe it will be bedeviled forever. One of the difficulties is that astronomers, like nearly all the scientists, first of all have a theory and they go out to prove it and when they do that, of course they generally pick out data which favour their hypothesis, even when they try not to. It is very important that we have machines like the PDS machine, the Cosmos machine and so on, from which you can select objects by using criteria which have been setted by the beginning and which the machine is looking at for you. And this is of enormous importance over the years to come.

The third thing which seems to me very exciting and obvious, but needs to be emphasized, is that the mind works in an evolutionary way. You very seldom have a brilliant idea which is right in one go. When one works on a problem, you usually approach it by a series of trials and errors, and if you have a sophisticated and easy to use data analysis facility, that is precisely what you can do. You can try a theory, you can disregard it; you can try a new one, you can compare the results of the two, you can store that and you can go on like that. I think is the way in which the human mind works. There has been a good example of that in the Conference when Don Wells showed what you can do with the VLA data; the gradual massaging of the data with all the software techniques is producing pictures with a dynamic range well beyond what the designers ever expected in a first place. And I mentioned in my talk in the beginning the difficulties which people experience in the past, for example in pioneering work on binary Galaxies, when no such help was available; the assumptions which were necessary to make, in the end ruined the conclusions completely. That is very important.

Fourthly, I want to mention briefly something that Fosbury brought up: there is only one reason really for an astronomer to be at a telescope these days and it is to make decisions; otherwise he may as well sit at home and get the results by some remote line. If he is going to make proper decisions, to do that job economically, he needs to see his data in an astrophysical form as possible. The raw data are insufficient. Fosbury showed us a good example of a very faint galaxy spectrum, where you could hardly distinguish the piece with the galaxy from the rest of the sky; you need to get that out, in order for the astronomer to make sensible decisions. I am sure to that extent we need more clever things going on at the front end.

Finally, I believe, when you go out observing the data, the data set you are going to bring back is only a millionth part of the data that exist in all the archival environment. I think really as the archival processing improves, one will be able to do a lot of astronomy without actually moving and doing observing. I think is sad at the moment, that other people data are not available to one. It is usually presented, sometime deliberately and very often accidentally, in a way in which it cannot be checked. I can think of a famous Radioobservatory a few years ago, which was producing a lot of wonderful maps for which nobody else really believed what was really going on because there was no way to check what procedure has to be used to turn raw data in astronomy. It was very important that the astronomers will be able to publish archival data in a form that everybody can have a look by himself and keep monitoring to a start.

<u>E.Groth</u>. I like to think back to the times, say before 1978, when all these Vaxes appeared on the scene. We were really in bad times as far as data processing was concerned. I think we made a lot of progress in the last few years getting these Vaxes and machines available to a lot of astronomers, but we have a long way to go. I personally find that processing takes so long that I feel unable to experiment. So I am looking forward , in the next five or ten years,

to some kind of machine that I can afford on University budget, which is very much faster than the one I have now. That is one thing.
There are different ways to look at data. The example that Friedman showed is a step in the right direction. Also these data cube that people have, they generally have to look at a slice of it; we need some clever way to look at all cube at once.
All that reminds me the two things I would like computer science people to do for astronomy: build us faster machines and find us a better way to look at the data.

<u>D.Wells.</u> Since the problem is posed on the potential of computer science for Astronomy, I have thought about the unsolved problems which I think computer science could help us and we can't do for ourselves; after all astronomers present in this Workshop are mostly computer experts and very knowledgeable about computer equipment and most ordinary computer techniques. So the question is about new things: "What is the potential of new technology, the future technology which we are unlikely to do for ourselves?". I think the main area is pattern recognition of various kind and this is particularly important in the long run because our data volume will rise rapidly in the future and the number of astronomers will not rise as rapidly. I just believe that we have to become more automatic. I do not mean to suggest by that , that machines should do astronomy instead of astronomers; I mean that we need more sophisticated tools in order to screen out portions of the data set that we judge less interesting and to find things that result unusual and worth of analysis. So I thought about what pattern recognition problems do we need to solve and which are sophisticated enough to justify outsiders in helping us. There are two problems which I see remaining as major items:one of them is galaxy morphology classification , which is currently done visually by people. They have to train their eyes and minds to see the patterns and proceed to the classification ; I think this is an extremely difficult problem to do with a machine , but the power of the machines and the software today is beginning to approach the point and I am optimistic. Five years ago I thought it was impossible ; now I am willing to think about. We may ask: "Why bother?" I think that the detection of unusual galaxy morphologies and their correlation with environment in Clusters both near and far, has got to have some astrophysics in it. And basically acquiring better statistics in such data allows one to do better analysis and it is desirable that this be an objective classification not a subjective one as done visually by people. So I think, if some Pattern recognition

person wants to help Astronomy, that is an area to work on.
A problem which I think is easier, but still is almost unsolved in Astronomy, is spectral classification. What I have in mind is for example objective prisma spectra as done by Schmidt telescopes in which one gets a medium resolution spectrum and wishes classify it of course by its temperature and luminosity. But what is by far more interesting often is the abundance of unusual elements, in other words spectrum anomalies. The problem is to recognize this automatically, construct the parameter space and analyze it. This is beginning to be done in recent years, but I think, is probably a useful problem to work on more.
In both of these cases, screening programmes should treat the data and construct the parameter space. At that point you are faced with the same statistical problems that Jerry Friedman showed us in cluster analysis: the recognition of anomalies in large data sets. It turns out, if one gets the million numbers, the million sample elements, probably most of the sample points are irrelevant; it is only one out of ten thousand, one out of hundred thousand, that has new astrophysics in it. It is a problem of very subtle cluster analysis in a noisy space and that is where the future is going to be.

J.Friedman. The first point I like to make is the one that was confusing me throughout much of this Conference: that is what astronomers refer to as data analysis. Actually they break up it into two parts. One, which we refer to as data reduction or data refining, takes the raw digitization and, processing it with complicated computer programmes, pattern recognition and other things, turns out variables of astronomical interest: this is called data reduction or refining. The other is the analysis of these variables and measurements of astronomical interest: that is the real data analysis, namely to look the refined data and try to see new and unexpected things. If we go to a library and look for material on data analysis, we find books describing the process of taking refined data and trying to look at them in various ways in order to discover phenomena and effects that are not suspected. I think that statistics can contribute to both areas, tha data reduction and the data analysis, although not so much in the data reduction because that tends to be very problem specific. There are some things that one can take from statistics and informatics: for example, spectral analysis is one area where classification techniques have been used very successfully, mostly in chemistry. For problems of signal to noise, smoothing and filtering, again there is a lot of theories and of algorithms which can be found in the statistics literature. In the other phase, that we call data analysis and is looking at the refined data , those data tend to have an enormous commonality over a wide variety of fields: for instance in the psychological field and the economical field you find many similar problems to what you have in Astronomy, like the clustering problems, for example. Many algorithms come from these fields as well as from statistics.

The second point I like to bring out is that I have had either the course or the honor to have been in two fields, the time in which the computer was just beginning to become important for them: High Energy physics in the mid 60's and Statistics where the computer is just now becoming important. The words I am hearing here are identical in spirit and in actuality to what was said in those two fields. People are beginning to realize that the computer exists and may have an impact, and exactly the same set of remarks that I have heard here I have heard in both of those instances: things like "computers are taking over and naturally it is terrible, programming is not science", ...; in the case of High Energy Physics, doing electronics was science, but programming was something else; in Statistics, doing science is proving theorems, not writing computer programmes. I heard the same kind of things here: that programming is not astronomy, astronomers should not have to program; we should shield the astronomers from having to program; doing programs somehow is not doing astronomy.

And that seems to be the reason for a lot of these packages: to shield the astronomers from the computer. Now, in the case of High Energy Physics people have realized that computing is important and that is something legitimate to do; that you are a physicist and all you do is to work with physics data on a computer, the same as design an electronic apparatus or a counter or a detection equipment for experiments. Another thing that I heard in the Workshop was that we need reliable software, we need friendly and easy to use software, we need transportable software. We may ask the question: "Why don't you have all those things?" One reason, and there are probably a lot of reasons, is that is expensive; but another reason may be found by examining the reward system. If someone turns out reliable, wonderful software, what reward does he get in your profession as compared to the reward he would get if he discovers a new astronomical phenomenon or does some more direct work? And again, in both the fields in which I had to deal with in the past at the time when computing was becoming important, that was the real problem. Computing was not science, there was no scientific reward in the field for doing computing. That is not true in High Energy Physics anymore, is still true in Statistics and it looks like . is still kind of true here. May be this is the reason why you don't have wonderful software, because you don't give the reward.

Students can't get thesis, by doing it assistant professors can't get tenure and so they stay away from it. The only people that can afford to do it are the older ones who have not to worry, but are less inclined to do it, because they are not grown up in the computer age. So I the one thing you have to come to grip with is a proper reward system for doing science on the computer as it is given for doing science at the telescope.

Another point I may make has to do with the notion of these large packages and of the command languages. What bothers me is that those command languages are yet another language you have to learn. There are already too many languages that we have to know to use the computer: one has to understand the language of the operating system, the programming language, the language of the test editor, the language of the debugging monitor; now you have to understand another command language in order to use the software and its packages. This can be worth if you really believe that you can shield most astronomers from ever having to do programming. I think this is a misguided effort. Programming to some people is not that hard: computers are one of the important tools to do science, and I think the better you know how to use them, probably the better science you can do.

<u>S.Levialdi.</u> Very globally one may state that the help Computer Science can give is to switch the emphasis from data processing to decision making and therefore free the astronomer so he can devote most of his time to higher intellectual work. The discussion here reminds me some discussions we had in the past in the field of biomedical automation.There were two lines of thought. One was to automate very simple things, very routine like procedures, those easy to automate; you could have faster results and perhaps aim towards reliability. The other approach was to automate very sophisticated, human decision making procedures. Eventually the first one won because it was easier to obtain automation. So, in other words, if one could draw a line between a low level task and a highlevel task and decide to automate the first one, perhaps this may relieve some of the burdens of the astronomers. Two requests were clearly stated out here: one was for fast processing, the other was for obtaining good visual facilities for observing data. In the first area we hope, although we have no certainty, that by mean of using many processors nicely orchestrated together, we might achieve figures of the order of 10^2 to 10^4 faster..So the speed up is between 100 and 10.000.Of course we have to pay for that; this means, we do not have operating systems orchestrating these processors yet, we do not have high level languages we do not have experience and reliable systems.But since that

ideas are not that new (they have started nearly 20 years ago or even earlier there was mention of parallelism in the Charles Babbage machine last century), we hope to achieve in the near future, and for the near future we mean one or two years, machines, reliable array processing machines that can produce this gain of 100 to 10.000. In the era of visual interaction I just want to mention some new ideas of a group in Japan, in Hiroshima, led by Professor Ichicawa. One is the notion of semantic zooming. The basic idea is that instead of increasing the enlargement of an image by standard zooming, you can represent your knowledge on a graph; at different levels on this graph you have different levels of information, like for instance for a territory you can start from a building, then have a square, then a city, then a region, then a country, a continent and so far. So you can climb this graph and represent on the screen different levels of generality of the information. I think this can be of some use in representing systems.

The other thing is the development of what is called visual language. This answers a bit the problem of not introducing new handbooks to read like for interpreting command languages and other things. Visual languages essentially are handled by icons, by symbols which clearly denote the actions which actually you want to produce on the system. So you could do image processing and other activities on your data not having to remember commands, but just associating icons with the operations.There is new open research on this field. Returning to the fact that it might exist some inertia in moving from one experienced system to another, and commenting on the reward which I think is an important point, I feel there is also a psychological "tie up" with the old system. If one has devoted a lot of effort in developing a program or a package of programs, and one has had to overcome a lot of problems, one inevitably is tied emotionally to his package, to one own package.This makes it difficult to change from a package to another and therefore there is a sort of tie up to one own way of representing images and even processing them. There might be some difficulty in overcoming this fact and this perhaps explains why there are so many different packages doing similar things.

To conclude I would like to say that instead of having on one side the astronomers and on the other side the computer scientists, perhaps we need a collaborative effort, a cooperative effort; what the result might be, there might be at least more talks and more cross work together. Otherwise if these fields diverge or if they just confront themselves, I doubt whether you could get results at all. Also from the experience on the biomedical field, we know that, unless the doctors and the biologists didn't get involved with computing and vice versa computer scientists got involved with medical problems, then the results were not very significant. I hope that this Workshop is one step toward this cooperative effort.

S.Tanimoto. I think that the comments made so far are quite helpful. I like to suggest that for the short term astronomers probably do not need computer scientists; I finally begin to realize that is probably true that most of the things you want from computing as astronomers, you can go out and buy: things like better software, more useful and friendly systems, and so forth. You can probably construct these systems yourselves after talking to professional programmers, to people who not necessarily must be at the fore front of the computer science research. On the other hand, as Don Wells pointed out, is interesting to ask the question of what new technology is likely to becoming along and might impact astronomy. He suggested pattern recognition as one area; it is probably true that there is some potential for pattern recognition to make an impact on astronomy. I would suggest that an area where astronomers and computer scientists really do have some common relative work to do is the formalization of astronomical knowledge.

There are a number of avantgarde efforts between computer scientists and scientists in other fields, and I mentioned during my talk the molecular genetics area, but there may be others along this line. It would seem to me that an interesting project where the two kind of scientists could really collaborate would be in trying to create a system that contains the representation for some aspect of astronomical theory. Let me give you a very simple example of a logical technique that is being studied now in artificial intelligence, that may or may not have applications in astronomy. It is called " non monotonic logic". In traditional logic one starts with a set of axioms and then tries to prove new things from the axioms; one property of such a logic is that generally if you add an axiom, that is you make a new assumption in addition to the ones you have already made, then you derive some new conclusions, you can make some new theorems. As you increase the set of axioms, the set of theorems also increases, i.e. there is a monotic relationship. In a non-monotic logic it is possible to introduce new axioms which actually reduce the number of theorems that you can prove from the system. An example of this can be the following situation: suppose your axioms are " All birds have feathers", "Robin is a bird","All birds can fly". These 3 axioms imply some other statements such as " A robin has feathers", "A robin can fly". If you add a new axiom such as " A bird weighs less than hundred pounds" or something like that, then you can infer more theorems. That is a monotonic situation. But suppose now that we change the system and make it a non monotonic one. Then we may change some of the axioms to have slightly different forms; for example: "In the absence of any information to the contrary, assume that any bird can fly". Now suppose that we introduce a new axiom: "An ostrich cannot fly". In the old system we would have a contradiction ; in the new system we simply make an exception in some sense, we go from the fault case to a special case and we reduce the number of theorems, but this does not pose any problem to the logic. Now it may well be that scientific theories really behave like non monotonic logic, that is the best theories seem to imply all these following things as new pieces of evidence come in and a lot of the old theories fall apart. Well, if we have a logic system which represents these theories in a non monotonic way, we will be able to cope with that radical changes. And this may be that kind of thing which can happen when you start merging astronomy with other branches of science. Anyway I like to suggest that this area of mathematical epistemology or knowledge representation may be an area which will be a good one for astronomers and computer scientists to work together.

A few contributions have been made from the floor. Some of them are synthetised in the following.

Rocca. Geophysics is using computers since the early 50's and in the beginning everybody thought that the computer could be used as an help in decision making. This did not happened. What happened in the other hand is that now computers are heavily used in geophysics (I may say that of the order of 3 - 4 billion dollars a year are spent in this area), but less that 1% of this figure is spent in decision making. Most of the computer use is as an help to instrumentation: in other words a deterministic use of the computer. You do not want a software programme to make decisions for you. You like the software do something that is completely repeatable and that is somehow an help to the instrumentation. An important point I would like also tomake is that the deterministic use of computers that does not mean that there is no use os statistics as an estimation technique to find deterministic parameters as for example the velocity of the earth, the propagation media, etc.; these parameters, estimated statistically are used as an input to the computer programme again to improve the results got by the data acquisition.
I am not sure that this may happen in astronomy too.

Simpson. What I see coming and I appreciate very much is better mass

storage and data base management systems. The Space Telescope, the
Gamma Ray Observatory, all these are generating a factor of 1000 or
more informations that we had in the past. The other tool that I want
which may or may not come from Computer Science, is a lot of dynamic
range in my ability to look at data. I got to be able to look at individual photons and make sure of what is going on on the right level
there and got to be able to look at the entire Galaxy and other galaxies.

Disney. One thing I would like to pick up is the Friedman remark about
the reward. This is absolutely right; nothing is done in this world
unless people feels that something can be taken out of it for himself.
Particularly in this field of data analysis, because is the young
people who does most of the work (most of the brilliant work seems
to be done by people in their 20's or early 30's who dont have permanent jobs). The snag is of course to put that message over to the
Directors of Observatories, who dont come to meetings like this.
I think is up to the senior astronomers around here to push this point
very hard. The problem, on the other side is that several graduate
students have the danger to become almost completely computer addicts
with very little reference to astronomy. It is a very difficult balance to achieve.

Groth. I have been first in a meeting like this, with a group of people interested in image processing, back in 1977; there were Landsat
people at that meeting and some astronomers. We didn't have too much
in common. We were unable to find edges to our galaxies, but the Landsat people could find the edges to their rivers and so. I think we
are still having the same problem of communication between astronomers and other image processing people.

Wells. When I suggested that Computer Science could contribute to
pattern recognition, I didn't bother to mention better computing hardware and laser discs and data base management systems, because I
assumed that all this is going to come anyway. I was mainly interested in very sophisticated, advanced kind of ideas we would not get
just going out and buying it. I want to come back to this pattern
recognition question. One thing I didn't mention was the reason of
the subject I picked , which was morphology of galaxies and classification of spectral data. The reason was that I know very well that
the expenditure of a large amount of money in the last decade or so
has permitted techniques of that kind to have an economical importance
in other fields. There are machines today that can automatically clas-

sify cells as seen through microscopes and apparently that is becoming economically important. I do not know the details, but I have visited such laboratories. Their data do not look exactly like ours, but I know that in order to classify them required a great technical effort by many people over a long period of time. Perhaps Astronomy cannot find such sort of things, while medicine has plenty of money. For spectral classification there are machines today that can automatically classify compound substances on the basis of gascromatography measurements. That is the kind of artificial intelligence application we are looking for. If one has the right people working on that class of problems, one makes progress. If there is an economic effect it is easy to get money to spend on the work of R. and D. It is not a priori the case for Astronomy.

J.Friedman. Just one comment about the need for collaboration between Computer scientists and Astronomers: I would like to emphasize that it should be a real collaboration, i.e. astronomers should give to the computer people their dues. For astronomers a piece of software is a tool to do science work; for the computer people is the end result, what they get their satisfaction from. If you want that computer people work on the problems of the astronomers, they have to give in exchange a proper recognition. I have heard often early in the week that Astronomers must stay in control and must keep things focused and must not let the computer scientists get off in their own world and produce fancy sophisticated things which astronomers cant use; but some room should be given to the computer people, otherwise the talented ones will not be interested and the end result will be very modest.

Levialdi. I am going back to the pattern recognition problem: perhaps in the future we might develop expert systems for astronomy. I think one good , positive aspect of our Workshop was that it may lead the way to a better formalisation of the problems of astronomy and the better they are formalised, then the better we can claim to put the astronomer knowledge into some kind of programme and this programme might help, I hope. So we can borrow techniques from artificial intelligence, we can borrow experiences from biomedical problems, cell analysis and so forth, and by doing this we might improve and help, as computers scientists, the work of the astronomers.

INDEX

ADAM (Astronomical Data Aquisition Monitor), 382
ADC (Atmospheric Dispersion Correctors), 380
Analysis functions, 327
Architecture triangle, 498
Astronomical data analysis, 431, 438

Background estimation, 58, 366, 373
Bootstrap sampling, 81, 83, 125
Brightness, 396

Calibration function, 326
Catalogs, 327
Classification, 70, 431
Cluster analysis, 29, 69, 72, 127, 364
Control strategies, 461
Convolution model, 368
COS-B satellite, 21, 23, 82, 120, 325
Cross-correlation, 510, 519

Deconvolution techniques, 128, 366, 368
Display systems, 6, 417
  DEC-VS11, 241
  Gould-DeAnza, 219
  IIS machines, 273, 418

Display systems (continued)
  SIGMA-ARGS, 249
Data compression, 425
Data structures, 497, 499
Differential Pulse Code Modulation (DPCM), 426

Echelle data, 323
Einstein X-ray satellite, 324, 325
Ellipticity analysis, 87
Exploratory data analysis, 46
EXPRESS, 472
Extended sources, 363, 401

Feature extraction, 347
Faint objects, 57
Finite mixture, 30
  by decision direct method, 33
  by maximum likelihood, 31, 36
  by method of moments, 39
FITS, 327, 454
Flux calibration, 413
FOC (Fant Object Camera), 165
Folding, 24, 83
Fourier descriptors, 402
Fourier transform, 15, 325, 329, 368

Gamma-ray pulsars, 21, 81, 123
GENESIS-2000, 521
Geometric corrections, 327

GKS, 136
Graph matching, 460
Growded fields, 439

Higher dimensional views, 49
HIPPARCOS, 99

Image analysis, 343, 349
  correction, 344
  detector CCD, 324, 390, 397
  display, 417
  enhancement, 344
  filtering, 434
  interpretation, 354
  processing, 241, 271, 323, 343, 411, 431, 479, 509, 510
Interactivity, 241, 271, 436, 481

Masking technique, 369
Maximum likelihood, 441
MIPS, 463
Mixture identification, 29
Morphological analysis, 401, 509
Multicomputer architectures, 461
  Cube, 465, 466
  MIMD, 465, 488, 493
  SFU, 485
  SIMD, 482, 465, 488
  Pyramid, 465
Multiple views, 51
Multivariate analysis, 46, 67, 69

Networks for data analysis,
  ASTRONET, 8, 211, 241, 410
  STARLINK, 8, 137, 249, 272
Noise reduction, 431

Optical data analysis, 135, 272 363

Parallel computers, 510, 520
Parallel machines,
  CLIP4, 502
  DDM1/2, 464
  MACSYM, 472
  MIT, 463
  MPP, 482
  PAPIA, 482, 492
  PASM, 489
  SISD, 482
Parallel processing, 459, 480
Parallel Programming languages,
  CAP4, 502
  PIXAL, 502
  Pascal PL, 502
Pattern recognition, 121, 124
Pipeline, 484, 520
Pyramidal computers, 520
Photometry, 363, 395
Photometric analysis, 364
Photometric corrections, 36
Programming languages,
  ADA, 225
  FORTRAN, 297, 314
  FORTRAN 8X, 315
  Pascal, 502
  Sheltran language, 297
  VSFORTRAN, 264
Projection pursuit, 50
Pulse processes, 75

Quasars, 389

Radioastronomy, 417
Radio data analysis, 135, 272
Radiopulsars, 16, 20

SAO catalog, 339
Segmentation, 432
Semantic analysis, 351
Systems for data analysis, 157
  AA, 227
  AIPS, 195, 250, 418, 422

Systems for data analysis (continued)
    CDCA, 223
    DWARF, 449
    FIPS, 165, 272
    GIPSY, 271, 418
    HERMES, 250, 282
    IHAP, 175, 306, 368
    ISIID, 229
    MIDAS, 175, 219, 272, 306
        317, 368, 451
    MOTHER, 249
    PANDORA, 272
    RIAIP, 241
    SAIA, 257
Smoothing, 329
Space Telescope, 191
Spatial filtering, 329

Spectrometry, 379
Statistical inference, 46
Structural analysis, 350
Supernova, 391, 397

Texture, 510
Texture extraction, 348
Three dimensional spectroscopy
    techniques, 379
  ASPECT, 379
  TAURUS, 379
Time dependent analysis, 387
Time series, 387
Transform coding, 48

Velocity field analysis, 93
Vision problems, 481
VLA, 195, 300, 422
VLSI, 460, 497

RAYMOND H. FOGLER LIBRARY
DATE DUE

**BOOKS ARE SUBJECT TO
RECALL AFTER TWO WEEKS**

MAY 20 1987

OCT 2 1987